# INFRARED SPACE INTERFEROMETRY: ASTROPHYSICS & THE STUDY OF EARTH-LIKE PLANETS

# ASTROPHYSICS AND SPACE SCIENCE LIBRARY

VOLUME 215

# INFRARED SPACE INTERFEROMETRY: ASTROPHYSICS & THE STUDY OF EARTH-LIKE PLANETS

Proceedings of a Workshop held in Toledo, Spain,
March 11–14, 1996

Edited by


C. EIROA

*Universidad Autónoma de Madrid, Spain*

A. ALBERDI

*Instituto de Astrofísica de Andalucía, CSIC, Granada, Spain and Laboratorio de Astrofísica Espacial y Física Fundamental, INTA, Madrid, Spain*

H. THRONSON

*NASA Headquarters, Washington D.C., U.S.A. and University of Wyoming, Physics Department, Laramie, Wyoming, U.S.A.*

T. DE GRAAUW

*Space Research Organisation Netherlands, Groningen, The Netherlands*

and

C. J. SCHALINSKI

*Institute of Space Sensor Technology, DLR, Berlin, Germany*



SPRINGER SCIENCE+BUSINESS MEDIA, B.V.

A C.I.P. Catalogue record for this book is available from the Library of Congress

ISBN 978-94-010-6300-5 ISBN 978-94-011-5468-0 (eBook)
DOI 10.1007/978-94-011-5468-0

---

*Printed on acid-free paper*

## TABLE OF CONTENTS

### 1. Scientific Case

## 2. Instrumental and Technical Cases

# FOREWORD

The past year has produced some of the most exciting results in the history of astronomy, particularly in the area of planets outside our solar system. Only a half-year before our meeting in Toledo, Spain, the first unambiguous detection of planet-sized masses orbiting main sequence stars were reported. Since that time, evidence for a new exoplanet has been reported almost at the rate of about once per month. Some of these objects are likely to turn out to be very low-mass stars, but something like half show characteristics – Jupiter-like mass and near-zero orbital eccentricity – which appear to be unique to planets.

Almost at the same time that giant planets were being discovered regularly, the two major space agencies, ESA and NASA, have identified searches for and detailed study of Earth-like planets as a major priority for the future. In ESA's "Horizon 2000 Plus" programme, an infrared interferometer has been proposed as a possible future Cornerstone mission. Similarly, scientists in the US produced the "Road Map for the Exploration of Neighboring Planetary Systems (ExNPS)", which provided NASA with a long-term plan which leads also to an infrared interferometer in space to study hypothetical Earth-like worlds beyond our Solar System. Such an observatory is designed to search for the thermal emission from a family of planets, using interferometric nulling to remove the contaminating light from the central star. The residual planetary light would be analyzed for the telltale signatures of molecules of water, carbon dioxide, and ozone. The presence of such molecules would be strong evidence for life on the planet.

The search for planets outside the Solar System is important also to the average citizen, who supports scientific exploration through taxes. All the participants at this meeting have had the experience of speaking with interested schoolchildren, of being stopped in buses or subways, or talking with a stranger in a long airline trip when the subject of astronomy comes up. In addition to the usual topics of black holes and quasars, over the past year our fellow citizens have become increasingly fascinated by the possibility of finding other worlds . . . and other life . . . beyond the Earth. Over the centuries, scientists have had to respond our curious fellow citizens by merely stating that perhaps there are other worlds around other stars, but we just did not know for sure. For the first time in human history, scientists and technologists are now able to say, "Yes, there are planets beyond the Solar System." More than that, as described at this conference, we might also be able to determine whether or not those distant worlds are inhabited.

This international conference originated during a small workshop on interferometry and planet detection in Paris in the spring of 1995. More elaborate planning was undertaken at a larger workshop hosted by the German Aerospace Research Establishment (DLR) in Berlin, in September, 1995. At that time, it was becoming clear that detection of planets outside the solar system was poised to become more than a theoretical exercise: new instruments, new techniques, and new facilities were going to turn one of the great mysteries of humanity into an area of active scientific research, solidly grounded in observational data. As part of our early discussion about this conference, it was clearly desirable that the first major European meeting on the direct detection of exo-planets be held at a unique location, perhaps reflecting the new scientific discipline that is just beginning. Toledo, Spain, with its cultural and historical traditions seemed ideal.

In Europe, we received support from ESA and we are grateful to the efforts by Dr Sergio Volonte. This meeting was supported in the US by NASA's Office of Space Science and by the University of Wyoming. We are grateful to Dr Michael Bicay at NASA Headquarters and to Dr Lee Schick at the University of Wyoming for their support. In Spain, many institutions recognized the importance of this meeting. We deeply appreciate the support provided by the Universidad Autónoma de Madrid, Sociedad Española de Astronomía, Consejo Superior de Investigaciones Científicas, Instituto Nacional de Técnica Aeroespacial and Comisión Interministerial de Ciencia y Tecnología. We are also grateful to Prof. Enric Banda, the Spanish State Secretary of Universities and Research, for his kind and warm welcome to all participants at the Toledo Conference.

The Editors (C. Eiroa, A. Alberdi, H. Thronson, T. de Graauw, C.J. Schalinski)

*Scientific Advisory Committee:* R. Angel, S. Beckwith, C. Eiroa, T. Encrenaz, T. Fukushima, T. de Graauw, A. Léger, J.M. Marcaide, J.M. Mariotti, A. Penny, R. Rodrigo, C.J. Schalinski, M. Shao, H. Thronson (Chairman), S. Volonté

*Local Organizing Committee:* A. Alberdi, C. Eiroa (Chairman), M. Guitart (Secretary), B. Montesinos, H. Thronson

UAM-LAEFF-IAA Workshop on

# INFRARED SPACE INTERFEROMETRY WORKSHOP

## Astrophysics & the Study of Earth-like Planets

11-14 March 1996
Toledo, Spain

**Participants:**

1. **Afonso, José**, University of Lisbon -Physics Department- (Lisbon, Portugal)
2. **Alberdi, Antxon**, Instituto de Astrofísica de Andalucía (Granada, Spain); Laboratorio de Astrofísica Espacial y Física Fundamental -INTA- (Madrid, Spain)
   e-mail: alberdi@laeff.esa.es
3. **Anselmi, Alberto**, Alenia Spazio (Torino, Italy),
   e-mail: aanselmi@to.alespazio.it
4. **Auh, Byung-Ryul**, Korea Astronomy Observatory (Taejeon, Korea)
5. **Bachem, Eberhard**, German Space Agency (DARA) (Bonn, Germany)
6. **Beichman, C**, Jet Propulsion Laboratory (Pasadena, California, U.S.A.),
   e-mail: chas@ipac.caltech.edu
7. **Bely, Pierre**, Space Telescope Science Institute (Baltimore, Maryland, U.S.A.),
   e-mail: bely@stsci.edu
8. **Blake, Rob**, The Open University-Department of Physics- (Milton Keynes, United Kingdom),
   e-mail: R.P.Blake@open.ac.uk
9. **Boss, Alan**, Carnegie Institution of Washington (Washington DC, U.S.A.),
   e-mail: boss@dtm.ciw.edu
10. **Briess, Klaus**, Institute of Space Sensor Technology -DLR- (Berlin, Germany),
    e-mail: klaus@asterix.ws.ba.dlr.de
11. **Casertano, Stefano**, Space Telescope Science Institute (Baltimore, Maryland, U.S.A.),
    e-mail: stefano@stsci.edu
12. **Castro-Tirado, Alberto**, Laboratorio de Astrofísica Espacial y Física Fundamental -INTA- (Madrid, Spain),
    e-mail: ajct@laeff.esa.es
13. **Claes, Peter**, European Space Agency-Villafranca Satellite Tracking Station (Madrid, Spain),
    e-mail: pclaes@iso.vilspa.esa.es
14. **Coudé du Foresto, Vincent**, Max Planck Institut für Astronomie (Heidelberg, Germany),
    e-mail: foresto@mpia-hd.mpg.de
15. **Coulter, Dan**, Jet Propulsion Laboratory (Pasadena, California, U.S.A.),
    e-mail: daniel.r.coulter@jpl.nasa.gov
16. **de Graauw, T**, Space Research Organisation Netherlands (Groningen, The Netherlands),
    e-mail: thijsdg@sron.rug.nl

17. **de Korte, Pieter**, Space Research Organisation Netherlands (Utrecht, The Netherlands),
e-mail: P.deKorte@sron.rug.nl

18. **de Winter, Dolf**, Universidad Autónoma de Madrid (Madrid, Spain),
e-mail: dolf@astro.uva.nl

19. **Dunkin, Sarah D.**, Department of Physics and Astronomy, University College London (London, England),
e-mail: skd@starlink.ucl.ac.uk

20. **Eiroa, Carlos**, Universidad Autónoma de Madrid (Madrid, Spain),
e-mail: carlos@astro1.ft.uam.es

21. **Encrenaz, Therese**, Observatoire de Paris (Meudon, France),
e-mail: tencrenaz@mesioa.obspm.fr

22. **Faucherre, Michel**, GRAAL/CNRS, Université de Montpellier II (Montpellier, France),
e-mail: faucherr@graal.univ-montp2.fr

23. **Ferlet, Roger**, Institut d'Astrophysique (Paris, France),
e-mail: ferlet@iap.fr

24. **Festou, Michel**, Observatoire Midi-Pyrenees (Toulouse, France),
e-mail: festou@obs-mip.fr

25. **Figueras, Francesca**, University of Barcelona -Dept. of Astronomy & Meteorology- (Barcelona, Spain),
e-mail: cesca@mizar.am.ub.es

26. **Fischer, Olaf**, Universitats Sternwarte-Astrophysikalisches Institut- (Jena, Germany),
e-mail: fischer@betty.astro.uni-jena.de

27. **Fontana, Walter**, University of Wien, Institute for Theoretical Chemistry (Wien, Austria),
e-mail: walter@stat.umn.edu

28. **Garzón, Francisco**, Instituto de Astrofísica de Canarias (Tenerife, Spain),
e-mail: fgl@iac.es

29. **Gautier, Daniel**, Observatoire de Paris -DESPA- (Meudon, France),
e-mail: daniel.gautier@obspm.fr

30. **Gilbreath, Charmaine**, US Naval Research Laboratory (Washington DC, U.S.A.),
e-mail: gilbreath@ncst.nrl.navy.mil

31. **Giménez, Alvaro**, Instituto Nacional de Técnica Aeroespacial (Torrejón de Ardoz, Madrid),
e-mail: ag@laeff.esa.es

32. **Glindemann, Andreas**, Max Planck Institut für Astronomie (Heidelberg, Germany),
e-mail: glindem@mpia-hd.mpg.de

33. **Goldsmith, Donald**, Interstellar Media (Berkeley, California, U.S.A.),
e-mail: dgoldsmith@astron.berkeley.edu

34. **Gómez, José Francisco**, Laboratorio de Astrofísica Espacial y Física Fundamental -INTA- (Madrid, Spain),
e-mail: jfg@laeff.esa.es

35. **Greenaway, Alan**, DRA Malvern (Malvern, United Kingdom),
e-mail: ahg%ccf.dnet@hermes.dra.hmg.gb

36. **Guillot, Tristan**, University of Arizona -Lunar and Planetary Laboratory- (Tucson, Arizona, U.S.A.),
e-mail: guillot@lpl.arizona.edu

37. **Harris, Alan**, DLR Institute for Planetary Exploration (Berlin, Germany),
e-mail: harris@terra.pe.ba.dlr.de

38. **Hawarden, Timothy G**, Joint Astronomy Centre (Hilo, Hawaii, U.S.A.),
e-mail: hawarden@jach.hawaii.edu

39. **Huygen, Eric**, Katholieke Universiteit Leuven -Instituut Sterrenkunde- (Heverlee, Belgium),
e-mail: rik@ster.kuleuven.ac.uk

40. **Johann, Ulrich**, Dornier SatellitenSysteme GmbH (Friedrichshafen, Germany),
e-mail: johann@eosys.dofn.de

41. **Jones, Barrie William**, The Open University-Department of Physics- (Milton Keynes, United Kingdom),
e-mail: b.w.jones@open.ac.uk

42. **Joubert, Martine**, CNES, (Paris, France),
e-mail: joubert@cst.cnes.fr

43. **Kaplan, Michael**, NASA Headquarters (Washington DC, U.S.A.),
e-mail: mskaplan@gm.ossa.hq.nasa.gov

44. **Kessler, Martin**, European Space Agency-Villafranca Satellite Tracking Station (Madrid, Spain),
e-mail: mkessler@iso.vilspa.esa.es

45. **Laskin, Robert**, Jet Propulsion Laboratory -CalTech- (Pasadena, California, U.S.A.),
e-mail: bob-laskin@343mac.jpl.nasa.gov

46. **Leger, Alain**, Université Paris-Sud -IAS- (Orsay, France),
e-mail: leger@iaslab.ias.fr

47. **Léna, Pierre**, Observatoire, Univ. Paris VII (Paris, France),
e-mail: lena@obspm.fr

48. **Liseau, René**, Stockholm Observatory (Saltsjobaden, Sweden),
e-mail: rene@astro.su.se

49. **Lund, Glenn**, Aeroespatiale -Centre Operationnel Satellites (Cannes La Bocca, France)

50. **Luri, Xavier**, University of Barcelona -Dept. of Astronomy & Meteorology- (Barcelona, Spain),
e-mail: xavier@mizar.am.ub.es

51. **Malfait, Koen**, Katholieke Universiteit Leuven, Instituut Sterrenkunde (Heverlee, Belgium),
e-mail: koenm@ster.kuleuven.ac.be

52. **Mampaso, Antonio**, Instituto de Astrofísica de Canarias (Tenerife, Spain),
e-mail: amr@iac.es

53. **Marcaide, Juan María**, Universidad de Valencia (Valencia, Spain),
e-mail: jmm@vlbi.matapl.uv.es

54. **Mariotti, Jean Marie**, Observatoire de Paris -DESPA- (Meudon, France),
e-mail: mariotti@obspm.fr

55. **Martín, Eduardo**, Instituto de Astrofísica de Canarias (Tenerife, Spain),
e-mail: ege@iac.es

56. **Mather, John C**, NASA Goddard Space Flight Center (Baltimore, Maryland, U.S.A.),
e-mail: mather@stars.gsfc.nasa.gov

57. **Mayor, Michel**, Observatoire de Genève (Sauverny Switzerland),
e-mail: mayor@scsun.unige.ch

58. **Mennesson, Bertrand**, University of Arizona -Steward Observatory- (Tucson, Arizona, U.S.A.),
e-mail: bmenness@as.arizona.edu

59. **Moitinho de Almeida, André**, Instituto de Astrofísica de Andalucía (Granada, Spain),
e-mail: andre@iaa.es

60. **Montesinos, Benjamín**, Instituto de Astrofísica de Andalucía (Granada, Spain); Laboratorio de Astrofísica Espacial y Física Fundamental -INTA- (Madrid, Spain),
e-mail: bmm@laeff.esa.es

61. **Moreira, Miguel**, University of Lisbon -Physics Department- (Lisbon, Portugal)

62. **Mundt, Reinhard**, Max Planck Institute für Astronomie (Heidelberg, Germany),
e-mail: mundt@mpia-hd.mpg.de

63. **Natta, Antonella**, Osservatorio di Arcetri (Firenze, Italy),
e-mail: natta@arcetri.astro.it

64. **Oro, Juan**, University of Houston-Dept. of Biochemical & Biophysical Sciences (Houston, Texas, U.S.A.)

65. **Pain, Thierry**, Aeroespatiale -Centre Operationnel Satellites (Cannes La Bocca, France)

66. **Paresce, Francesco**, European Southern Observatory (Garching bei Munchen, Germany)
e-mail: fparesce@eso.org

67. **Parodi, GianCarlo**, BCV Progetti (Milano, Italy),
e-mail: gian@bcv.sublink.org

68. **Penny, Alan**, Rutherford Appleton Laboratory (Oxon, United Kingdom),
e-mail: ajp@ast.star.rl.ac.uk

69. **Perrin, Guy**, Observatoire de Paris-Meudon (Meudon, France),
e-mail: perrin@hplyot.obspm.fr

70. **Petro, Larry**, Space Telescope Science Institute (Baltimore, Maryland, U.S.A.)
e-mail: petro@stsci.edu

71. **Piña, Robert**, University of Florida -Dept. of Astronomy- (Gainesville, Florida, U.S.A.),
e-mail: rpina@astro.ufl.edu

72. **Quirrenbach, Andreas**, Max Planck Institute für Extraterrestrische Physik (Garching, Germany),
e-mail: quirrenbach@mpe-garching.mpg.de

73. **Rabbia, Yves**, O.C.A. (Grasse, France)

74. **Ridgway, Stephen**, Kitt Peak National Observatory (Tucson, Arizona, U.S.A.),
e-mail: ridgway@noao.edu

75. **Rodrigo, Rafael**, Instituto de Astrofísica de Andalucía (Granada, Spain),
e-mail: rodrigo@iaa.es

76. **Rodríguez Espinosa, J.M.**, Instituto de Astrofísica de Canarias (Tenerife, Spain),
e-mail: jre@ll.iac.es

77. **Rogers, John**, Cambridge University -Radio Astronomy Department- (Cambridge, United Kingdom),
e-mail: john@mrao.cam.ac.uk

78. **Rouan, Daniel**, Observatoire de Paris-Meudon (Meudon, France),
e-mail: rouan@obspm.fr

79. **Rowan-Robinson, M**, Astrophysics Group, Blackett Lab., Imperial College (London, England),
e-mail: mrr@ic.ac.uk

80. **Sandler, David**, University of Arizona & ThermoTrex Corp. (San Diego, California, U.S.A.),
e-mail: dsandler@cts.com

81. **Saraceno, Paolo**, Istituto di Fisica dello Spazio Interplanetario -CNR- (Frascati, Italy),
e-mail: saraceno@ifsi.fra.cnr.it

82. **Scaramuzzi, Franco**, Istituto di Astrofisica Spaziale (Frascati, Italy)
e-mail: scaramuzzi%efr419@frascati.enea.it

83. **Schalinski, Cornelius J.**, Institute of Space Sensor Technology -DLR- (Berlin, Germany),
e-mail: schalinski@dlr.de

84. **Schneider, Jean**, Observatoire de Paris (Meudon, France),
e-mail: schneider@obspm.fr

85. **Schilling, Govert**, Free-Lance Science Writer (Utrecht, The Netherlands)
e-mail: goverts@knoware.nl

86. **Shao, Michael**, Jet Propulsion Laboratory (Pasadena, California, U.S.A.),
e-mail: mshao@huey.jpl.nasa.gov

87. **Shibai, Hiroshi**, The Institute of Space and Astronautical Science (ISAS) (Kanagawa, Japan),
e-mail: shibai@astro.isas.ac.jp

88. **Simeoni, Denis**, Aeroespatiale -Centre Operationnel Satellites (Cannes La Bocca, France)

89. **Simon, Richard**, National Radio Astronomy Observatory (Charlottesville, Virginia, U.S.A.),
e-mail: rsimon@nrao.edu

90. **Stanton, Richard**, Jet Propulsion Laboratory -Caltech- (Pasadena, California, U.S.A.),
e-mail: r.stanton@jpl.nasa.gov

91. **Telesco, Charles**, University of Florida -Dept. of Astronomy- (Gainesville, Florida, U.S.A.),
e-mail: telesco@astro.ufl.edu

92. **Thronson, Harley**, University of Wyoming-Physics Department- (Laramie, Wyoming, U.S.A.); NASA Headquarters (Washington D.C., U.S.A.),
e-mail: Harley.Thronson@hq.nasa.gov

93. **Torra, Jordi**, University of Barcelona -Dept. of Astronomy & Meteorology- (Barcelona, Spain),
e-mail: jordi@mizar.am.ub.es

94. **Tytler, David**, University of California - San Diego, (La Jolla, California, USA)

95. **Vázquez, Roberto**, Instituto de Astrofísica de Andalucía (Granada, Spain),
e-mail: vazquez@iaa.es

96. **Voit, Mark**, Space Telescope Science Institute (Baltimore, Maryland, U.S.A.),
e-mail: voit@stsci.edu

97. **Volonté, Sergio**, European Space Agency -Headquarters- (Paris, France),
e-mail: svolonte%esa.bitnet@listserv.gmd.de

98. **Waelkens, Christoffel**, Institut voor Sterrenkunde (Leuven, Belgium),
e-mail: christoffel@ster.kuleuven.ac.be

99. **Walmsley, Malcolm**, Osservatorio di Arcetri (Firenze, Italy),
e-mail: walmsley@arcetri.astro.it

100. **Waters, Rens**, University of Amsterdam -Astronomical Institute- (Amsterdam, The Netherlands),
e-mail: rensw@astro.uva.nl

101. **Wesselius, Paul R.**, SRON (Groningen, The Netherlands),
e-mail: p.r.wesselius@sron.rug.nl

102. **Wooden, Diane**, NASA Ames Research Center (Moffett Field, California, U.S.A.),
e-mail: wooden@ssa1.arc.nasa.gov

103. **Woolf, Neville**, University of Arizona -Steward Observatory (Tucson, Arizona, U.S.A.),
e-mail: nwoolf@as.arizona.edu

104. **Yanagisawa, Masahisa**, University of Electro-Communications (Tokyo, Japan)
e-mail: yanagi@alfven.ee.uec.ac.jp

105. **Yun, Joao**, University of Lisbon -Physics Department- (Lisbon, Portugal),
e-mail: yun@astro.cc.fc.ul.pt

# PART I

# SCIENTIFIC CASE

# THE FORMATION OF PLANETARY SYSTEMS

A. P. BOSS
*Department of Terrestrial Magnetism, Carnegie Institution of Washington*
*5241 Broad Branch Road, NW, Washington, DC 20015-1305 USA*

**Abstract.** The orbital characteristics of the planets imply their formation in the rotating disk of gas and dust that earlier produced the sun. Observations of star-forming regions suggest that such protoplanetary disks are common around young stars, and hence that planetary systems are abundant in the galaxy. Our understanding of the star formation process is fairly robust, in large part because of our ability to study pre-collapse clouds, protostars, young stellar objects, and pre-main-sequence stars in nearby star-forming regions, and to revise and extend the theory of star formation accordingly. While we are achieving a similar understanding of the large scale structure of protoplanetary disks, the absence of examples of other newly-formed or mature planetary systems (prior to the discoveries of 1995-96) has greatly hampered our ability to test and improve upon theories of planet formation. The recent discovery of several giant planets and brown dwarf stars orbiting nearby stars marks the advent of a new era, *the era of the discovery of extrasolar planetary systems.* Current theories of the formation of terrestrial and giant planets and of brown dwarf stars are discussed in the light of these discoveries. The extremely small separation (0.05 AU) between 51 Peg and its $\sim 1/2$ Jupiter-mass companion implies that if 51 Peg B formed as a giant planet, then it must have migrated inward to its present location following its formation. Gravitational interaction with an exceptionally long-lived disk is the likely cause of this migration.

**Key words:** planet formation, brown dwarfs, giant planets, terrestrial planets

## 1. Introduction

Embarking on an ambitious search for extrasolar planetary systems would not be prudent without a reasonable chance of success. *Why do we expect planets to exist around other stars in our galaxy?* The reason is that we have strong evidence that the planets of our solar system formed as a part of the same process that led to the formation of the sun. Because stars like the sun are found throughout our galaxy, and because we can observe the formation of new generations of solar-type stars in nearby star-forming regions, we conclude that the same natural processes that led to the formation of our solar system some 4.6 billion years ago have most likely also led to the formation of planetary systems around many other stars in our galaxy.

The fact that the planets all revolve about the sun in the same direction as the sun rotates, on roughly circular orbits, and in a common orbital plane only slightly inclined to the sun's equatorial plane, led Laplace to hypothesize that the planets and sun formed from a single rotating, flattened cloud of gas. The Laplacian hypothesis has easily



*C. Eiroa et al. (eds.),*
*Infrared Space Interferometry: Astrophysics & the Study of Earth-Like Planets, 3–8.*

withstood the scientific advances of the intervening two centuries, and today still forms the underlying foundation of all research on solar system formation.

## 2. Star Formation

The recognition that the youngest stars are always found embedded within dense clouds of interstellar gas and dust largely proves the stellar half of Laplace's hypothesis. Astronomical observations of nearby (150 pc) star-forming regions such as Taurus-Auriga and Rho Ophiuchus have resulted in a sharply improved understanding of the star formation process. Numerous examples, or at least good candidates, now exist for nearly all phases of the star formation process – pre-collapse cloud cores, protostars surrounded by infalling molecular cloud material, young stellar objects driving strong bipolar outflows at the same time that accretion of gas is occurring, and optically visible, pre-main-sequence stars, still surrounded by gas and dust. These observations continue to guide and validate our theoretical understanding of the star-forming process.

Contemporary star formation begins with interstellar clouds of molecular hydrogen containing dust grains formed by previous generations of stellar evolution and envelope ejection. Molecular clouds contain "cloud cores", where the gas density is higher than that of the cloud as a whole, and these cloud cores are the site of subsequent star formation. The tendency of cloud cores to contract because of their self-gravity is nearly balanced by the resistance to contraction of the magnetic fields and turbulent motions inferred to be present by the large nonthermal line widths of the clouds. Magnetic fields slowly leak out of cloud cores, however, and dense cloud cores begin to contract at an ever-increasing rate, until supersonic velocities are obtained and the dynamic collapse phase is well underway. This collapse phase is halted once the center of the cloud becomes optically thick to infrared radiation and the cloud's temperature rises because of compressional heating. First an outer core forms, of size 10 AU, and soon thereafter a final protostellar core forms with a size several times the sun's radius. The gas and dust of the cloud core continue to fall supersonically inward and accrete onto the growing protostar. In a rotating cloud, a protostellar disk is formed, through which most of the cloud's gas must pass in order to be accreted by the central protostar. The cloud may form a binary or higher order stellar system at this point by fragmenting into two or more protostellar cores while dynamic collapse is still underway – as a consequence, stars are formed on highly eccentric orbits. The lowest mass stars thought to form from fragmentation have masses on the order of 0.01 $M_{\odot}$. Such stars, with masses less than the minimum of about 0.08 $M_{\odot}$ needed to sustain hydrogen fusion on the main sequence, are termed *brown dwarf stars.*

Initially, the disk mass is comparable to or greater than that of the protostar, and the disk evolves primarily through self-gravitating, spiral density waves. Such a protostellar disk would resemble the spiral galaxies that Laplace and William Herschel mistook two centuries ago as evidence for solar systems in the process of formation elsewhere in the universe. As the protostar's mass increases, gravitational instability of the disk becomes less important, and magnetic fields may become the primary agent of disk evolution. The protostar ejects bipolar flows from a very early stage, focused jets of high velocity gas that blow outward along the rotation (symmetry) axis of the disk in both directions. In combination with the eventual exhaustion of infalling matter, the energetic early stellar wind and radiation field halts the mass accretion process and clears away the remaining debris within $\sim 10^6$ yrs. The protostar becomes an optically visible, pre-main-sequence star, surrounded by a circumstellar disk radiating in the infrared.

## 3. Planet Formation

Our present understanding of planet formation is necessarily more purely based on theory than is our grasp of star formation, given the fact that until late 1995 we had only one example of a planetary system against which to test our theories. Our understanding of star formation would be similarly hampered if the only young star we could observe was T Tauri. Young stars like T Tauri often have several indicators of the presence of a circumstellar disk, such as excess infrared emission, and obscuration of the half of a bipolar jet that is hidden behind the disk. The masses inferred for these disks are often on the order of 0.01 to 0.1 $M_{\odot}$, which is the range inferred to have been necessary to produce our solar system. The astronomical evidence for these circumstellar disks is one of the best reasons that we believe planetary systems are not rare, and that the planetary half of Laplace's hypothesis is also correct.

Planetary formation starts with the growth of dust grains residing in the disk, where the space density of grains is high enough for appreciable mutual collisions to occur. Starting from sizes on the order of a micron or less, the grains grow through sticking following collisions to sizes on the order of 10 km. These planetesimals are massive enough to begin interacting with each other through gravity, and subsequent growth of the planetesimals occurs through a runaway accretion phase, where one planetesimal rapidly grows larger by accreting most of its smaller neighbors. After about $10^5$ yrs, the runaway planetesimals have grown to become about 2000 km in diameter, and are termed planetary embryos. The planetary embryos are formed on orbits that are initially nearly circular and coplanar. Long-range gravitational interactions between the planetary embryos results in the growth of the eccentricity of their orbits. During this time period, about $10^6$ to $10^7$ yrs after the formation process started, most of the gas and dust remaining in the disk is removed by the combination of stellar wind and radiation. The planetary embryos begin to have close encounters that increase their eccentricity, and they collide with each other to form the final planetary system. This final period of collisional growth requires about $10^8$ yrs in the inner solar system, and is punctuated by giant impacts, such as that believed to have led to the formation of the Earth's moon.

The collisional accretion process is known to be highly stochastic, with planets as massive as about twice the Earth's mass being a possibility in a system which could just as easily end up looking like our solar system (Wetherill, 1996). As long as Jupiter forms prior to this final phase, the asteroid region will be cleared of planetary embryos, and the resulting planets will be confined to the inner 2 AU from the sun; few will form much closer to the sun than 1/2 AU. If no Jupiter-like planet forms, then Earth-like planets may form out to 3 or 4 AU or even beyond. Increasing or decreasing the disk surface density has a proportionate effect on the masses of the resulting solid planets.

In the outer solar system, it is generally believed (Pollack, 1984) that planetary embryos must grow to sizes on the order of 10 Earth masses through runaway accretion, in order to accrete hydrogen and helium envelopes prior to removal of the remainder of the disk gas. Jupiter and Saturn apparently grew fast enough ($< 10^6$ yrs) to accrete significant gaseous envelopes, acquiring most of their mass by gravitational capture of gas from the disk. Uranus and Neptune grew too slowly ($\sim 10^8$ yrs) and retained little disk gas; their masses (10 to 15 Earth masses) are thought to represent the masses of the intermediate planetary embryos, formed largely of rock and ice. Our understanding of planet formation in the outer solar system is not as well developed as it is for the inner solar system – much depends on the invocation of runaway accretion to 10 Earth mass objects, an invocation that so far rests more on hope than calculation. In fact, in the one detailed calculation that has been performed (by G. W. Wetherill), the same physics that successfully reproduces the inner planets and their nearly circular orbits leads to massive planetary embryos moving on highly eccentric orbits in the outer solar system (Lissauer et al., 1995), a situation very unlike our solar system. An alternative idea for giant planet formation is the rapid formation of giant gaseous protoplanets (GGPP) through gravitational instability of the gaseous portion of the disk (Cameron, 1978). This process could form giant planets extremely quickly, as required, but may not be able to account for the ice/rock cores of Jupiter and Saturn. Considering that evidence for ice/rock cores inside *extrasolar* Jupiters may well be unobtainable, the GGPP mechanism might still be invoked to explain giant planets found elsewhere in our galaxy.

Stepping back, one can draw inferences about the location of giant planet formation from the need to have icy planetesimals in order to grow planetary embryos through collisional accumulation, or from the need to have a disk cold enough for GGPP formation. In the former case, this requires that the ice condensation point fall around 5 AU, in order to allow the giant planets to grow icy cores, yet prevent the inner planets from following suit. The calculated thermal structure of the solar nebula appears to be quite consistent with this requirement, and further suggests that the ice condensation point is determined more by the disk's properties than by the protostellar mass, implying that the ice condensation point should not be a strong function of the mass of the final main-sequence star (Boss, 1995). Thus we might expect to find giant planets formed at distances only slightly smaller than 5 AU, even for K and M dwarf stars much less massive than the sun, assuming that planets do not migrate significantly after their formation (see below).

The upper limit on the mass of a giant planet is not well understood, but masses on the order of 10 Jupiter masses are thought to be the minimum mass for brown dwarf stars (which do not have cores formed through a prior phase of collisional accumulation of icy planetesimals). The falling surface density of the disk dooms the accumulation process to failure in the outermost regions, where at most a population of comet to Pluto-sized bodies might form or be transported to following close encounters with the giant planets.

## 4. Extrasolar Planets and Brown Dwarfs

Within just a few months, the revolutionary discoveries of 1995-96 turned the field of extrasolar planet research from one governed primarily by theoretical concerns to one constrained by the hard facts of observational reality. The discovery by Mayor and Queloz (1995) of an object with a mass about 1/2 that of Jupiter orbiting the solar-type star 51 Pegasi launched the revolution. Nearly simultaneously, strong spectroscopic evidence for the existence of a brown dwarf star (Gliese 229 B) was also presented (Oppenheimer et al., 1995; Nakajima et al.,

1995). Marcy and Butler (1995) soon thereafter confirmed the reality of 51 Pegasi B, and announced two more candidates of their own, multiple-Jupiter-mass objects orbiting the solar-type stars 47 Ursae Majoris and 70 Virginis. While the former object appears to be a giant planet on a nearly circular orbit (like 51 Pegasi B), the latter appears to another brown dwarf star, given its high orbital eccentricity and what we understand about star and planet formation.

The biggest surprise about these discoveries was the location of 51 Pegasi B only 0.05 AU from its primary, a separation that cannot be primordial, but must result from inward orbital migration (Lin et al., 1996), probably through gravitational interactions with long-lived disk gas (Boss, 1996). Considering that the radial velocity technique used to discover 51 Pegasi B is most sensitive to short period orbits, perhaps the tiny separation is not all that astonishing. More importantly, these discoveries have confirmed the fundamental validity of our understanding of the planet formation process – other planetary systems really do exist. Laplace would have been pleased.

## References

Boss, A.P.: 1995, *Science* **267**, 360
Boss, A.P.: 1996, *BAAS* **27**, 1379
Cameron, A.G.W.: 1978, *Moon Planets* **18**, 5
Lin, D.N.C., Bodenheimer, P., Richardson, D.C.: 1996, *Nature*, in press
Lissauer, J.J., Pollack, J.B., Wetherill, G.W., Stevenson, D.J.: 1995, in Neptune and Triton, D.P. Cruikshank (Ed.), Univ. Arizona, p. 42
Marcy, G.W., Butler, R.P.: 1995, *BAAS* **27**, 1379
Mayor, M., Queloz, D.: 1995, *Nature* **378**, 355
Nakajima, T., Oppenheimer, B.R., Kulkarni, S.R., Golimowski, D.A., Matthews, K., Durrance, S.T.: 1995, *Nature* **378**, 463
Oppenheimer, B.R., Kulkarni, S.R., Matthews, K., Nakajima, T.: 1995, *Science* **270**, 1478
Pollack, J.B.: 1984, *ARAA* **22**, 389
Wetherill, G.W.: 1996, *Icarus* **119**, 219

# HIGH RESOLUTION SPECTROSCOPY OF VEGA-LIKE STARS

S.K. DUNKIN and M.J. BARLOW
*Dept. of Physics and Astronomy, University College London, Gower Street, London WC1E 6BT, England*

S.G. RYAN
*Anglo-Australian Observatory, PO box 296 Epping, NSW 2121 Australia*

**Abstract.** We have studied the photospheric abundances and circumstellar lines of a sample of Vega-like stars in an effort to improve the information available on this class of star.

King (1994) recently drew attention to the possible link between Vega-like stars and the photospheric metal-depleted class of A-stars, the $\lambda$ Bootis stars. In the 6 Vega-like stars studied here for depletions, none showed the pattern of underabundance observed in $\lambda$ Bootis stars. However, depletions of silicon and magnesium found in two of the sample could be linked to the presence of silicate dust grains in the circumstellar environment of those stars.

Circumstellar gas absorption lines have been positively identified in three stars from our sample. Individual cases show evidence either of high-velocity outflowing gas, variability in the circumstellar lines observed or evidence of circumstellar gas in excited lines of Fe II. No previous identification of circumstellar material has been made for two of the stars in question.

**Key words:** Vega-like stars, abundances, circumstellar lines

## 1. Introduction and Observations

During the course of the *IRAS* mission, excess flux above that expected for a black body was found for four bright main-sequence stars: Vega ($\alpha$ Lyr) (Aumann et al. 1984), $\alpha$ PsA (Fomalhaut), $\epsilon$ Eri and $\beta$ Pic. This excess emission has been attributed to circumstellar dust in the form of a disk, toroid or envelope, somewhat similar to that postulated for our own Solar System. These four stars have become known as the prototypes of the "Vega-like" phenomenon. Since then, searches of the IRAS database (e.g. Walker & Wolstencroft 1988 and references therein) have expanded the known membership of the class significantly.

Echelle spectra of 13 Vega-like stars selected mainly from the list of Walker & Wolstencroft (1988) were obtained from two sites: the Anglo-Australian Telescope and the Observatorio Astronomico Nacional, Mexico in 1993. The resolving power of the data was R($=\lambda/\Delta\lambda$)$\sim$44,000 and 15,000 respectively, with wavelength coverage extending from approximately 3300Å to 9000Å. Full details of the observations and sample can be found in Dunkin, Barlow & Ryan (1996a,b).



*C. Eiroa et al. (eds.),*
*Infrared Space Interferometry: Astrophysics & the Study of Earth-Like Planets,* 9–11.

## 2. Photospheric Abundances

The accretion of dust-depleted gas from the circumstellar environment onto the stellar surface has been proposed as an explanation of the $\lambda$ Bootis phenomenon (Venn & Lambert 1990). $\lambda$ Bootis stars are chemically peculiar A-stars, exhibiting underabundance in elements such as magnesium, calcium and iron. King (1994) investigated the possible link between the $\lambda$ Bootis phenomenon and Vega-like stars, with inconclusive results. We investigated such a link by carrying out an abundance analysis of a sample of A-type Vega-like stars.

Model spectra were produced using the LTE-spectrum synthesis program UCLSYN (Smith 1992), using atomic data from Kurucz (1995). In our full analysis of four Vega-like stars having low enough v sin $i$'s and partial analysis of two, we found no depletions similar to those found in $\lambda$ Bootis stars, or in the interstellar medium. We did find underabundances in two particular elements in two of our stars. HD 169142 has a depletion of 0.86 dex (relative to solar) of Si and 0.56 dex in Mg, whilst HD 139614 also has a Si depletion of 0.52 dex below solar. Silicon and magnesium are the two main constituents of silicate dust grains. Such grains are indeed present around HD 169142 (from the presence of the 18$\mu$m silicate feature, Sylvester et al 1996); HD 139614 has yet to be studied for the presence of this feature in its spectrum. However, the absence of depletion of other elements such as Fe, Ca and Ti is puzzling, as in the dust-depletion model these would be expected to be even more depleted than Mg and Si.

## 3. Circumstellar Gas

The presence of dust in the circumstellar environment of a Vega-like star should also imply the presence of gas in the region. This gas can be searched for by looking at their characteristic narrow absorption lines, easily distinguishable from the broad, shallow stellar lines upon which they are superimposed. Distinguishing between IS and CS lines is a difficult task, and often the CS lines can only be positively identified if they are variable, or at very high velocities relative to the star, or show narrow absorption in excited lines which are not present in interstellar sightlines.

We studied the lines of Ca K (3933.663Å) and Na D (5889.951Å and 5895.924Å) for evidence of narrow absorption lines. The majority of our sample had some evidence for these lines, but most could not be positively identified as circumstellar. We found that three stars had lines which were highly likely to be of circumstellar origin. 51 Oph has previously been studied (i.e. Lagrange-Henri et al. 1990) and, as it is also a shell star, was thought likely to have some component of this narrow line emanating from its circumstellar environment. We confirmed this by discovering the presence of narrow absorption components in excited lines of Fe II (4549 and 4583Å). These lines are not expected to appear in the interstellar medium, so must be originating from the circumstellar environment of 51 Oph. HD 35187 has also been previously studied by Grady et al. (1996), who classify it as a Herbig Ae star. Their spectrum of the Na doublet is distinctly different from ours, in that another narrow absorption line has appeared in our spectrum in the time between the two observations. This variability does not occur in the interstellar medium on these timescales, and we therefore ascribe a circumstellar origin to this variable narrow absorption line in HD 35187. HD 144432 is obviously an active star, judging by its prominent P Cygni profiles in the Na D and Ca K lines. In addition to two probably interstellar lines visible in the the Na D profile, a weaker, high velocity component is also seen. Its high velocity of approximately −84 km/s would suggest that its origin is circumstellar, rather than interstellar. This component is also visible in the Ca K line, after the observed profile has been divided through by a standard model for its spectral type.

## References


Aumann, H.H., Gillett, F.C., Beichman, C.A., de Jong, T., Houck, J.R., Low, F.J., Neugebauer, G., Walker, R.G., Wesselius, P.R.: 1984, *ApJ* **278**, L23

Dunkin, S.K., Barlow, M.J., Ryan, S.G.: 1996, submitted

Dunkin, S.K., Barlow, M.J., Ryan, S.G.: 1996, submitted

Grady, C.A. et al.: 1996, *A&A*, in press

King, J.R.: 1994, *MNRAS* **269**, 209

Kurucz, R.L.: 1995, ASP Conference Series **78**, 205

Lagrange-Henri, A.M., Ferlet, R., Vidal-Madjar, A., Beust, H., Gry, C., Lallement, R.: 1990, *A&AS* **85**, 1089

Smith, K.C.: 1992, Ph.D. Thesis, University of London, Chapter 5

Sylvester, R.J., Skinner, C.J., Barlow, M.J., Mannings, V.: 1996, *MNRAS* in press

Venn, K.A., Lambert, D. L.: 1990, *ApJ* **363**, 234

Walker, H.J., Wolstencroft, R.D.: 1988, *PASP* **100**, 1509

# INFRARED OBSERVATIONS OF PLANETARY ATMOSPHERES

T. ENCRENAZ
*DESPA, Observatoire de Paris, F-92195 Meudon.*

**Abstract.** Infrared spectra of planetary atmospheres show a large variety of features. They are dominated by the signatures of their most spectroscopically active constituents: $CO_2$ in the case of the terrestrial planets, and $CH_4$ in the case of the giant planets. Minor species are also detectable, in particular: CO and $H_2O$ in the terrestrial planets with, in addition, $O_3$ in the Earth atmosphere; $NH_3$ and $PH_3$ in Jupiter and Saturn; several hydrocarbons in all giant planets and Titan. In order to search for spectral signatures in exoplanets, the 6-15 $\mu$m range is especially appropriate. Assuming that all sources of noise (apart from the detector noise) can be completely removed, the required integration time for detecting an Earth-like planet at 5 parsec (assuming the performances of the ISO detectors) is about two weeks for a 5-m telescope (R=100, S/N=5). Improving the sensitivity of the detectors would reduce these integration times accordingly.

**Key words:** planetary atmospheres

## 1. Introduction

There are two classes of planets in the Solar system. In the vicinity of the Sun, the terrestrial planets (Mercury, Venus, the Earth and Mars) are characterized by a small size, a high density and a small number of satellites; apart from Mercury (too small and too hot to keep a permanent atmosphere), their atmospheres are primarily composed of $CO_2$ and $N_2$ (in the case of the Earth atmosphere, $CO_2$ has been trapped in $CaCO_3$ under the oceans and $O_2$ has appeared as a consequence of the development of life). At larger distances from the Sun, the giant planets (Jupiter, Saturn, Uranus and Neptune) are characterized by a large mass and volume, a low density, a large number of satellites and a system of rings; their atmospheres are dominated by hydrogen and helium. Pluto, the last planet, shows more similarities with the small bodies of the outer Solar system (outer satellites, outer asteroids, Trans-Neptunian objects).

The classification into two classes of planets can be simply understood as a result of the condensation sequence which took place in the protosolar nebula as it cooled down after its gravitational collapse into a disk. At small heliocentric distances (Rh< 2 AU), the temperature is such than only metals and silicates can condense; this material will form the cores of the terrestrial planets. At larger heliocentric distances,



*C. Eiroa et al. (eds.),*
*Infrared Space Interferometry: Astrophysics & the Study of Earth-Like Planets*, 13–23.

the temperature is low enough for most of the elements (C, N, O...) to condense into ices ($H_2O$, $CH_4$, $NH_3$, $CO_2$...). As the C, N, O elements are, after H and He, the most abundant in the Universe, this material allows to build large cores, whose mass is about ten times the terrestrial mass. This mass becomes sufficient to bind the surrounding gas gravitationally, so that the outer planets are able to accrete the surrounding subnebula, mostly composed of hydrogen and helium (Mizuno 1980). This leads to the formation of very massive, voluminous, low density planets. This simple scenario accounts well for the two classes of planets observed in the Solar system; we can expect to find the same classification in any stellar system which would have been formed according to the same scenario.

The chemical composition can be also simply understood, on the basis of thermochemical models of the primordial nebula. Under thermochemical equilibrium, the abundances of carbon-bearing and nitrogen-bearing molecules can be estimated according to the following reactions:

$$CH_4 + H_20 <-> CO + 3\ H_2$$
$$2\ NH_3 <-> N_2 + 3\ H_2$$

These reactions evolve toward the right-hand sides at high temperatures, and the left- hand side at low temperature. As a consequence, C and N are expected to be found in their reduced form in the giant planets: this explains the presence of $CH_4$ in all giant planets and $NH_3$ in Jupiter and Saturn (in the case of Neptune, the situation might be more complicated, as the presence of $N_2$ is not completely excluded presently). In contrast, in the case of the terrestrial planets, the formation of CO, $N_2$ and $H_2$ is favored; hydrogen escapes, and CO, in turn, reacts with $H_2O$ to form $CO_2$; this qualitatively accounts for the observed atmospheric composition of the terrestrial planets. The true story, however, is likely to be more complicated than this model; the rapid cooling of the nebula might lead to the kinetic inhibition of the CO and $N_2$ reduction (Lewis & Prinn 1980). The thermochemical equilibrium reactions provide no more than a qualitative explanation of the observed species; the same characteristics can be reasonably expected in any outer stellar system comparable to our own.

With respect to these general schemes, the detection of a planet at a very close distance (0.05 AU) from 51 Peg (Mayor & Queloz 1995) is puzzling. According to the above scenario, massive stars are not expected to form at such a small distance from the central star, because most of the mass would be in gaseous form and there would not be enough material to build a large core. The possible origin of this object is still a matter of debate. Concerning the composition of its atmosphere, we might expect to find $H_2O$; carbon, would be expected in the form of CO rather than $CH_4$.

## 2. Infrared spectra of planetary atmospheres

As any Solar-system object, planets are characterized by a two-component spectrum. When a solar photon is received by a planet, it can be either reflected or absorbed and converted into thermal heat, characterized by an effective temperature $T_e$. In the first case, the photon is observed in the solar reflected component which follows a blackbody continuum at solar temperature (T= 5770 K) peaking at 0.5 $\mu$m; in the second case, a thermal emission is observed, which roughly follows a blackbody curve at the $T_e$ temperature. $T_e$ ranges from about 50 K (Neptune, maximum at 60 $\mu$m) to 500 K (Mercury, maximum at 6 $\mu$m).

The effect of the planetary atmosphere is to introduce spectroscopic features on these two blackbody spectra. In the case of the reflected sunlight component, atmospheric signatures appear as absorption features in front of the solar continuum. From the depth of the lines, the integrated abundance of the absorber can be derived. In contrast, in the thermal component, the situation is more complicated. The infrared outgoing flux is mostly a function of the thermal atmospheric profile. As a consequence, molecular signatures can appear in emission or in absorption, according to the atmospheric region in which they are formed. In the case of the giant planets, if the lines are formed in the lower tropospheric part (where the temperature decreases as the altitude increases) the lines are in absorption; if they are formed in the upper stratospheric region (where the temperature increases again with altitude) the lines appear in emission. At some wavelengths, a combination of both effects can be observed; in this case, a spectral resolution higher than 100 is necessary to separate the different spectral signatures. From an analysis of the outgoing flux, using a radiative transfer model, it is possible to retrieve information on the thermal profile and on the vertical distributions of the atmospheric constituents.

## 3. The terrestrial planets

In spite of their similarities regarding their chemical composition, the atmospheres of Venus and Mars show two extreme cases, with surface

Table I. Chemical Atmospheric Composition of the Terrestrial Planets

| Molecule | Venus | Earth | Mass |
|---|---|---|---|
| $CO_2$ | 0.965 | 0.00035(1) | 0.953 |
| $N_2$ | 0.035 | 0.781 | 0.027 |
| $O_2$ | - | 0.209 | 0.0013 |
| $O_3$ | - | $< 10^{-6}$ | $< 10^{-6}$ |
| $H_2O$ | $3 \cdot 10^{-5}$ | $3 \cdot 10^{-3}(2)$ | $3 \cdot 10^{-4}(3)$ |
| $CO$ | $4 \cdot 10^{-5}$ | $10^{-5}(4)$ | $7 \cdot 10^{-4}$ |
| $SO_2$ | $2 \cdot 10^{-4}$ | - | - |
| $CH_4$ | - | $10^{-6}$ | - |
| Ar | $7 \cdot 10^{-5}$ | 0.009 | 0.016 |
| Ne | $10^{-5}$ | $1.8 \cdot 10^{-5}$ | $2.5 \cdot 10^{-6}$ |

(1) Trapped in $CaCO_3$; (2) in oceans;
(3) variable with season; (4) stratospheric.

pressures of 97 bars (Venus) and 0.007 bar (Mars), and surface temperatures of 730 K (Venus) and 230 K (Mars). With a surface pressure of 1 bar and a surface temperature of about 290 K, the Earth stands as an intermediate case. As a consequence, in spite of the similarities of the absorbers (mostly $CO_2$, CO and $H_2O$), there are large differences between the spectra produced by the hot and dense atmosphere of Venus, the cold and tenuous Martian atmosphere, and the oxygen-rich atmosphere of the Earth (see Hunten et al. 1983 and Kieffer et al. 1992 for reviews on Venus and Mars respectively). The atmospheric composition of the terrestrial planets is given in Table 1.

In the case of Venus, the spectrum is entirely dominated by $CO_2$. In the reflected sunlight component, minor species have been detected (CO, HCl, HF). The thermal component is observed beyond 7 $\mu$m, but also at lower wavelengths, in the spectral windows left between the $CO_2$ absorption bands, in some hot spots observable on the dark side. In this case, the radiation comes from the lowest atmospheric levels, where the temperature is high enough for the thermal radiation to be detectable, even at $\lambda = 1$ $\mu$m. In these windows, high-resolution ground-based spectroscopy has allowed the study of minor species of the lower atmosphere, such as CO, $H_2O$, HDO, $SO_2$ and OCS (Bezard et al. 1990, 1993; Fig.1). At longer wavelengths, the thermal emission corresponds to the cloud emission at about 130 K. Around 15 $\mu$m, the spectrum is dominated by the strong n2 band of $CO_2$, which provides a retrieval of the thermal atmospheric profile.

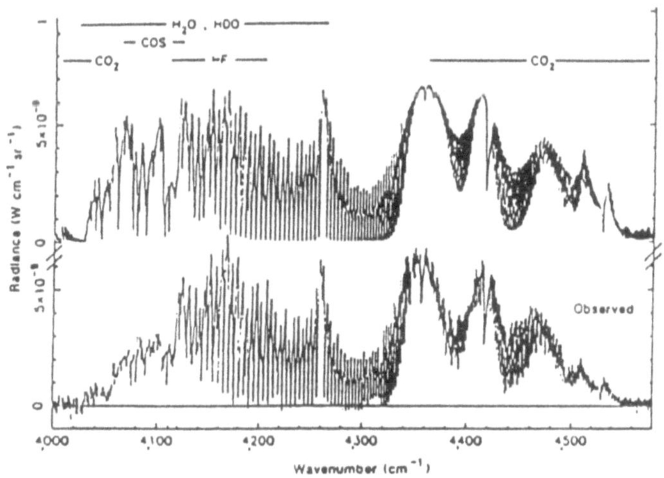


*Figure 1.* Thermal emission from the Venus night-side, in the 2.3 $\mu$m window. Top: synthetic model; bottom: observed spectrum. Several minor species ($H_2O$, HDO, HF, OCS are identified. The figure is taken from Bézard et al. (1990).

In the reflected spectrum of Mars, absorption features are all due to $CO_2$, plus weak signatures of $H_2O$ and CO. In the thermal regime, the continuum refers to the surface emission; at first approximation, this continuum is a blackbody at the surface temperature. Atmospheric signatures are the $CO_2$ n2 band at 15 $\mu$m and the broad silicate signature at 8-12 $\mu$m due to aerosols. An important caracteristic of the thermal spectrum of Mars is that it is very dependent upon the surface temperature. In the equatorial region, for instance, the surface temperature is warmer than the atmosphere and the $CO_2$ band is seen in absorption. In contrast, in the polar region covered with $CO_2$ and $H_2O$ frost, the surface temperature is colder than the atmosphere and the $CO_2$ band appears in emission (Hanel et al. 1972; Fig. 2).

The spectrum of the Earth, also dominated by $CO_2$ and $H_2O$ signatures, is again different from those of Mars and Venus. Absorption features due to other minor species ($O_3$, $N_2O$, $CH_4$) are also identified, in abundances which show strong variations over the location. The 4.3 $\mu$m and 15 $\mu$m bands of $CO_2$ can be used to retrieve the thermal profile from 0 to 50 km. At long wavelengths, the spectrum of the Earth atmosphere distinctively shows the $O_3$ signature at 9.6 $\mu$m (Hanel et al. 1992).

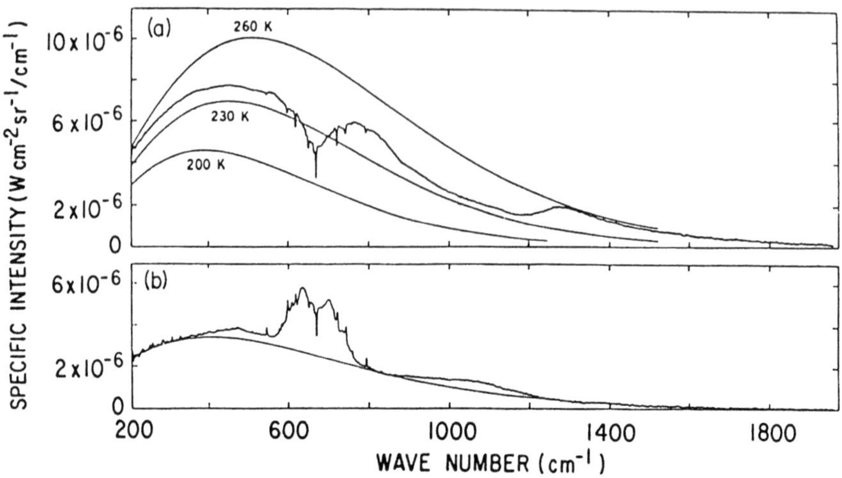


*Figure 2.* The infrared spectrum of two regions of Mars, recorded near the center and in the south polar cap. At the pole, the surface is colder than the atmosphere, so that the 15 $\mu$m $CO_2$ band is shown in emission. The figure is taken from Hanel et al. (1972).

The Earth is the only planet showing the ozone signature (Fig.3), and there is good indication that this signature could be considered as a reliable indicator of the development of life. Indeed, oxygen is present on Mars but at a level of 0.13% of the total pressure, and has not been detected spectroscopically. Rosenqvist & Chassefière (1995) have investigated the various processes which could lead to the formation of oxygen in a dense primitive atmosphere; their conclusion is that these processes could not generate more than a few millibars of $O_2$. As a consequence, they concluded that an abundance of oxygen larger than 10 mb in an exoplanet would be a serious indicator for life. $CH_4$, although present in the Earth' atmosphere, cannot be considered as an indicator for life, as methane is expected to be, generally, the preferred molecular form of carbon at relatively low temperatures, as illustrated by its presence in giant planets, Titan, and the brown dwarf Gl 229B (Oppenheimer et al. 1995).

## 4. The Giant Planets

The reflected spectrum of the giant planets is mostly dominated by methane absorption, with, in addition, a contribution duc to the rotational pressure-induced spectrum of hydrogen and, in the case of Jupiter

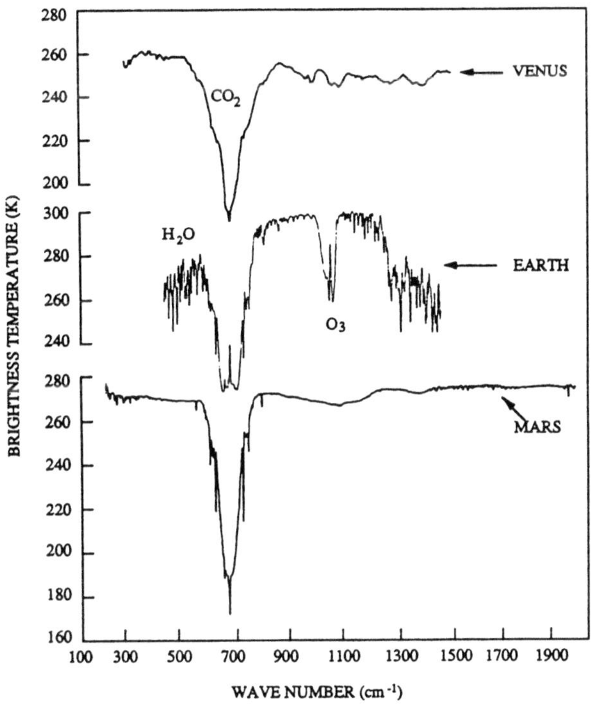


*Figure 3.* Fourier-transform spectra of Venus, the Earth and Mars in the infrared range, between 100 and 2000 cm$^{-1}$ (5 - 100 $\mu$m). The figure is taken from Hanel et al. (1992).

and Saturn, absorptions by $NH_3$ and $PH_3$. The near- infrared spectrum of Jupiter also shows thermal emission lines due to $H_3+$, especially in the auroral regions of the planet (Drossart et al. 1989). In the thermal component, the continuum is due to the pressure-induced spectrum of hydrogen. Various signatures are superimposed over this continuum, including emissions by methane and various hydrocarbons and, in the case of Jupiter and Saturn, absorptions by $NH_3$, $PH_3$, CO, $GeH_4$, $AsH_3$ (see Encrenaz et al. 1995 for a review). It has to be noted that CO and HCN have been detected in Neptune's stratosphere, which has been interpreted as the possible presence of $N_2$ in Neptune (Gautier et al. 1995). The atmospheric composition of the giant planets is given in Table 2.

The 5-50 $\mu$m spectrum of Jupiter and Saturn has been measured with the IRIS-Voyager spectrometer (Hanel et al. 1979, 1981; Fig.4) with a resolution of 4.3 cm$^{-1}$ (R=200 at 10 $\mu$m). The spectra show that a spectral resolution of at least 100 is indeed necessary for an

Table II. Chemical Atmospheric Composition of the Giant Planets

| Molecule | Jupiter | Saturn | Uranus | Neptune |
|---|---|---|---|---|
| $H_2$ | 1 | 1 | 1 | 1 |
| $HD$ | $6 \cdot 10^{-5}$ | $1 \cdot 10^{-4}$ | - | - |
| He | 0.11 | 0.03 | 0.18 | 0.23 |
| $CH_4$ | $2 \cdot 10^{-3}$ | $4 \cdot 10^{-3}$ | $2 \cdot 10^{-2}$ | $4 \cdot 10^{-2}$ |
| $CH_3D$ | $3 \cdot 10^{-7}$ | $4 \cdot 10^{-7}$ | $10^{-5}$ | $3 \cdot 10^{-5}$ |
| $NH_3$ | $3 \cdot 10^{-4}$ | $2 \cdot 10^{-4}$ | - | - |
| $C_2H_2$ | $3 \cdot 10^{-8}$(*) | $7 \cdot 10^{-8}$(*) | $10^{-7}$ | $1 - 9 \cdot 10^{-7}$ |
| $C_2H_6$ | $2 \cdot 10^{-6}$ | $3 \cdot 10^{-6}$ | - | $3 \cdot 10^{-6}$ |
| $C_3H_4$ | $3 \cdot 10^{-9}$(**) | - | - | - |
| $C_3H_8$ | $6 \cdot 10^{-7}$ | - | - | - |
| $C_2H_4$ | $7 \cdot 10^{-9}$(**) | - | - | - |
| $C_6H_6$ | $2 \cdot 10^{-6}$(**) | - | - | - |
| $H_2O$ | $10^{-6}$(*) | - | - | - |
| $CO$ | $1.5 \cdot 10^{-9}$(+) | $2 \cdot 10^{-9}$(+) | - | $6 \cdot 10^{-7}$(++) |
| $PH_3$ | $5 \cdot 10^{-7}$ | $1.5 \cdot 10^{-6}$ | - | - |
| $GeH_4$ | $7 \cdot 10^{-10}$ | $4 \cdot 10^{-10}$ | - | - |
| $AsH_3$ | $3 \cdot 10^{-10}$ | $2.4 \cdot 10^{-9}$ | - | - |
| $HCN$ | - | - | - | $3 \cdot 10^{-10}$(++) |
| $H_3^+$ | (++) | (++) | - | - |

Mixing ratios only indicate orders of magnitude
(*) variable; (**) in the north polar region;
(+) detected in the troposphere; (++) detected in the stratosphere.

unambiguous determination of the various emitters and absorbers.

Titan, Saturn's largest satellite, is the only satellite surrounded by a thick atmosphere. The major constituent is $N_2$, followed by $CH_4$. Its surface pressure is 1.5 bar, and its surface temperature is 94 K (see Encrenaz 1990 and Coustenis 1995 for a review).

Titan's reflected spectrum is entirely dominated by $CH_4$, with an absorption band due to CO (Lutz et al. 1983). Emission by methane is also visible in the thermal component, together with many emission features due to various hydrocarbons and nitriles, and also to $CO_2$. Many of them were discovered by the IRIS spectrometer of the Voyager mission (Hanel et al. 1981). The presence of $CO_2$ and CO could be the result of chemical reactions involving infalling meteoritic oxygen (Samuelson et al. 1983). The $N_2$ atmosphere, the surface pressure and the detection of complex molecules are arguments in favor of a possible similarity between Titan and the early Earth atmosphere. However, the very low temperature of Titan is likely to slow down any chemical

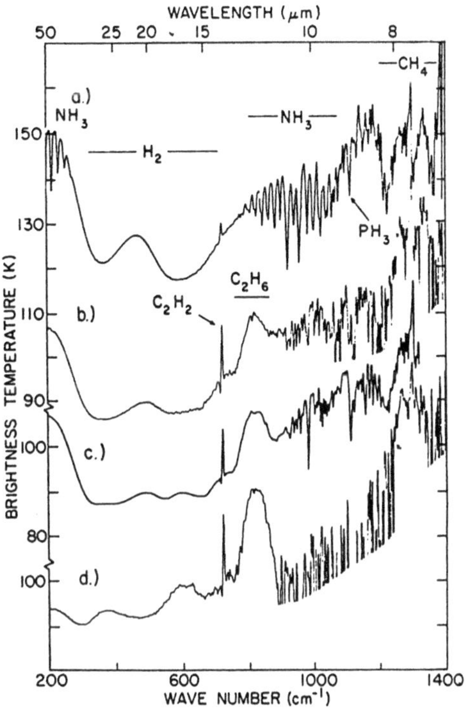


*Figure 4.* Spectra of Jupiter and Saturn in the 7-50 μm range, as observed by the IRIS-Voyager spectrometer: (a) Jupiter NEB; (b) Saturn mi-latitude; (3) synthetic spectrum of Saturn, mid-latitude; (d) Saturn south pole. The figure is taken from Hanel et al. (1981).

evolution considerably.

## 5. Detection of molecular species in exoplanets

What are the signatures which can be expected in the planets of external stellar systems? From the study of the planets we know, the following conclusions can be drawn. Terrestrial planets will exhibit $CO_2$, $H_2O$ and CO signatures; in addition, an Earth- like planet with life development will show the $O_3$ signature. Giant planets will show $CH_4$ and hydrocarbons signatures. Two spectral ranges are especially favorable to search for these species: the 2-5 μm range and the 6-15 μm range. As pointed out by Angel (1990), the 6-15 μm range is more favorable because this is the spectral range where the flux ratio Planet/Star is maximum. As mentioned above, a resolving power of 100 or more would be adequate for a reliable identification of the molecular species. A rough estimate of the expected signals is given below.

1) If we consider a Jupiter-size planet at 5 parsec, with an effective temperature of 1300 K, the expected signal is about 0.1 Jy at 2 $\mu$m, and 1 mJy at 100 $\mu$m. This case is comparable to the brown dwarf Gl 229 B recently discovered, for which a near-IR spectrum has been recorded (Oppenheimer et al. 1995). The observation of this type of object is indeed within the capabilities of a large ground-based telescope for the near- IR range, and within the capabilities of ISO in the 10 $\mu$m region.

2) If we consider a Jupiter-like planet at 5 parsec, with an effective temperature of 130 K ("true giant planet"), the maximum expected signal is about 30 mJy at 30 $\mu$m; the flux at 10 $\mu$m is expected to be 3 mJy.

3) If we consider an Earth-like planet at 5 parsec, with an effective temperature of 290 K ("true Earth-type planet"), the maximum signal is about 5 mJy at 10 $\mu$m.

For both cases 2 and 3, a search of molecular species requires spectroscopy in the 10 $\mu$m region with a spectral resolving power of about 100. Assuming instrumental performances comparable to the ones of the ISO spectrophotometer PHT-S, calculations show that, in the case of an Earth-like planet at 5 parsec, for a S/N of 5, an integrating time of about 1 week is required for a 5-m telescope, or about 1 hour for a 16-m telescope. Improving the detector sensitivity would allow to reduce the integration time. It should be noted that, in this calculation, all other sources of noise (star contamination, zodiacal light) are assumed to be completely removed.

## References

Angel, R.: 1990, in The next generation: A 10-16-meter UV-visible-IR successor to the Hubble Space Telescope, University of Arizona Press.

Bézard, B., de Bergh, C., Crisp, D., Maillard, J.P.: 1990, *Nature* **345**, 508

Bézard, B. et al.: 1993, *Geophys. Res. Letters* **20**, 1587

Coustenis, A.: 1995, *Earth Moon and Planets* **67**, 95

Encrenaz, Th.: 1990, *Rep. Prog. Physics* **53**, 793

Encrenaz, Th.: 1995, *Earth, Moon and Planets* **67**, 77

Gautier, D., Conrath, B., Owen, T., de Pater, I., Atreya, S.K.: 1995, in Neptune and Triton, D. Cruikshank et al. (Eds.), Univ. of Arizona Press.

Kieffer, H.H., Jakosky, B.M., Snyder, C.W., Matthews, M.S.: 1992, in Mars, University of Arizona Press.

Hanel, R. et al.: 1972, *Science* **175**, 305

Hanel, R. et al.: 1979, *Science* **204**, 972
Hanel, R. et al.: 1981, *Science* **212**, 192
Hanel, R. A., Conrath, B.J., Jennings, D.E., Samuelson, R.E.: 1992, in Exploration of the Solar system by infrared remote sensing, Cambridge University Press.
Hunten, D.M., Colin, L., Donahue, T.M., Moroz, V.I.: 1983, in Venus, University of Arizona Press.
Lewis, J.S., Prinn, R.G.: 1980, *ApJ* **238**, 357
Lutz, B.L., de Bergh, C., Owen, T.: 1983, *Science* **220**, 1374
Mayor, M., Queloz, D.: 1995, *Nature* **378**, 355
Mizuno, H.: 1980, *Prog. Theor. Phys.* **64**, 544
Oppenheimer, B.R., Kulkarni, S.R., Matthews, K., Nakajima, T.: 1995, *Science* **270**, 1478
Rosenqvist, J., Chassefière, E.: 1995, *Plan. Space Sci.* **43**, 3
Samuelson, R.E. et al.: 1983, *Journal of Geophys. Res.* **88**, 8709

## 6. Questions

*A. Léger* : Do you find evidence for non-thermodynamical equilibrium gaseous mixtures in planetary atmospheres? To which extent does it ruin the argument that out of equilibrium gaseous mixture is a strong criterion for life on the planet?
*T. Encrenaz* : In the case of giant planets, disequilibrium species have been observed as an effect of vertical motions (e.g. $PH_3$). However, this kind of effect cannot apply on a telluric planet with a surface. Rosenquist & Chassefière (1995) have investigated whether a large amount of $O_2$ could be built by any mechanism and concluded negatively. Thus, it seems that the presence of $O_2$ in large amounts ($> 10$ mb) is a serious indicator for life.

*D. Tytler* : In some cases, when exo-planet orbit is perpendicular to the plane of the sky (and exo-zodiacal disk is edge-on) we might get spectra of both day and night sides of planets at different points in planet's orbit. How important would this be, and what would you learn with both spectra?
*T. Encrenaz* : Observing day-side and night-side spectra of a planet is important to separate the reflected component from the thermal flux. This is specially important in the NIR below 5 $\mu$m (see the case of Venus in particular). Separating the two components properly is essential for a reliable retrieval of the T(P) structure and the vertical distribution of the minor atmospheric constituents. Note that about 7 $\mu$m the thermal component is likely to dominate in all cases.

# DETECTION OF PLANETS VIA MICROLENSING

R. FERLET
*Institut d'Astrophysique de Paris, CNRS*
*98bis Bd. Arago, 75014 Paris, France*

**Abstract.** We briefly present the gravitational microlensing phenomenon and the different on-going surveys. We then discuss its application to search for extrasolar planets.

**Key words:** microlensing, extrasolar planets

## 1. The microlensing phenomenon

The genesis of the gravitational lens concept goes all the way back to Newton (bending of light rays), then Soldner in Munich in 1804 (angular deflection by the Sun), the General Relativity of Einstein, Lodge and Eddington in 1919 (multiple images by the Sun) – also the year of a solar eclipse which spectacularly confirmed the Einstein theory – Chwolson in 1924 (ring of light from a star) and finally Einstein in 1936 who calculated the microlensing effect by a distant massive object in the vicinity of the line of sight toward a background light source. It is interesting to note that Einstein was very pessimistic about the possibility of detecting such an effect.

It is Paczyński (1986) who first proposed to apply this effect to stars in the Magellanic Clouds in searching for massive astrophysical compact halo objects (MACHOS). With masses below the thermonuclear threshold $\sim 0.08\ M_\odot$ but above the evaporation limit infered to be $\sim 10^{-7}\ M_\odot$, MACHOS provide a plausible form of the baryonic dark matter needed in the halo of the Milky Way.

In the special case of perfect alignment between the observer, an LMC star at 50 kpc and a 1 $M_\odot$ deflector located 10 kpc from the Sun, the observer should see the so-called Einstein ring with a radius $R_o \sim 1.4\ 10^9$ km. If the alignment is not perfect, two images are created. For a 1 $M_\odot$ deflector lying at a distance $R_o$ from the sight line in the plane perpendicular to it (or equivalently if the impact parameter equals 1), the angular separation is then 5 mas. Note that this effect should be called "millilensing" instead of microlensing in the case of non cosmological distances. One of the image (the faintest) is within the Einstein ring, while the other (the brightest) is outside in the opposite direction.



C. Eiroa et al. (eds.),
*Infrared Space Interferometry: Astrophysics & the Study of Earth-Like Planets*, 25–30.

Such an angular deflection is out of reach from present instrumental capabilities in the visible. However, since the surface brightness is conserved, there is also an apparent magnification of light: by a factor 1.34, or $\sim$ 0.3 mag, for an impact parameter unity. This effect is detectable, even for faint stars in the Magellanic Clouds! As the deflector is moving, the microlensing effect will show up as a smooth event of increase brightness of the source star, the resulting light curve having specific characteristics which should allow to distinguish it from intrinsically variable stars: symmetry in time, achromaticity, unicity and same magnitude before and after the event.

The probability of deflection at a given time is the solid angle fraction occupied by the Einstein circles of the deflectors. For a standard spherical halo of $\sim$ 5 $10^{11}$ $M_\odot$ entirely made of MACHOS, the "optical depth" $\tau$ for having a microlensing event with a magnification factor larger than 1.34 is of the order of 0.5 $10^{-6}$. Deflectors of larger masses will give rise to longer events but rarer, implying to survey millions of stars, while for smaller masses, events will be shorter but more numerous, implying to survey less stars but at higher frequencies.

## 2. The observational surveys

Two groups are presently monitoring LMC stars. The french EROS (Expérience de Recherche d'Objets Sombres) collaboration started in 1990 at ESO–La Silla in Chile with i) $5° \times 5°$ Schmidt plates in two colors every two nights or so which provide about 5 million usable stars, and ii) a mosaic of 16 CCDs at the focus of a 0.4 m telescope, providing more than $10^5$ stars in a $1° \times 0.4°$ field in the LMC bar, again in two colors and up to about 50 times per night. The Australian–US MACHO collaboration started in 1992 from Mount Stromlo in Australia at a 1.27 m telescope with an 8 CCDs camera covering a $0.5° \times 0.5°$ field in two colors simultaneously which provide about 9 million stars through multi-field operation.

In September 1993, the EROS team isolated two light curves compatible with a microlensing event (Aubourg et al. 1993), and the MACHO team one (Alcock et al. 1993). Since then, MACHO has identified six more events. All these events lasted several tens of days. The duration of a microlensing event is a function of the mass of the deflector, its transverse velocity, and its geometrical position with respect to the observer. The present mass estimates point to brown dwarfs. As a conclusion, less than 20% of a spherical standard halo consist of baryonic dark matter under the form of objects in the mass range $10^{-7}$–$10^{-2}$ $M_\odot$; at both ends, the fraction can still go up to 50%. Definitively, the Galactic halo is not dominated by such objects (Aubourg et al. 1995) and could apparently be consistent with expectation for a Universe whose primary component is cold (non-baryonic) dark matter.

A third group started in 1992 to search for microlensing events, but only toward the Galactic bulge through Baade's window. OGLE (Optical Gravitational Lensing Experiment), a polish–US collaboration, operates a 1 m telescope from Las Campanas in Chile. and also in 1993 published its first event (Udalski et al. 1993). They have now detected eighteen events. Also monitoring the bulge, the MACHO group has now identified near 100 candidates, well distributed over the HR diagram. Another french project, DUO (Disk Unseen Objects), started in 1993 a Schmidt survey of the Galactic bulge from ESO in Chile. Monitoring 15 million stars, they have detected 13 microlensing candidates (Alard 1996). To be complete (see e.g Ferlet & Maillard, 1996), one should mention two other projects (one french called AGAPE, one US) which are underway, both searching for microlensing events toward the Andromeda galaxy through pixel monitoring, and a New-Zealand–Japan microlensing project called MOA which is in the process of construction.

The observed microlensing optical depth toward the Galactic bulge seems to be far above the prediction $\sim 8.5\ 10^{-7}$ (Kiraga & Paczyński, 1994) computed with known stellar populations of the bulge. This might strengthen plenty of earlier evidences that the bulge is in fact barlike, elongated toward us (Paczyński et al. 1994; Zhao et al. 1995). Nevertheless, a more heavy ("maximum") disk could still be an alternative explanation.

The phenomenon of gravitational microlensing has now been convincingly proved to be a manageable important new tool for addressing a wide variety of scientific questions, from star formation to galaxy evolution and cosmology. It is clear also that all the projects will end up with huge unprecedented catalogues of all kinds of variable stars, of extreme astrophysical potentiality (see e.g Ferlet & Maillard, 1996). A further crucial step is to look for deviations from the standard microlensing light curve expected from lens point-mass. The first such events have already been reported. For instance, the MACHO group published the longest event (about 100 days; Alcock et al. 1995) ever yet recorded, during which the light curve has begun to be distorted by the effect of the motion of the Earth. This parallax event provides thus new information which enables to break the degeneracy of the problem (more parameters defining the system than observables).

Even if error bars on the mass estimates are still very large, most of the lenses toward the Galactic bulge are very likely low mass stars in the disk. Binary lens events are then expected, and OGLE, DUO and

MACHO have each observed such light curves which evidently differ from the standard one. Thus, OGLE event 7 (Udalski et al. 1994) seems to be due to a system with a mass ratio of 1.02. Although the number of parameters needed to completely constrain the system is large, the model for the DUO event 2 (Alard et al. 1995) has made a prediction which has been verified through very high spatial resolution CFHT observations. So there is some confidence in this model which implies a mass ratio of 0.33. Therefore, if the primary of the lens is a very low mass star, it is not impossible that the companion is a brown dwarf.

## 3. The detection of planets

If the lensing star is surrounded by planets, the well-defined smooth light curve may also be significantly altered. This method for discovering planetary systems in the Galactic disk has been first suggested by Mao and Paczyński (1991), and explored in detail by Gould and Loeb (1992) and more recently by e.g. Bennett and Rhie (1996). The Einstein ring of a Jupiter mass object orbiting a disk lens star at a few kpc from the Sun toward the Galactic bulge is roughly $10^{-3}$ that of a solar mass object. There will be perturbation of the point lens light curve when the Einstein ring of the planet will get near one of the unperturbed image positions due to the lens itself. This defines the lensing zone, whose $\pm$ 5% magnification contours in the source plane has roughly the shape of an elongated box for a Jupiter-like planet. The actual excess planetary signature will then depends on the trajectory of the source star across this boxlike structure whose width scales with the square root of the mass ratio.

The probability for detecting a planet given that the source passes within one Einstein radius of the parent star is about 0.17 for a Jupiter-like orbital radius of the planet, assuming that the minimum detectable perturbations are 5%. It means that a solar-like planetary system could be detected about 20% of the time, the largest contributor being by far Jupiter. The distribution of lensing time scales still for a Jupiter-mass planet peaks around one day, assuming the standard flat rotation curve for the Galactic disk. The lensing time decreases to few hours for an Earth-like planet. From the observations, it is straightforward to derive the planet/star mass ratio and the projected planet/star separation normalized to the Einstein radius. Without additional information or statistical sample, it will be difficult to directly estimate the mass of the planet and its distance to the parent star and to the observer.

Advances in data processing allow the OGLE, MACHO and EROS projects to detect in real time on-going microlensing events and therefore immediately alert the community, providing thus the opportunity to look in detail at the fine structure of the light curves, especially at caustic crossings. The PLANET (Probing Lensing Anomalies NETwork) group has performed a four weeks pilot campaign in June 1995 with three telescopes around the world and a 22 hours coverage per day (Sackett et al. 1996). A new campaign will last much longer in 1996 with a fourth telescope in order to monitor more events at a better accuracy and a better sampling time. According to the actual detection rate of microlensing events toward the bulge, about 200 events are expected per 6 months with the new EROS II upgraded experiment (1m telescope and 2 × 8 CCDs 2K × 2K) now operational in Chile.

The target (lens) stars of a programme searching for extrasolar planets are the most common, typically low mass ones everywhere within about 4–7 kpc from the Sun toward the bulge. The target planets are of all masses, including Earth's. The target orbital radii range from about 0.5 to 8 AU, depending on the mass of the parent star. If each target star has a planet in this lensing zone, with one 1m telescope able to do more than five measurements per hour at 1% photometry down to 20th mag stars, about 9 Jupiters, 3 Neptunes and 1 Earth are expected in 6 months.

However, an important parameter has to be taken into account, namely the finite size of the source stars which are mainly giants when monitoring the bulge. The center to limb darkening is a source of chromaticity in gravitational microlensing, and microlens imaging provides an opportunity to study the surfaces of normal red giants (Loeb & Sasselov, 1995). Another aspect of resolving the microlensed star is the possibility to break the degeneracy in the microlensing light curve by determining the angular radius of the Einstein ring and the proper motion of the lens. Recent detailed calculations by e.g. Bennett and Rhie (1996) reach roughly similar conclusions: more than 2% of all Earth mass planets in the lensing zone can be detected, if one requires a minimum deviation of 4%, which translates into about two Earth mass planet discovered per year (with a 24 hours coverage) if only a third of all lenses have such a planet.

Planetary lenses seem now well modeled and the finite source effects easily computed. Microlensing presently appears as one of the most powerful tool for detecting extrasolar planets on large scale, and in particular Earth mass ones, provided optical photometric follow-up of alerts are adequately operating. The use of near infrared photometry would be most important for measuring proper motion and thus break degeneracy. However, much theoretical work is still needed to quantify the measurability of the involved parameters and determine the optimal strategy for planetary searches.

## References


Alard, C., Mao, S., Guibert, J.: 1995, *A&A* **300**, L17

Alard, C.: 1996, in Variable Stars and the Astrophysical Returns of Microlensing Surveys, 12th IAP Colloquium. Editions Frontières, in press.

Alcock, C. et al.: 1993, *Nature* **365**, 621

Alcock, C. et al.: 1995, *ApJ* **454**, L125

Aubourg, E. et al.: 1993, *Nature* **365**, 623

Aubourg, E. et al.: 1995, *A&A* **301**, 1

Bennett, D., Rhie, H.S.: 1996, in Variable Stars and the Astrophysical Returns of Microlensing Surveys, 12th IAP Colloquium. Editions Frontières, in press.

Einstein, A.: 1936, *Science* **84**, 506

Ferlet, R., Maillard, J.P.: 1996, Editors of Variable Stars and the Astrophysical Returns of Microlensing Surveys, 12th IAP Colloquium. Editions Frontières, in press.

Gould, A., Loeb, A.: 1992, *ApJ* **396**, 104

Kiraga, M., Paczyński, B.: 1994, *ApJ* **430**, L101

Loeb, A., Sasselov, D.: 1995, *ApJ* **449**, L33

Mao S., Paczyński, B.: 1991, *ApJ* **374**, L37

Paczyński, B.: 1986, *ApJ* **304**, 1

Paczyński, B. et al.: 1994, *ApJ* **435**, L113

Sackett, P. et al.: 1996, in Variable Stars and the Astrophysical Returns of Microlensing Surveys, 12th IAP Colloquium. Editions Frontières, in press.

Udalski, A. et al.: 1993, *Acta Astron.* **43**, 289

Udalski, A. et al.: 1994, *ApJ* **426**, L69

Zhao, H.S., Spergel, D., Rich, M.: 1995. *ApJ* **440**, L13

# DETECTION OF PLANETARY SPECTRAL FEATURES THROUGH CIRCUMSTELLAR DUST - A MONTE CARLO SIMULATION

O. FISCHER and W. PFAU
*Astrophysikalisches Institut und Universitäts-Sternwarte, Schillergäßchen 2, 07745 Jena, Germany*

**Abstract.** One of the methods envisaged to detect planets outside of our solar system assumes that certain spectroscopic features at 10 and below 20 $\mu$m wavelength are typical of planets. Their detection would then be an unambiguous sign of the presence of Earth-like planets. In these spectral regions, there might be interference with other features being generally typical of circumstellar material: the silicate features of dust. In order to get an estimate of this effect, we used our 3D Monte Carlo radiation transport code to model the spectrum of a Keplerian disk with embedded artificial planets encircling a central star of solar luminosity and temperature. The disk structure and the grain model resemble real properties in protoplanetary disks. The model was calculated for a set of different optical depths at various positions of the observer with respect to the configuration, the detectability of the embedded planetary sources is discussed.

**Key words:** radiative transfer, monte carlo, circumstellar dust

## 1. Introduction

Spectroscopic observations should provide information about the existence and properties of Earth-like and Jupiter-like planets around nearby solar-type stars. Especially the infrared (IR) part of the extrasolar planet's spectrum is of interest for such observations because of the enhanced fluxes at these wavelengths. IR spectra of solar system planets (see Burke 1992) show prominent absorption and emission features due to atmospheric constituents. Besides pointing to the existence of an atmosphere, these features could help to find out whether atmospheres of extrasolar planets can be divided into $CO_2$-dominant and $CH_4$-dominant ones, as Earth-like and Jupiter-like planets are. The detection of ozone would be a strong indicator for Earth-like biological activity.

We limited our investigations to spectral features of atmospheric constituents (the $O_3$ $\nu_1$ and $\nu_3$ bands at $\lambda = 9.6$ $\mu$m and the $CO_2$ $\nu_2$ band at $\lambda = 15.0$ $\mu$m), which interfere with circumstellar dust features in the debris disk (the silicate bands at $\lambda = 10$ $\mu$m and 18 $\mu$m). For assumed extrasolar planets with $CO_2$ and $O_3$ bands these features would be washed out or shifted into the noise from radiation scattered or reemitted by dust particles in the line of sight to the planets.



*C. Eiroa et al. (eds.),*
*Infrared Space Interferometry: Astrophysics & the Study of Earth-Like Planets,* 31–36.

## 2. Model

To evaluate the maximum amount of dust which does not influence the atmospheric bands at different viewing angles, Monte Carlo simulation of radiative transfer was performed for a configuration consisting of an Earth-like and a Jupiter-like planet in a debris dust disk around a solar-type central star. Distances and emissions of the planets are assumed similar to those of Earth and Jupiter (planetary photon fluxes were taken from Burke 1992). For our investigation we performed monochromatic radiative transfer calculations at 11 wavelength points in the range $\lambda = 8.5 \ldots 11.0$ $\mu$m for the $O_3$ band and at 16 wavelength points in the range $\lambda = 12.6 \ldots 18.4$ $\mu$m for the $CO_2$ band.

### 2.1. Monte Carlo simulation of radiative transfer

The Monte Carlo simulation of radiative transfer aims at the construction of a stochastic model in which the expected values of the intensities have to be determined. This model consists of the random walk of weighted photons. The random path is determined by the random quantities: starting-point, free path length and propagation direction. We started weighted photons from four sources: the central star, the two planets, and the thermally emitting dust (photon fluxes of the sources are shown in Fig. 1). During the course of the random walk of a weighted photon, the initial intensity (weight) will change as the result of scattering events and absorption. After the last scattering (or without any scattering event), the weighted photon becomes "observable". In our 3D model, we can "observe" the configuration from 480 positions around it. Therefore, the full solid angle of all possible observer directions is divided into 480 equal sized solid angle intervals of $\pi/120$. The centre of each cone is determined by the angles $\theta_{obs}$ und $\phi_{obs}$. The photons are accumulated in circles of different size (0.173 ... 1.73 AU in diameter) centred on the lines of sight to the planets (comparable with beam sizes). More details about the Monte Carlo simulation of radiative transfer can be found in Fischer et al. (1994, 1995).

### 2.2. Circumstellar dust

Assumptions about the dust particles of the disk were made on the basis of observational hints from Vega phenomenon dust as it was detected around $\alpha$ Lyr, $\alpha$ PsA and, as a particularly clear case, for $\beta$ Pic.
The dust was modelled as orbiting around the central star in a Keplerian disk (Shakura & Sunyaev 1973). We assumed a disk with a radius $r_D = 100$ AU, a thickness $z_D = 10$ AU and the midplane density at the

disk edge in the range of $\rho_D = 10^{-18} \ldots 10^{-19}$ g/cm$^3$. Ground based and HST spectroscopic studies (Lagrange-Henri et al. 1989) of the $\beta$ Pictoris debris disk have provided hints of a central hole in it. We assumed the inner edge of the disk to be at a radius of 0.25 AU. Using these parameters we get total disk masses in the range of $10^{-8} \ldots 10^{-7}$ $M_\odot$ ($3 \cdot 10^{-5} \ldots 3 \cdot 10^{-4}$ $M_\oplus$ dust mass), corresponding to masses typical of disks around G-type main-sequence stars (see André 1994). The mid-plane density of the solar system interplanetary dust of $\approx 10^{-19}$ kg/m$^3$ at 1 AU (Leinert & Grün 1990) is about $10^2 \ldots 10^3$ times lower than in our less evolved disk. As a first attempt we modelled the emission from the mid-size grains by $T \propto r^{-0.4}$ (Artymowicz et al. 1989). For a distance of 1 AU we took $T = 150$ K.

The $\beta$ Pic spectrum indicates the existence of micron-sized silicate particles (Knacke et al. 1993). For the dust population in our debris disk model we used grains with radii $a = 0.1 \ldots 10$ $\mu$m and a distribution $n(a) \propto a^{-2}$. For the dust composition we applied the main components of the interplanetary dust particle (IDP) population proposed by Sandford (1988), which give quit well a fit to the $\beta$ Pic spectrum – 60% olivines ($MgFeSiO_4$) and 40% pyroxenes ($Mg_{0.8}Fe_{0.2}$ $SiO_3$). The optical constants are taken from Dorschner et al. (1995) who prepared silicate glasses as laboratory analogues of circumstellar silicate dust.

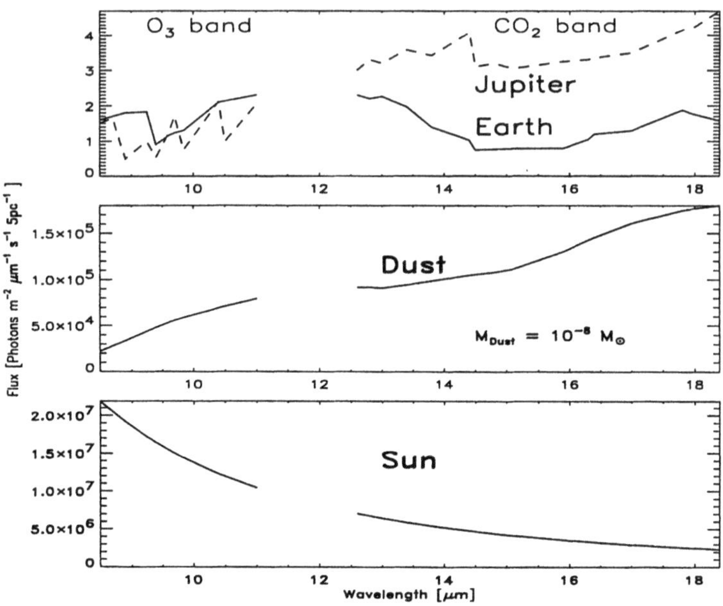


*Figure 1.* Model sources of emission.

## 3. Results

We performed the radiative transfer, isolating the effects of scattered stellar radiation and the thermally emitted (and scattered) dust radiation from the planet radiation. The last one was taken without transfer because it is negligibly influenced as compared to the effects called before.

Despite the large number of (weighted) photons started from the star (for each wavelength $2 \cdot 10^7$), the statistical error is still high because of the very low probability of scattering in the disks used for the calculations. The error increases (if the number of collected weighted photons decreases) going from Earth to Jupiter , from edge-on to pole-on positions, and from large to low beam sizes (see Fig. 2). For the interpretation we have to notice this "noise effect".

The effects of the beam size and observer's position with respect to the disk are shown in Fig. 2. Close to the pole-on view no absorption acts and the curves for scattered stellar light show the distribution of $C_{\mathrm{sca}}$ (see Fig. 1). For larger inclination angles $\theta_{\mathrm{Obs}}$ this light becomes increasingly absorbed. As a consequence, the minimum in the source emission close to 9 $\mu$m is partly filled and shifted to the right. In case of the thermal dust contribution, the curves for the Earth are similiar to the emission from the whole dust. The course of the Jupiter curves differs especially in the $O_3$ band range.

## 4. Discussion

Our results show that the discovery of atmospheric bands of planets through circumstellar dust of a debris disk strongly depends on the dust mass, i. e. the age of the central star. Beam size and the observer's position with respect to the disk also influence the contrast of the band features to the "background".

- In all cases, a band detection is most likely for pole-on candidates.
- It is important to know what additional emission component can mask the band features most – the thermal dust emission or scattering. For Earth-like planets photons from the thermal dust radiation dominate the whole wavelength range of investigation. Therefore, the $O_3$ $\nu_1$ and $\nu_3$ bands (9.6 $\mu$m) are easier to be detected than the $CO_2$ $\nu_2$ band (15 $\mu$m). For Jupiter-like planets more scattered stellar photons are collected at wavelengths shorter than about 14 $\mu$m.

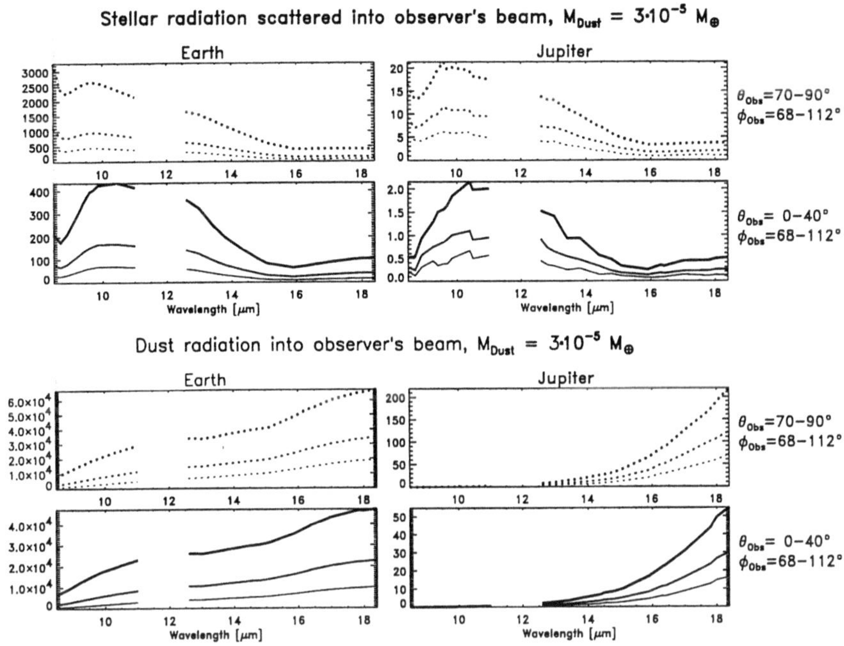


*Figure 2.* Photon flux portions of scattered stellar radiation and thermally dust radiation for different beam sizes (with decreasing line thickness: 1.7, 1.4, 1.0 AU) at different "observer positions" between the pole-on and edge-on view normally to the line connecting the central star with the planets: (dotted lines: $\theta_{Obs} = 0 \ldots 40^\circ, \phi_{Obs} = 68 \ldots 112^\circ$, solid lines: $\theta_{Obs} = 70 \ldots 90^\circ, \phi_{Obs} = 68 \ldots 112^\circ$). To get a better statistics, we summarized the results for different "observational directions".

- To get photon fluxes for the $O_3$ bands from an Earth-like planet comparable with the fluxes from the thermally emitting dust, beam sizes significantly smaller than 0.1 AU are necessary for the $3 \cdot 10^{-4}$ $M_\oplus$ dust disk. For the $3 \cdot 10^{-5}$ $M_\oplus$ dust disk a beam size of 0.1 AU would allow a detection for configurations close to the pole-on view (Fig. 2). In case of the Jupiter-like planet, beam sizes smaller than about 1 AU would enable a detection of atmospheric features in the range up to about 12 $\mu$m already for the $3 \cdot 10^{-4}$ $M_\oplus$ dust disk. However, edge-on disks demand (especially because of the scattered light) much smaller beams. A 1 AU beam is small enough to discover a Jupiter-like planet for any disk tilt (see Fig. 2) investigating the interesting spectral range around 15 $\mu$m (see Fig. 1) and possible features of $CH_4$-dominated atmospheres at shorter wavelengths ($CH_4$ $\nu_4$ at 7.7 $\mu$m, $C_2H_6$ $\nu_9$ at 12.2 $\mu$m, $C_2H_2$ $\nu_5$ at 13.7 $\mu$m).

- With the help of the curves in Fig. 2, we can derive how the band profiles change by the inference of the "background light". So we predict that the right wing of the $O_3$ band is raised by thermal dust emission. The same effect is expected for Jupiter's spectral features around 15 $\mu$m. Bands of Jupiter-like planets in the range up to about 12 $\mu$m are also influenced by the scattered light. Especially a possible $CH_4$ $\nu_4$ band at 7.7 $\mu$m can be distorted by the dust continuum.

- Taking into account a correlation between the age of the central star and the circumstellar dust mass (André 1994, $t^{-2}$ variation), Jupiter-like planets should be detectable by their spectral characteristics up to about $\lambda = 15$ $\mu$m using beam sizes $< 1$ AU around G-type stars with ages larger than $10^9$ yr. Earth-like planets can be discovered with 0.1 AU beams around G-type stars older than $5 \cdot 10^9$ yr.

In a forthcoming investigation we shall include additional details, such as a different dust population (particles with ice mantles) and gaps in the disk along the planetary orbits.

## References

André, P.: 1994, in Circumstellar dust disks and planet formation. Editions Frontieres, Ferlet, R., Vidal-Madjar, A. (Eds.), p. 115

Artymowicz, P., Burrows, C., Paresce, F.: 1989, *ApJ* **337**, 494

Artymowicz, P.: 1994, in Circumstellar dust disks and planet formation. Editions Frontieres, Ferlet, R., Vidal-Madjar, A. (Eds.), p. 47

Burke, B. F. (ed.): 1992, Toward Other Planetary Systems (TOPS), A Report by the Solar System Exploration Division, NASA print, Fig. 4.3, p. 78

Dorschner, J.,Begemann, B., Henning, Th., Jäger, C., Mutschke, H.: 1995, *A&A* **300**, 503

Fischer, O., Henning, Th., Yorke, H.: 1994, *A&A* **284**, 187

Fischer, O.: 1995, *Reviews in Modern Astronomy* **8**, 103

Knacke, R. F., Fajardo-Acosta, S. B., Telesco, C. M., Hackwell, J. A., Lynch, D. K., Russell, R. W.: 1993, *ApJ* **418**, 440

Lagrange-Henri, A. M., Beust, H., Vidal-Madjar, A., Ferlet, A.: 1989, *A&A* **215**, L5

Leinert, Ch., Grün, E.: 1990, in Physics of the inner heliosphere I, Schwenn, R., Marsch, E. (Eds.), p. 236

Sandford, S. A.: 1988, in Dust in the Universe, Bailey, M. E. & Williams, D. A. (eds.), Cambridge University Press, p. 193

Shakura, N. I., Sunyaev, R. A.: 1973, *A&A* **24**, 337

# EVOLUTION AND SPECTRA OF EXTRASOLAR GIANT PLANETS

T. GUILLOT
*Lunar and Planetary Laboratory, University of Arizona, Tucson, AZ 85721*

M.S. MARLEY
*Dept. of Astronomy, New Mexico State University, Las Cruces, NM 88003*

D. SAUMON *
*Lunar and Planetary Laboratory, University of Arizona, Tucson, AZ 85721*

R.S. FREEDMAN
*Sterling software, NASA Ames Research Center, Moffett Field, CA 94035*

**Abstract.** We present evolution models predicting both the radius and effective temperature of the newly discovered extrasolar giant planets (EGPs). Theoretical spectra of 47 UMa b and 70 Vir b, two moderately hot EGPs whose atmospheres are dominated by absorption of water and methane, are compared to similar calculations for the brown dwarf Gl 229 B, and observations of Jupiter. On this basis, we predict that a wide variety of EGPs should be bright in the 4-5 $\mu$m spectral region.

**Key words:** planetary systems – planets, evolution – planets, theoretical spectra – stars: individual (51 Peg, $\upsilon$ And, 55 Cnc, Lal 21185, 47 UMa, $\tau$ Boo, 70 Vir, HD 114762, Gl 229)

## 1. Introduction

With nine indirect detections of extrasolar giant planets around nearby stars, two direct detections of brown dwarf candidates, and the direct detection of Gl 229 B, the first confirmed brown dwarf, all that in less than a year, the research of the formerly elusive EGPs and brown dwarfs has entered a new era. Observations have broken into the realm of extrasolar non-stellar compact objects, and as time goes by, we will be able to detect fainter and fainter objects. Although the *direct* detection of Earth-like planets is a tremendous technological challenge, jovian extrasolar planets may soon fall under our scrutiny. In order to aid in the formulation of search strategies of giant planets around nearby stars, we detail the evolution and spectra of these objects.

* Permanent address: Dept. of Physics and Astronomy, Vanderbilt University, Nashville, TN 37235



*C. Eiroa et al. (eds.),*
*Infrared Space Interferometry: Astrophysics & the Study of Earth-Like Planets*, 37–46.

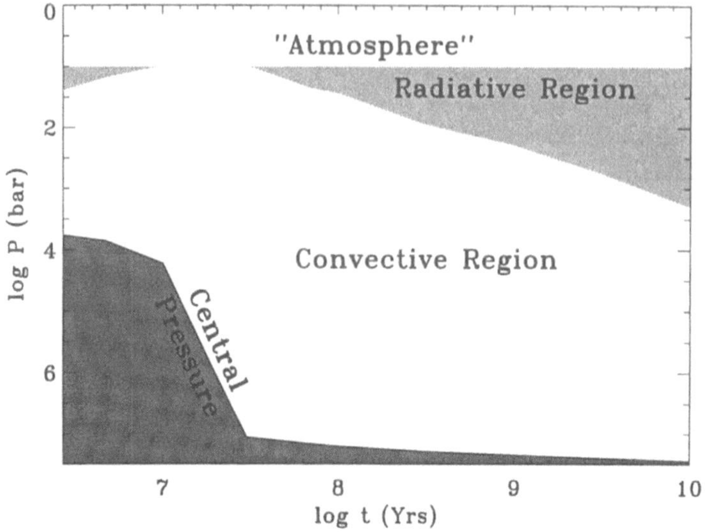


*Figure 1.* Evolution of the interior of a 1 $M_J$ planet at 0.05 AU from a star similar to 51 Peg a, assuming an albedo of 0.35 (see the corresponding evolution track in Guillot et al. 1996). The stellar heat flux is supposed to be absorbed at 10 bar. At slightly smaller pressures, the atmosphere is likely to be convective, because of the strong stellar heat flux and because of higher opacities in the infrared than at visible wavelengths. Deeper, a radiative region appears when the intrinsic luminosity of the planet becomes small enough.

## 2. Evolution of EGPs

Giant planets and brown dwarfs are believed to begin their life as extended objects, with radii of a few times Jupiter's radius ($R_J$). The subsequent loss of gravitational potential energy, added to the contribution due to stellar insolation accounts for most of the luminosity of these objects (in brown dwarfs, one part of the energy is also due to thermonuclear reactions; radioactivity is always negligible).

As shown by Guillot et al. (1996), two different phases in the evolution of EGPs can be distinguished: (1) when their internal luminosity (caused by the cooling and contraction of the planet) is not negligible compared to stellar insolation, they closely follow the Hayashi, fully-convective evolution track. (2) When $T_{\rm eff} \approx T_{\rm eq}$ ($T_{\rm eq}$ is the effective temperature of an illuminated planet with no intrinsic luminosity), the atmosphere cannot cool significantly, and the planet evolves away from the Hayashi track at almost constant effective temperature. Whereas most of the planet remains convective, a small part becomes radiative due to the low intrinsic luminosity. As the planet cools and contracts, the intrinsic luminosity still becomes smaller with the consequence that the radiative region steadily grows (Fig. 1). Note that this phenomenon occurs earlier for planets under significant stellar heating, but is otherwise unavoidable for any giant planet.

We extend the calculations of Guillot et al. (1996) to the newly discovered giant planets. Our results are summarized in Table 1. For these objects, the radius is a very weak function of the mass. Thus, we predict radii very close to that of Jupiter, independently of the planet considered. The largest deviations are for small masses ($M < M_J$), either when hot (the radius of 51 Peg b could be 45% larger than that of Jupiter), or when cold (the radius of Saturn, M=$0.34 M_J$, is about 15% smaller than that of Jupiter). Except for objects which suffer important insolation from a luminous star, the effective temperature of a planet is a strongly increasing function of its mass, for a given age.

We assume that the orbital radius of the planet is constant. Yet, our calculations show that hydrogen-helium gaseous planets could not form in close planetary orbits because tidal interactions with the star would disrupt the distended protoplanet and because the high temperatures would prevent refractory elements from condensing in the early nebulae. The period during which planets can spiral inward is short however (Lin, Bodenheimer & Richardson 1996), and has little effect on the subsequent evolution of the planet (see Guillot et al. 1995).

Another important assumption entering these calculations is that EGPs are mainly formed with hydrogen and helium. This is verified for Jupiter and Saturn for which the mass of the non-hydrogen and helium elements is less than 25% (9% in Jupiter) (Guillot et al. 1994). This is also verified for the recently discovered brown dwarf companion of Gl 229, whose observed spectrum is in agreement with theoretical spectra assuming solar composition (Marley et al. 1996a). Moreover, forming giant planets with mostly heavy elements (ices/rocks) seems difficult: a first scenario would involve a massive protostellar disk whose hydrogen and helium (but not the heavier elements) would be swept away by the stellar wind before proto-giant planets reach a critical mass of about $\sim 1-10 M_\oplus$. Thus, they would not be capable of capturing the surrounding hydrogen and helium (Mizuno 1980) but might continue to grow by accretion of cometary-like material. This scenario cannot apply to EGPs in close orbit because they would not be able to spiral inward (Lin, Bodenheimer & Richardson 1996) after the disk has been swept away. Another possibility is a stripping of the planet by evaporation. This evaporation requires very short star-planet separations (within 0.1 A.U. at least, probably less –see Guillot et al. 1996). In this case, and if the escape of hydrogen leads to a significant mass loss, the enhanced abundance of heavy elements would tend to decrease significantly the radius. Estimations of the decrease of planetary radius with increasing $Y$ or increasing core mass can be found in Saumon et al. (1996). For

Table I. Newly discovered giant planets and brown dwarfs*

| Star | $t$ (Gyr) | Pl. | $a$ (AU) | $M \sin i$ ($M_J$) | $T_{eq}$ (K) | $R$ ($R_J$) | $T_{eff}$ (K) |
|---|---|---|---|---|---|---|---|
| 51 Peg (G2V) | 6-10 | b | 0.05 | 0.46 | 1270 | 1.21-1.46 | 1190-1375 |
| $\upsilon$ And (F8V) | 1-6 | b | 0.06 | 0.6 | 1220 | 1.18-1.70 | 1140-1330 |
| 55 Cnc A (G8V) | 1-10 | b | 0.11 | 0.8 | 680 | 1.05-1.19 | 637-739 |
| | | c? | >6 | >5 | <100 | <1.13 | >172 |
| Lal 21185 (M2V) | 1-10 | b | 2.3 | 0.9 | 45 | 1.01-1.09 | 64-163 |
| | | c? | ~7 | >1? | ~26 | <1.1 | >63 |
| 47 UMa (G0V) | 4-8 | b | 2.1 | 2.39 | 185 | 1.06-1.09 | 182-250 |
| $\tau$ Boo A (F7V) | 1-6 | b | 0.046 | 3.87 | 1390 | 1.09-1.34 | 1300-1500 |
| 70 Vir (G4V) | 6-10 | b | 0.43 | 6.6 | 360 | 0.98-1.07 | 344-430 |
| HD 114762 (F9V) | 1-6 | b | 0.41 | 10 | 440 | 0.94-1.08 | 425-770 |
| Gl 229 (M1V) | 1-5 | B | >44 | 30-55 | <10 | 0.86-1.00 | 890-1030 |

* Some of these observations have yet to be confirmed. The spectral type of the stars (given in parenthesis), as well as their age and luminosities have been estimated from various sources including Lang K.R, *Astrophysical Data: Planets and Stars*, Springer-Verlag, New-York, (1992), the SIMBAD database (operated at CDS, Strasbourg, France), and other references given hereafter. Following the nomenclature used for binary stars, indirectly detected companions are designed with small letters, beginning with b, and directly detected ones with capital letters. As an example, the 55 Cnc system is believed to possess 2 stars (A and B) and 2 planets orbiting the brightest star (Ab and Ac). The equilibrium temperature is defined as in Guillot et al. (1996), and calculated using a Bond albedo of 0.35. This can be scaled for any albedo $A$ using the following formula: $T_{eq}(A) = T_{eq}(0.35)[(1 - A)/(1 - 0.35)]^{1/4}$. The quantities $R$ (radius at the 10 bar level) and $T_{eff}$ have been calculated as described in Guillot et al. (1996). The uncertainty on these values reflects possible variations of the Bond albedo of the planet ($0.1 \leq A \leq 0.5$), of $M$ due to $\sin i$ (except for 51 Peg b, and Lal 21185 b&c, we assumed $0.5 \leq \sin i \leq 1$) and of the age of the planet (as indicated). Note that these calculations assume pure hydrogen/helium composition (we use $Y = 0.30$, slightly higher than the solar value in order to mimic the presence of a solar abundance of heavy elements), and use the interpolated Saumon-Chabrier equation of state (Saumon, Chabrier & Van Horn 1995). In the case of Jupiter, this would tend to overestimate the radius by about 5%. (We used $R_J = 69800$ km and $M_J = 1.8986 \times 10^{30}$g).

*51 Peg*: see Mayor & Queloz (1995) (see also the note on $\upsilon$ And). The inclination of the rotation plane of the star (assumed equal to that of the planet) is constrained by spectroscopic measurements of the projected rotation rate of the star and chromospheric activity. The mass of 51 Peg b is thus such that $0.46 \leq M/M_J \leq 0.7$ [François et al. (1996)].

*$\upsilon$ And*, *55 Cnc A* and *$\tau$ Boo A*: see communications by G. Marcy and P. Butler (IAU colloquium 161, Capri, 1-5 July 1996) and http://cannon.sfsu.edu/~williams/planetsearch/planetsearch.html/.

*Lal 21185*: see Gatewood (1996). Astrometric measurements allow for a direct estimate of $M$, without the $\sin i$ ambiguity of Doppler measurements.

*47 UMa*: Butler & Marcy (1996).

*70 Vir*: Marcy & Butler (1996).

*HD 114762*: Latham et al. (1989) (see also the note on $\upsilon$ And).

*Gl 229*: Nakajima et al. (1995); estimations of age, mass ($M$ instead of $M \sin i$), radius and effective temperature from Marley et al. (1996).

example, a 10% decrease of the radius of a 1 $M_J$ planet requires an increase of $Y$ by $\sim 0.15$, or $M_{\rm core} \sim 50 M_\oplus$ instead of $M_{\rm core} = 0$. As a limiting case, a 1 $M_J$ rocky planet would have a radius of about 1/3 $R_J$ (Guillot et al. 1996).

## 3. Spectra of EGPs

First estimations of the spectra of EGPs based on reliable evolution calculations were presented by Burrows et al. (1995). Using the same method, Saumon et al. (1996) calculated the detectability of a wide variety of planetary systems, exploring most of the space of parameters (type of star, age, mass of the planet, orbital distance), and studying the consequences of different assumptions, especially concerning chemical composition and rotation. The major drawback of these studies is the assumption that EGPs radiate as blackbodies. In the case of Jupiter, significant departures from the blackbody emission are due to the absorption of methane and ammonia (see Saumon et al. 1996).

The recent detection of Gl 229 B, the first genuine brown dwarf (Nakajima et al. 1995) gave us an observational constraint for the calculation of more detailed theoretical spectra of these objects. Using a radiative transfer code originally developed for studying the atmospheres of the giant planets, Marley et al. (1996a) were able to theoretically reproduce the observed spectrum of Gl 229 B, to identify absorption of water and methane in the 1-2.5 $\mu$m region, and to estimate the effective temperature ($890 \leq T_{\rm eff} \leq 1030$ K), and mass ($30 \leq M/M_{\rm Jup} \leq 55$) of this object. It is important to notice that the presence of $H_2$ is crucial to reproduce the observations, although its effect on the spectrum is more subtle than that of $H_2O$ or $CH_4$. Very similar results were found by Tsuji et al. (1996) and Allard et al. (1996), using radiative transfer methods and data previously applied to stars. Differences arise from the fact that these authors assume that metal oxides and hydride, such as TiO, VO and FeH, are present in the atmosphere of the brown dwarf, whereas Marley et al. assume that they condense deep in the atmosphere. This last hypothesis is confirmed by the fact that no absorption lines of these molecules could be identified in the observed spectrum. However, a problem arises with the *visible* spectrum of Gl 229 B: water and methane are unlikely to absorb efficiently enough in this spectral region to explain the observations. But even though the presence of a large abundance of grains can be ruled out (Tsuji et al. 1996), clouds could still explain the discrepancy.

Using the method and data that successfully reproduced the spectra of Uranus ($T_{\rm eff}$=75 K –Marley et al. 1996b) to Gl 229 B ($T_{\rm eff}$=960 K

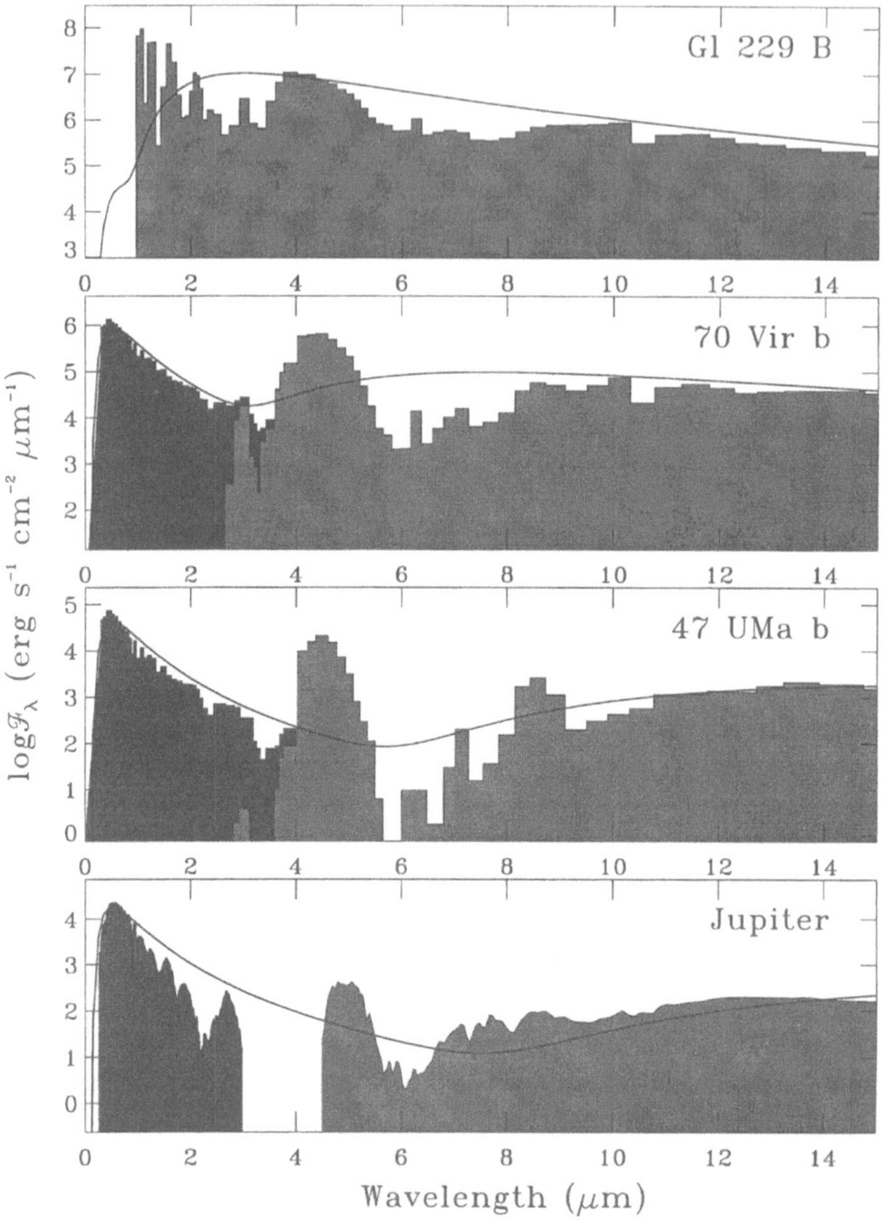


*Figure 2.* Theoretical spectra of Gl 229 B (Marley et al. 1996a) ($T_{\rm eff}$ = 960 K), 70 Vir b ($T_{\rm eff} \approx$ 380 K) and 47 UMa b ($T_{\rm eff} \approx$ 170 K), compared to an observed spectrum of Jupiter ($T_{\rm eff}$ = 124.4 K). The reflected part of the flux is shown in dark grey. The internal flux is shown in light grey. The line represents the flux of a blackbody at the same effective temperature, including the reflected stellar flux. All the fluxes are given at the surface of the object considered. Although most of these calculations are preliminary, the presence of a strong flux in the 4-5 $\mu$m region, is a robust conclusion, confirmed both by observations of Jupiter and Gl 229 B.

–Marley et al. 1996a), we calculated the first spectra of EGPs having effective temperatures between these two extrema. These spectra include both the thermal and the reflected components of the heat flux emitted by the planet. They are still preliminary because we neglect the condensation of molecular species in the atmosphere and the consequent decrease of their chemical abundance, because the absorbed stellar flux is treated separately from the thermal flux and includes absorption of methane, but not water, and because the spectral resolution is low. Note that contrary to our evolution calculations, the albedo of the planet is *consistently calculated* by taking into account scattering and absorption of the incoming stellar light in the atmosphere. Solar composition is assumed. Two examples for 70 Vir b and 47 UMa b are shown in Fig. 2 and compared to a low-resolution spectrum obtained for Gl 229 B, and an observed spectrum of Jupiter. We assumed that an arbitrary, purely scattering cloud of unit optical depth was present at the 1 bar level. Thus, contrary to Tsuji's assumption our cloud particles or grains are *not* well mixed in the atmosphere, but rather form discrete layers, as observed on Jupiter. Other models calculated with no cloud present (i.e. incoming light reflected only because of Rayleigh scattering by $H_2$) show qualitatively the same results, although absorption features tend to be deeper. While the albedos calculated with a cloud are close to $\sim$0.4, removing the cloud lowers it to about $\sim$0.2.

In all our calculations, we find that the emitted flux is significantly larger than the blackbody flux in the 4-5 $\mu$m region, which therefore emerges as a favorable spectral region for the detection of EGPs. This region is bright for very different effective temperatures (from 100 to 1000K) because both methane and water are the main absorbers in this region. Methane opacity has a minimum around 5 $\mu$m. Water opacity has a minimum around 3.5 $\mu$m. Other molecules also play a role, especially when water condenses out: the absorption of ammonia is minimum at 4.5 $\mu$m, which coincides with a peak of phosphine ($PH_3$) absorption. The small shift of the peak for Jupiter compared to the other planets (around 5 $\mu$m instead of 4.5 $\mu$m) is due to the fact that water condenses so that its abundance is small in a significant part of the atmosphere.

## 4. Conclusion

We have presented evolution calculations for a variety of newly discovered extrasolar planets, with the assumption that they are mainly formed with hydrogen and helium. These calculations show that with effective temperatures ranging from about 100 to 1500 K, these objects will be unlike what we see in our solar system, especially in the visible where cloud formation will play a significant role. Their infrared spectra are dominated by absorption of water, methane and ammonia. We calculated the first theoretical spectra of these objects, showing the presence of a favorable window in the 4 to 5 $\mu$m region. These predictions should be most valuable in aiding in the formulation of search strategies around nearby stars. They also provide tools for a direct interpretation of the future spectra (determination of effective temperature, gravity, and composition). Clearly, much more will be learned and more surprises are to be expected from future direct detections of extrasolar giant planets and brown dwarfs.

**Acknowledgements:** This study was made possible with the help of Adam Burrows, William Hubbard and Jonathan Lunine.

## References

Allard, F., Hauschildt, P.H, Baraffe, I., Chabrier, G.: 1996, *ApJ* **465**, L123
Burrows, A., Saumon, D., Guillot, T., Hubbard, W.B., Lunine, J.I.: 1995, *Nature* **375**, 299
Butler, R.P., Marcy, G.W.: 1996, *ApJ* **464**, L147
François, P., Spite, M., Gillet, D., Gonzalez, J.-F., Spite, F.: 1996, *A&A* **310**, L13
Gatewood, G.: 1996, *BAAS* **28**, 885
Guillot, T., Chabrier, G., Morel, P., Gautier, D.: 1994, *Icarus* **112**, 354
Guillot, T., Chabrier, G., Gautier, D, Morel, P.: 1995, *ApJ* **450**, 463
Guillot, T., Burrows, A., Hubbard, W.B., Lunine, J.I., Saumon, D.: 1996, *ApJ* **459**, L35
Lin, D.N.C., Bodenheimer, P., Richardson, D.C.: 1996, *Nature* **380**, 606
Latham, D.W., Mazeh, T., Stefanik, R.P., Mayor, M., Burki, G.: 1989, *Nature* **339**, 38
Marcy, G.W., Butler, R.P.: 1996, *ApJ* **464**, L15
Marley, M.S., Saumon, D., Guillot, T., Freedman, R.S., Hubbard, W.B., Burrows, A., Lunine, J.I.: 1996a, *Science* **272**, 1919
Marley, M.S., McKay, C.P., Pollack, J.B.: 1996b, *Icarus*, in press
Mayor, M., Queloz, D.: 1995, *Nature* **378**, 355
Mizuno, H.: 1980, *Prog. Theo. Phys.* **64**, 544
Nakajima, T., Oppenheimer, B.R., Kulkarni, S.R., Golimowski, D.A., Matthews, K., Durrance, S.T.: 1995 *Nature* **378**, 463
Saumon, D., Chabrier, G., Van Horn H.M.: 1995, *ApJS* **99**, 713
Saumon, D., Hubbard, W.B., Burrows, A., Guillot, T., Lunine, J.I., Chabrier, G.: 1996, *ApJ* **460**, 993
Tsuji, T., Ohnaka, K., Aoki, W., Nakajima, T.: 1996, *A&A* **308**, 29

## 5. Questions

*P. Saraceno* : Will the limit mass for deuterium burning change for planets close to stars?

*T. Guillot* : No. Deuterium burning always occur when the planet (or probably brown dwarf in fact) is hot, at early stages of its evolution for low mass objects. This is a period when the contribution due to stellar insolation is negligible. Therefore, the deuterium burning limit is about $13\,M_J$, independently of the orbital radius.

*A. Quirrenbach* : Which assumptions about abundances, and chemistry have gone into your models?
*T. Guillot* : We used solar abundance everywhere (Anders & Grevesse 1989). The chemical equilibrium between $NH_3$ and $N_2$ and between $CH_4$ and CO was taken into account. Unlike other studies (Tsuji et al., Allard et al.) we did *not* include chemical species that condense deep in the atmosphere, as Fe, Ti, V. These elements would form FeH, VO, TiO, whose features are not observed in the spectra of Gl 229 B. So far, we cannot tell whether *e.g.* the C/O ratio is solar: we need better opacity data for $CH_4$ and high resolution observations.

*T. Encrenaz* : It should be noted that the calculated thermal spectra of the newly-found planets are strongly function of the thermal profile. Depending upon the presence of a temperature inversion or its absence, a given signature, like $CH_4$, can appear in emission or in absorption.
*T. Guillot* : That's right. Our models take that into account. However, it is very difficult to estimate the magnitude of this thermal inversion, which depends on the presence of aerosols, cloud particles,...etc. Fortunately, we find that the gross features of the spectra (*e.g.* the $4\,\mu$m peak) do not depend that much on the thermal profile.

*D. Tytler* : You can infer the effective temperature from the spectrum. Can you tell other properties of a planet, such as its mass?
*T. Guillot* : In the case of Gl 229 B, the effective temperature is quite precisely constrained by our theoretical spectra. Unfortunately, this is not the case for the gravity. The effect of gravity on the spectra is more subtle than that of the effective temperature. Briefly, increasing the gravity tends to widen the peaks, but only slightly, so that the constraint on this quantity is weak. This is why we can't infer a precise mass for Gl 229 B. Since the age of the primary is larger than 1 Gyr, the mass is between 30 and 55 $M_J$.

*D. Tytler* : What type of data ($\lambda$, spectral resolution, signal to noise) is needed to get the mass more precisely?
*T. Guillot* : We can think of two kinds of observations: spectra at $10\,\mu$m would be very interesting because $NH_3$ strongly absorbs in this spectral region, and the $NH_3/N_2$ equilibrium happens to be quite sensitive to changes of the temperature/pressure of the atmosphere. Another

## 2. When is a Planet Habitable?

If the search for exo-life is restricted to chemistry-based ones, the concept of habitable planet results, as well as a possibility of detecting it from distance.

Macromolecules can react rapidly, only if they are in *solution* . Solid state reactions are too slow and large molecules cannot stand the temperatures required for them to be in the gas phase. As pointed out by A. Brack (1993), among the different possible solvents, liquid water is a special one. It has a very high dielectric constant, $\epsilon = 80$, that allows salt ionisation and, most important, has the capability of building H-bonds with dissolved molecules. The latter property allows very specific conformation of macromolecules by attraction of their hydrophilic groups ($OH$, $CO$, $COOH$...) and repulsion of the hydrophobic ones ($CH$, $CH_3$ , aliphatic chains...). These specific conformations of macromolecules are favourable to specific chemical reactions as "key-keyhole" ones which are valuable to build reproducible complex structures with a rich information content. Alternative solvents as liquid hydrocarbons, alcohols, liquid $NH_3$ are less favourable than water because of a lower $\epsilon$ or/and lack of H-bonding.

Chemically, $H_2O$ has some activity, i.e. hydrolysis, that can be important to select between different chemical pathways by destroying the products of some of them (Brack 1993). However, its activity remains moderate as opposed, for instance, to liquid $NH_3$ that attacks essentially any organic compounds.

Last but not least, T. Owen (1980) has pointed out that water is also appropriated for a subtle reason: it is indirectly *more resistant to UV* because some of its photolysis products, $O_2$ and $O_3$, protect it from further attack.

The definition of the habitable zone (HZ) around a star results: it is the region where a planet can sustain liquid water at its surface, at least during some time of its local year.

J.Kasting et al. (1993) have further proposed the concept of continuously habitable zone (CHZ) requiring that such a situation lasts long "enough" for life to evolve towards elaborated structures. This excludes massive stars which have a short life and possibly small stars where habitable planets are so close to them that their spinning and orbital rotations are phase locked by tidal forces, which may lead to unfavourable conditions for life appearance and development, although detailed climate models have to settle this point (Williams & Kasting 1996). The authors concluded that mid to early *K stars and G ones* ($0.78 M\odot < M\star < 1.1 M\odot$) were optimal candidates for having a CHZ.

# LIFE SIGNATURES ON EXOPLANETS

A. LÉGER
*Institut d'Astrophysique Spatiale, CNRS, bat 121*
*Université Paris-Sud - F-91405 Orsay, France*
*e-mail: leger@iaslab.ias.fr*

**Abstract.** We show that, within our present understanding of what could be alien life and knowledge of planetary physics, the *simultaneous* detection of $O_3$ and $H_2O$, in the atmosphere of a planet located in the *outer part* of the habitable zone of its star, is a good criterion for remote detection of primitive life.

**Key words:** exobiology, exoplanets, extrasolar planets

## 1. What could be Alien Life?

The search for a good tracer of alien life requires first a general definition of what is life. Biophysicists and biochemists agree to define a living being as a system that:
(1) contains information (negentropy),
(2) is able to replicate itself and
(3) undergoes few random changes in its information package that allow a Darwinian selection of the fittest.

Searching for such beings, several lightyears away from us, seems *a priori* a hopeless task. We show that, fortunately enough, this is not the case.

Trying to avoid being biased by the nature of life on Earth, one can consider different possibilities for the support of its information. Organised physical structures as those of the magneto-optic or semiconductor memories of our computers could be thought of, but the building of such systems by natural processes alone appears unlikely. A more attractive possibility is a chemical support: the coding by a sequence of chemical entities (chain cells) in a *linear macromolecule* which behaves as a message written with letters. A replication process can be imagined, provided that cells have homo or hetero affinities - of course, we have in mind the example of DNA but this can happen with quite different macromolecules ·



*C. Eiroa et al. (eds.),*
*Infrared Space Interferometry: Astrophysics & the Study of Earth-Like Planets,* 47–54.

## 2. When is a Planet Habitable?

If the search for exo-life is restricted to chemistry-based ones, the concept of habitable planet results, as well as a possibility of detecting it from distance.

Macromolecules can react rapidly, only if they are in *solution* . Solid state reactions are too slow and large molecules cannot stand the temperatures required for them to be in the gas phase. As pointed out by A. Brack (1993), among the different possible solvents, liquid water is a special one. It has a very high dielectric constant, $\epsilon = 80$, that allows salt ionisation and, most important, has the capability of building H-bonds with dissolved molecules. The latter property allows very specific conformation of macromolecules by attraction of their hydrophilic groups ($OH$, $CO$, $COOH$...) and repulsion of the hydrophobic ones ($CH$, $CH_3$ , aliphatic chains...). These specific conformations of macromolecules are favourable to specific chemical reactions as "key-keyhole" ones which are valuable to build reproducible complex structures with a rich information content. Alternative solvents as liquid hydrocarbons, alcohols, liquid $NH_3$ are less favourable than water because of a lower $\epsilon$ or/and lack of H-bonding.

Chemically, $H_2O$ has some activity, i.e. hydrolysis, that can be important to select between different chemical pathways by destroying the products of some of them (Brack 1993). However, its activity remains moderate as opposed, for instance, to liquid $NH_3$ that attacks essentially any organic compounds.

Last but not least, T. Owen (1980) has pointed out that water is also appropriated for a subtle reason: it is indirectly *more resistant to UV* because some of its photolysis products, $O_2$ and $O_3$, protect it from further attack.

The definition of the habitable zone (HZ) around a star results: it is the region where a planet can sustain liquid water at its surface, at least during some time of its local year.

J.Kasting et al. (1993) have further proposed the concept of continuously habitable zone (CHZ) requiring that such a situation lasts long "enough" for life to evolve towards elaborated structures. This excludes massive stars which have a short life and possibly small stars where habitable planets are so close to them that their spinning and orbital rotations are phase locked by tidal forces, which may lead to unfavourable conditions for life appearance and development, although detailed climate models have to settle this point (Williams & Kasting 1996). The authors concluded that mid to early *K stars and G ones* ($0.78M\odot < M\star < 1.1M\odot$) were optimal candidates for having a CHZ.

## 3. How to detect Life by Remote Sensing?

Operational tools for remote detection of life appear if one considers the possible nature of the macromolecules supporting the living being's information (Owen 1980). To allow replication, labile bonds are favourable and possibly necessary. Then, *carbon* based chemistry, organic chemistry, appears to be the most powerful. Carbon can be easily oxidised ($CO_2$) or reduced ($CH_4$) which allows a great variety of chemical species. This property is unique. Even Si, the element most similar to C in Mendeleev periodic table, makes much stronger bonds with O than with H and has a poorer chemistry.

This richness of C chemistry is confirmed by the study of a medium where physical conditions (pressure, temperature...) are quite different from those in the laboratory, the Interstellar one: among 112 species identified in 1996, 84 contain carbon and only 8 silicon.

Consequently, restricting our quest for life to carbon chemistry based ones is probably not too severe and T. Owen (1980) concluded that our present understanding of life requirements and planetary conditions provides us with some real support for *carbon-water* chauvinism.

It is thought that carbon was mostly fully oxidised ($CO_2$) in the primitive atmosphere of terrestrial planets, $10^8$ yrs after their formation (Gautier 1992). If a carbon based life has developed at a large scale on a planet, it will have required a large amount of organic molecules and made it from the only abundant raw material, $CO_2$. This implies the reduction of the latter by some process, as photosynthesis or the use of planet internal heat, according to a reaction scheme:

$$CO_2 + H_2O + energy \rightarrow (CH_2O) + O_2,$$

that releases *free oxygen.* As this gas is very reactive and oxidises iron or sulphurs contained in planet rocks, if not continuously replenished, it would disappear rapidly from an atmosphere, specially if plate tectonic and volcanism are present which bring fresh reducing material from the planet interior to its surface. On Earth, this would happen in 2 $10^7$ yrs (Broecker & Teng 1982).

*The* simultaneous and massive *presence of free $O_2$ ($P > 10$ mbar)* and *$H_2O$ in an exo-planet atmosphere appears to be a criterion for the presence of C based life.*

To qualify this criterion, it must be shown that *abiotic* processes cannot provide free oxygen as well. This is a key question that has already been addressed but requires additional studies in view of its primary importance. Photodissociation of $CO_2$ has been considered

(Rosenqvist & Chassefière 1995) but only a small amount of $O_2$ ($< 5$ mbar) can be produced under the most favourable conditions. A more plausible possibility is $H_2O$ photodissociation by the stellar UV, followed by H escape from the atmosphere. On terrestrial planets, i.e. rocky ones with temperature close to 300 K, this process is strongly reduced by the presence of a cold trap at the tropopause which blocks most of the ascending water vapour and makes it fall back to the planet surface as rain or snow before it has reached the UV-rich upper regions. Moreover, the escape of atomic H is limited by the diffusion in the dry stratosphere and depends upon interactions with the stellar wind (Chassefière 1996). As a result, 99.9999% of breathable oxygen on Earth is produced by photosynthesis and only 1 ppm comes from $H_2O$ photodissociation, although the latter species is quite abundant (Walker 1977).

However, an efficient production of abiotic $O_2$ seems possible when a planet undergoes a runaway greenhouse effect, as Venus did. The tropopause cold trap no longer exists and the H escape is favoured, the more as the planet has no magnetic field to protect it from its star's wind (Chassefière 1996). Such a process has been studied (Owen 1980, Kasting 1995). The authors'conclusion is that the detection of $O_2$ in the atmosphere of a planet located close to its star would be of little relevance but, if the planet is located at the *outer part* of its HZ, such a detection would be *meaningful.*

In addition, T.Owen has pointed out that if $O_2$ in a planetary atmosphere comes from $H_2O$ photolysis during a greenhouse runaway, it would have used up these species whose bands would no more be present in the planet's spectrum.

Therefore, the *simultaneously* presence of $O_2$ and $H_2O$ appears to be a *good criterion* for revealing a biological activity.

As early as 1975, J. Lovelock (1975) has proposed that a good criterion for life on an exo-planet is the presence of atmospheric gases very far from a thermodynamical equilibrium, e.g. $O_2$ and $CH_4$ , as they exist on Earth. This idea has also been used by C. Sagan et al. (1993), when showing that an observer, using Galileo probe instruments, would have revealed life on Earth. We do not think that, this is an "*absolute* " criterion either, because a planet is a thermodynamically open system and purely abiotic processes can be produced out of equilibrium products, e.g. OH and H as the result of $H_2O$ photodissociation. Accurate planetary models has to be worked out in order to *validate* or *falsify* this criterion as any other.

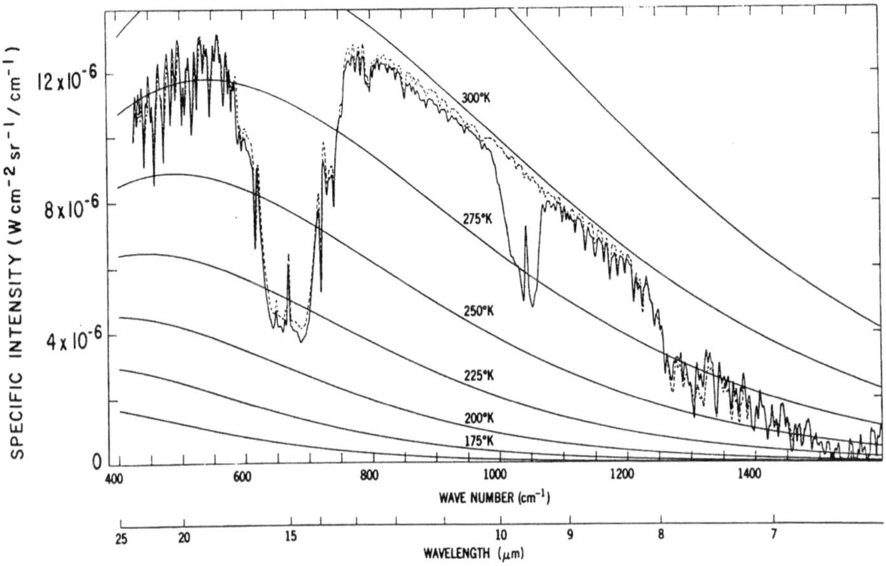


*Figure 1.* Earth emission as measured from space over the Pacific Ocean near Guam by Nimbus 4 (Kunde et al. 1974). The dotted curve is a model calculation that does not include the ozone component. Note the strong $CO_2$ band at $15\mu m$, which is present in the spectra of all solar terrestrial planets that have an atmosphere, the $O_3$ band at $9.6\mu m$ and the structure shorter than $8\mu m$ due to $H_2O$ . The latter two spectral features are specific to the Earth and reveal that the planet is habitable ($H_2O$) and that a photosynthetic activity takes place on it at a large scale ($O_3$). (Courtesy of Hanel et al. 1992).

## 4. Ozone, a better tracer than Oxygen

Considering the massive presence of $O_2$ and $H_2O$ as a good criterion to evidence exo-life, T. Owen (1980) has proposed to search for the spectroscopic signatures of the former gas in the visible (A and B-band at 760 and 720 nm, respectively). In practice, this proposal is not very useful because it faces a huge difficulty: in the visible, the star is much brighter than the planet ($5\ 10^9$ for Sun / Earth) and the spectrum of the latter would be extremely difficult to obtain.

R. Bracewell (1978) has pointed out that this contrast is more favourable in the IR ($7\ 10^6$ at $10\mu m$ for Sun/Earth) than in the visible and R. Angel, A. Cheng and N. Woolf (1984) showed that the mid IR region is of special interest because it contains *many* valuable spectroscopic signatures: $15\mu m$ ($CO_2$) whose presence or absence would indicate a major similarity or difference with the atmospheres of solar terrestrial planets, $6-8\mu m$ ($H_2O$) telling whether it is habitable or not and $9.6\mu$ m ($O_3$) whether $O_2$ is present, as the former species is made from the latter. The IR spectrum of the Earth illustrates the informative power of this part of the spectrum (Fig. 1). Observing the Solar terrestrial planet spectra would indicate that Venus, the Earth and Mars contain $CO_2$ but the Earth alone is habitable ($H_2O$) and inhabited ($O_3$).

In addition, the ozone $9.6\mu m$ band intensity is a very sensitive tracer of $O_2$. The dependence of the former gas abundance upon the latter one is not linear but logarithmic (Paetzold 1962, Kasting et al. 1985, Léger et al. 1993). On Earth, a small amount of $O_2$ (10 mbar) would lead to a $O_3$ abundance only 1.7 times smaller than the present one (Léger et al. 1993). The $O_2$ bands would be much more reduced (by a factor 20), making the detection of the latter species more difficult than that of the former. As an illustration, an outer observer of the Earth, with the capability of measuring these bands at their present level with a signal to noise ratio of 12, would have detected life on Earth (at 7 sigma) for 2 Gyr if observing the $O_3$ band, but only for 0.5 Gyr if observing the $O_2$ ones (Kasting 1993).

The *simultaneously* presence of $O_3$ and $CH_4$ seems to be an *excellent criterion* for the presence of life. $CH_4$ can be detected in this part of the IR spectrum (a tiny sharp band at 1300 $cm^{-1}$ in Fig. 1) but it requires a much higher spectral resolution (Res = 500) than the detection of $H_2O$ and $O_3$ (Res = 20). It should be considered in a next generation mission.

Space missions dedicated to this remote sensing of life indices have been described (Léger et al. 1996, Angel & Woolf 1996) and are presently under study by ESA ( DARWIN mission) and NASA (Planet Finder).

## References

Angel, J.R., Cheng, A.Y., Woolf, N.J.: 1984, *Nature* **322**, 341

Angel, J.R., Woolf, N.J.: 1996, *ApJ*, in press

Bracewell, R.N.: 1978, *Nature* **274**, 780

Brack, A.: 1993, *Origin of Life and Evol. of the Biosphere* **23**, 3

Broecker, W.F., Teng, T.H.: 1982, Tracers in the Sea, *Lamont-Boherty Geol. Observ.*, Columbia U. Press

Chassefière, E.: 1996a, *Icarus*, in press

Chassefière, E.: 1996b, *Icarus*, submitted

Gautier, D.: 1992: in Frontiers of Life, Trân Thanh Vân J.& K. et al (Eds.), Editions Frontiéres

Hanel, R.A., Conrath, B.J., Jennings, D.E., Samuelson, R.E.: 1992, Exploration of the Solar System by IR Remote Sensing, Cambridge Univ. Press

Kasting, J.F.: 1993, *Science* **259**, 920

Kasting, J.F.: 1995, *Planet. Space Sci.* **43**, 11

Kasting, J.F. et al.: 1985, *J. Geophys. Res.* **90**, 10497

Kasting, J.F., Whitmire, D.P., Reynolds, R.T.: 1993, *Icarus* **101**, 108
Kunde, V.G. et al.: 1974, *J. of Geoph. Research* **79**, 777
Léger, A., Mariotti, J.M., Mennesson, B., Ollivier, M., Puget, J.L., Rouan, D., Schneider, J.: 1996, *Icarus*, in press
Léger, A., Pirre, M., Marceau, F.: 1993, *A&A* **277**, 309
Lovelock, J.E.: 1975, *Proc. R. Soc. Lond. B* **189**, 167
Owen, T.: 1980, in Strategies for the Search for Life in the Universe, Papagiannis (Ed.), Reidel, p. 177
Paetzold, H.K.: 1962, *Mem. Soc. Roy. Sci., Liége* **7**, 452
Rosenqvist, J., Chassefière, E.: 1995, *Planet. Space Sci.* **43**, 3
Sagan, C., Thompson, W.R., Carlson, R., Gurnett, D., Hord, C.: 1993, *Nature* **365**, 715
Walker, J.C.:1977, Evolution of the Atmosphere, Macmillan, N.Y.
Williams, D.M., Kasting J.F.: 1996, Habitable Planets with High Obliquities, *LPS* **27**, 1437

## 5. Questions

*O. Fischer* : What do you think about the age of the central star which is probably correlated with the amount of debris dust around it, in connection with a possible detection of the $O_3$ bands?
*A. Léger* : Old stars, say older than 0.5 Gyr, are better candidates for life detection for two reasons : (i) they have probably less interplanetary dust which makes the spectroscopy of planets easier, (ii) possible life would have more time to spread all over the planet and make major atmospheric changes. As an example, $O_3$ would have been detectable on Earth by remote spectroscopy only for the last 2 Gyrs.

*O. Fischer* : Did you include the effect of beam size in your approximation of detectable life-giving planets?
*A. Léger* : Yes, the SNR calculations include the actual planet emission which fills only a very small fraction of the telescope FOV.

*V. Coudé du Foresto* : Ozone is a good tracer of $O_2$ but it also comes in much lower concentration in the atmosphere. Why does this not make it more difficult to detect?
*A. Léger* : To make a good detection, one needs a band with an optical depth, $\tau$, of the order of unity. $\tau$ depends upon the product of the abundance of the carrier by the transition oscillator strength. The latter is much stronger for the $9.6\mu m$ $O_3$ band (allowed transition) that for the 760 nm $O_2$ band (forbidden one). The net result is basically a compensation between the two factors and a band with $\tau \simeq 1$ for both species, although $O_2$ is many orders of magnitude more abundant than $O_3$.

*A. Penny* : If ozone is found, then what will be the best way to see if any of it is on land, rather than in the oceans? A simple way would be to measure the colour in the visible as the planet rotates. One would deconvolve the blue from the oceans, the brown from land without life, and the green for plant covered land. Are there other ways?
*A. Léger* : Ozone forms from $O_2$ which is uniformly distributed both in latitude and longitude as any long life gas in a planet atmosphere. As ozone needs stellar photons to form, there may be a variable abundance distribution, e.g. a day/night effect. This would not trace the place where the photosynthetic activity takes place.

Imaging a extra-solar planet in the visible, as you suggest, would be extremely costly because a minimum of 124x124 pixels is needed with the corresponding 4 orders of magnitude increase in the telescope collecting area if we want the same SNR. In addition, in the visible we loose 3 orders of magnitude in the planet/star contrast, with respect to the situation in the IR. Personally, I think that improving spectroscopy, e.g. increasing the *resolving power* by a factor of 100 would cost much less (100 times) in term of collecting area and would be *extremely informative* giving access to trace gases as $CH_4$, whose presence simultaneously with that of $O_2$ is considered as a sharp criterion for evidencing life (out of thermodynamical equilibrium system) or possibly CFC - like gases which, as far as we presently know, cannot be produced by natural process and would indicate a technological activity. Clearly, there is room for future "Super-DARWIN" missions.

Presently, I do not see a way to distinguish whether life is present in oceans or on continents.

# MOLECULAR GAS PRODUCTION IN THE $\beta$ PICTORIS DISK

R. LISEAU and P. ARTYMOWICZ
*Stockholm Observatory, S–133 36 Saltsjöbaden, Sweden*
*e-mail: rene@astro.su.se and pawel@astro.su.se*

**Abstract.** Based on CO (2–1) observations we find that the rate of steady state production of gaseous CO in the inner 400 AU of the $\beta$ Pic disk is extremely low. Model calculations suggest the SiO production rate due to grain-grain collisions to be higher by orders of magnitude. However, because of insufficient angular resolution the predicted SiO line emission escaped detection. A new generation of interferometers is required to test the Keplerian disk hypothesis for $\beta$ Pictoris.

**Key words:** $\beta$ Pic, stellar evolution, circumstellar matter, mm–radio lines

## 1. Introduction

The circumstellar disk around the nearby star $\beta$ Pic is unique among other similar objects (Vega-type stars), because of both its exceptionally large IR excess in the IRAS bands, and its resolved images in the visible and near infrared. Dust grains, probably dominated by bright, anhydrous silicates are spread out to distances $r \sim 1000$ AU from $\beta$ Pic (60″) in what looks like a thin disk seen nearly edge-on.

### 1.1. Previous 1.3 mm Observations of the Dust

The lower limit on the total mass is comparable to the mass of the Earth, whereas the upper limit of about $10^3\,M_\oplus$ is less reliable (Table 1). A similar lower limit on the mass of the *dust alone* in the disk was obtained by Chini et al. (1991) from observations of $\beta$ Pic in the dust continuum at 1.3 mm with the 15 m SEST (24″ beam = 400 AU). The dust temperature is of order 100 K. For equal amounts of hydrogen gas and dust, i.e. for $R_{\rm H}\,(\equiv m^{\rm H}_{\rm gas}/m_{\rm dust}) = 1.0$, one finds from the model of $\beta$ Pic by Chini et al. average values $n({\rm H_2}) \geq 3\,10^5\,{\rm cm^{-3}}$ and $N({\rm H_2}) \geq 3.6\,10^{21}\,{\rm cm^{-2}}$ for the hydrogen volume and column density respectively.

### 1.2. 1.3 mm Observations of the Gas: CO ($J$=2–1)

The above result indicates that excitation of the lower rotational levels of CO would be in LTE, since $n({\rm H_2}) \gg n_{\rm crit}$, where the critical density



C. Eiroa et al. (eds.),
*Infrared Space Interferometry: Astrophysics & the Study of Earth-Like Planets*, 55–57.

for the (2–1) line is about $2\,10^4\,\mathrm{cm}^{-3}$. Assuming the gas and the dust to be at the same temperature, $T_{\mathrm{kin}} \sim 100\,\mathrm{K}$, the above $H_2$ column density leads to the prediction of the intensity of the CO(2–1) line from the $\beta$ Pic disk, viz. $I(\mathrm{CO}_{2-1}) > 20\,\mathrm{K\,km\,s^{-1}}$, taking into account the fractional beam filling of the Chini et al. source ($8.4\,10^{-2}$) and using the interstellar medium value of the CO abundance with respect to $H_2$, $X(\mathrm{CO})_{\mathrm{ISM}} = 8\,10^{-5}$. From SEST observations (23″ beam) we find $I(\mathrm{CO}_{2-1}) < 0.058\,\mathrm{K\,km\,s^{-1}}$, implying that the LTE column density of CO molecules $N(\mathrm{CO}) < 10^{15}\,\mathrm{cm}^{-2}$ ($f_{\mathrm{beam}} = 0.084$). Therefore, in the inner 400 AU of the $\beta$ Pic disk $\frac{N(\mathrm{CO})/N(\mathrm{H_2})}{m_{\mathrm{gas}}/m_{\mathrm{dust}}} < 0.003\ \frac{X(\mathrm{CO})_{\mathrm{ISM}}}{R_{\mathrm{H}}}$ and we conclude that any steady-state production of CO gas in the disk around $\beta$ Pic would proceed at only a minute rate. However, the above $N(\mathrm{H_2})$ is a gross overestimate and we conclude further that $R_{\mathrm{H}} \ll 1$, i.e. that the $\beta$ Pic disk is virtually devoid of hydrogen gas.

## 2. SiO Gas Production from Grain-Grain Collisions

We study the process of dust grain collisions and evaporation in the $\beta$ Pic system. Mostly shattered and subsequently blown out of the system by radiation pressure, the grains are also partially converted to $SiO_2$, SiO, Si and other gas species in grain-grain collisions occurring at relative velocities exceeding $12\,\mathrm{km\,s^{-1}}$. This process and other, less efficient, sources produce a tenuous gaseous atmosphere of the particulate disk. The gas-phase chemistry of silicon is driven by the photochemical cascade from $SiO_2$, through SiO, to Si and $Si^+$, and the gas-grain processes. In addition, synthesis of SiO from Si and $O_2$ may be important.

The equilibrium gas production, chemistry, and recondensation on grains in the inner part of the $\beta$ Pic disk is computed in a closed-box model (with no radial flow of gas) including a moderately large chemical reaction network. For given density and temperature profiles, the mass of SiO gas is obtained (Table 1).

### 2.1. SiO Gas Observations: SiO($v$=0, $J$=2–1)

In the lower rotational transitions of SiO, these models predict optical depths much larger than unity. The radiative transfer through the disk has been treated in a Sobolev approximation, assuming Keplerian rotation of the disk. For chemical models with $p = 1$ and 0 (Table 1), the intensity of the $\mathrm{SiO}_{v=0}(2\text{–}1)$ line from the $\beta$ Pic disk is predicted to be $I(\mathrm{SiO}_{2-1}) \sim 2600 - 4300\,\mathrm{K\,km\,s^{-1}}$. From SEST observations (60″ beam) we find the $1\,\sigma$ upper limit $I(\mathrm{SiO}_{2-1}) < 0.0013\,\mathrm{K\,km\,s^{-1}}$.

The beam dilution factor of the model SiO source at $\beta$ Pic is $f_{\rm beam} \sim 7.7\,10^{-6}$, so that the models predict a line intensity about $15-25$ times the observed upper limit.

This result could indicate the presence of mass flows in the $\beta$ Pic disk. However, the theoretical line, being relatively broad, is of low signal in each resolution element of the observations. Further, given the considerable latitude of parametric values in both the chemical and the radiation transfer models, the apparently indicated 'disagreement' is as yet not conclusive. Further progress requires observations at significantly higher spatial resolution. At long wavelengths, observations at the subarcsec level will have to await the advent of a new generation of interferometers.

Table I. Parameters for the $\beta$ Pic System and Modeled SiO Gas Abundances

| | | | |
|---|---|---|---|
| Star | | | |
| | Distance | $d$ (pc) | 16.4 |
| | Radial Velocity | $v_{\rm hel}$ (km s$^{-1}$) | $+21 \pm 3$, (LSR: $+2.3$) |
| | Spectral Type | | AV 5 |
| | Effective Temperature | $T_{\rm eff}$ (K) | 8 200 |
| | Mass | $M_\star$ ($M_\odot$) | 1.8 |
| | Radius | $R_\star$ ($R_\odot$) | 1.7 |
| Disk | | | |
| | Mass | $M_{\rm disk}$ ($M_\oplus$) | $\gtrsim 1$, $\lesssim 10^3$ |
| | Radius | $R_{\rm disk}$ (AU) | $10^3$ |
| | Inclination | $i$ (°) | 3 ('nearly edge-on') |
| | Scale Height | $h(r)/r$ | $4\,10^{-2}$ (opening $\sim 2°$) |
| | Temperature Distribution | $T(r)$ (K) | $100\,(r/20\,{\rm AU})^{-0.5}$ |
| | Optical Depth (10 $\mu$m) | $\tau_{\rm Silicate}(r)$ | $10^{-3}\,(r/20\,{\rm AU})^p$ |
| | Outer Evaporation Radius | $r_{\rm ev}$ (AU) | 14.5 (14.5)[a] |
| | Grain Mass ($r < r_{\rm ev}$) | $M_{\rm gr}$ (g) | $3.0\,10^{23}$ ($2.1\,10^{23}$)[a] |
| | SiO Mass ($r < r_{\rm ev}$) | $M_{\rm SiO}$ (g) | $9.7\,10^{22}$ ($3.1\,10^{22}$)[a] |
| | SiO Volume Density | $\log n({\rm SiO})$ (cm$^{-3}$)[b] | 3.4 (3.9)[a] |
| | SiO Column Density | $N_{\rm SiO}$ (cm$^{-2}$) | $2\,10^{17}$ ($7\,10^{17}$)[a] |

Notes to the table: [a] For the exponents of the adopted radial distribution of the optical depth in the 10$\mu$m silicate feature, viz. $p = 0$ and $p = 1$ (in parenthesis).
[b] Average volume density at the characteristic evoporation radius, $0.5\,r_{\rm ev}$.

## References

Chini, R., Krügel, E., Shustov, B., Tutukov, A., Kreysa, E.: 1991, *A&A* **252**, 220

# PLANETS IN CM DRACONIS: A MULTI-SITE PHOTOMETRIC SEARCH

E.L. MARTÍN and H. DEEG
*Instituto de Astrofísica de Canarias*
*38200, La Laguna, Tenerife, Spain*

M. CHEVRETON and J. SCHNEIDER
*Observatoire de Meudon, 92195, Meudon Cedex, France*

L. DOYLE and J. JENKINS
*SETI Institute, MS 245-3, NASA Ames Research Center, CA 94035, USA*

E. PALAIOLOGOU
*University of Crete, P.O. box 1527, Heraklion, Crete, Greece*

W. LEE
*Korean Astronomical Observatory, Whamdong Yuseonggu, Daejon 305-348, Korea*

**Abstract.** We present preliminary results of the international collaboration "TEP" aimed at monitoring photometrically the eclipsing binary CM Draconis. Planetary companions with more than ~3 Earth radii should produce detectable occultations at the accuracy of our measurements. The observations started in spring 1994 and are still going on. Eight observatories have participated so far: Haute Provence (France), Lick (USA), La Palma (Spain), Rochester (USA), Skinakas (Greece), Taejon (Korea), Teide (Spain) and Wise (Israel). We have obtained a homogeneous dataset by using CCD detectors with standard broad-band filters (V and R) and maintaining always common reference stars between all the observatories. So far, we have obtained about 500 hours of effective integration time on CM Dra, distributed in ~ 16000 CCD frames. We have analysed about 40% of the 94–95 database, and have found encouraging, albeit inconclusive, results.

**Key words:** extrasolar planets, eclipsing binaries

## 1. Introduction

The TEP (Transits of Extrasolar Planets) project is a current effort with a realistic chance to detect Earth-sized planets. It is a photometric method, which depends on the continuous monitoring of a small, main sequence, eclipsing binary. Planets around short period binary systems are thought, from angular momentum considerations, to orbit in a plane very close to that of the binary (Schneider and Chevreton 1990, AA, 232, 251). The transit of a planet across the binary components will then produce a characteristic brightness variation, which we attempt to detect (Deeg et al. 1996, AA Tr., in press).

The most suitable system for these observations is CM Dra, a well known (Lacy, 1977, ApJ, 218, 444) system with dM components. CM



*C. Eiroa et al. (eds.),*
*Infrared Space Interferometry: Astrophysics & the Study of Earth-Like Planets,* 59–61.

Dra is unique in its small size ($R_*$=0.25, 0.23 $R_\odot$), low temperature (~3100 K), nearly edge-on inclination of 89.8 degress, and proximity of 16.9 pc. These properties are advantageous for several reasons: Due to the small size of CM Dra (total surface area: 12% of the sun), the transit of an Earth-sized planet across one binary component would cause a brightness drop of 0.1%; a Neptune-sized body would cause a drop of 1.2%. These drops are an order of magnitude larger than the transits of similar planets across stars of solar size. Second, in systems of low luminosities and temperatures it is expected, that terrestrial planets may form in a region much closer to the central star. If planets comparable to those of the solar system have formed around CM Dra at distances of similar equilibrium temperature, then the orbital period of a "Venus-equivalent" is 11 days, of an "Mars-equivalent" it is 33 days and for a "Jupiter" it would be 7 months. The nearly edge-on inclination of CM-Dra will lead to the near certain occurrence of an inner-planet transit in a continuous observational campaign of a few weeks, and to a larger than 95% probability of detection in a non-continous campaign lasting 3-4 months. Non-detection after several months of observation would also be of astrophysical interest, as the absence of planets around this system can then be concluded with high certainty.

Of photometric concern may be the occurrence of starspots, which are expected on the low mass components of CM Dra. They can be readily separated from planetary transits, as they are tidally locked to the binary rotation, and starspot cycles are much longer than the duration of a transit. CM Dra has rarely flares (only two in our database). The project is also expected to produce significant spin-offs in the knowledge about flare rates and surface granulation on low mass binaries and should lead to a much better characterization of the orbital elements of CM Dra.

## 2. Observations and Results

To obtain photometric observations of CM Dra with the required precision, the TEP working group was formed. A preliminary observational campaign of CM Dra was performed in May and June 1994 with sites in France, the Canary Islands, and the USA. A total of about 50 observatory-nights data were collected. At all sites, CCD cameras were employed as they allow the simultaneous measurement of CM Dra and several comparison stars with a high duty cycle and allow the adcquisition of new data every few minutes, needed to trace the characteristic light curves produced by a transiting planet.

Transit signals of planets with 2.6 Earth radii (halfway in area between Earth and Neptune), for example, would produce a drop in brightness of about 0.6%, a precision in photometry we have achieved in our 1994-1995 observations using 1-meter class telescopes. For the detection of smaller planets (an Earth-sized planet would cause a drop of 0.08%), whose signals are hidden in noise, cross-correlation techniques have been developed (Jenkins et al. 1996, Icarus, in press): A model of the eclipsing binary convolved with all possible planetary transit signatures can be used as a matched filter to detect planetary transit signals below the observational noise level. Sufficient continous observational coverage of CM Dra will allow detections or non-detections with high confidence, and provide the first bona fide evidence on the formation of terrestrial-sized planets around another main-sequence star system.

In addition, jovian-mass planets may also be detected by causing a periodic shift in the epoch of the binary eclipse minima due to the motion of the binary-giant planet barycenter (Doyle et al. 1996, in preparation). A Jupiter mass planet at 5.2 AU would cause a shift of 11 sec and this should be easily detectable in the photometric data. As CM Dra is the lowest mass eclipsing binary known, the project is also expected to produce significant spin-offs in the knowledge about flare rates and surface granulation of low mass stars, and will lead to a better characterization of the orbital elements of CM Dra. Of particular interest is an improved determination of the radii, leading to a determination of the specific brightness of the system.

A few preliminary results have been obtained from analysis of about 40% of our database. We note here the following: i.) a delay $\geq$100 seconds in the timing of our eclipses with respect to the prediction of Lacy (1977), ii.) several dips in the light-curves which are in some cases compatible with planetary occultation expectations, iii.) a very low flare occurrence.

In 1996, observations are scheduled in Greece, France, Israel, Spain and USA. The TEP collaboration is open to newcomers that would like to join forces with us and contribute to a more extensive coverage of CM Dra. The coordinates of CM Dra make it a suitable target from northern hemisphere observatories: RA: 16.34.21.4 Dec:+57.09.35.5 (Eq 2000.0, Epoch 1992.28). The steps needed to obtain photometric data from observations of CM Dra taken for the TEP project can be obtained by emailing Hans Deeg (hdeeg@iac.es) or Eduardo Martin (ege@iac.es). An automatic reduction procedure is available within the IRAF environment.

# EPICURUS WAS RIGHT: OTHER WORLDS EXIST!

*A year rich in discoveries.*

M. MAYOR and D. QUELOZ
*Geneva Observatory*
*CH-1290 Sauverny*
*Switzerland*

**Abstract.** The discovery of the first planet orbiting a solar type star acted on all groups working in that field as a tremendeous stimulation. Nine months after the discovery of 51 Peg B, five planetary companions have been detected by radial velocity techniques. The unexpected orbital elements of most of these jovian planets give a first evidence of the diversity of planetary systems which is still far to be fully understood.

## 1. From July 1995 to March 1996

More than 2000 years ago the Greek philosopher Epicurus in a letter to Herodote already mentioned his belief that other worlds exist (Epicurus, translation M. Conche 1992). A few years ago a planetary system with three telluric planets has been discovered around a neutron star (Wolszczan & Frail 1992; Wolszczan 1994). Despite this remarkable discovery, the possibility to find a planet around a solar–type star was eagerly expected. After many false alarms it seems now that we have detected the very first planet orbiting a solar–type star (Mayor, Queloz 1995a). Two weeks after the announcement of the discovery of the companion to 51 Peg (Mayor, Queloz 1995b), the result has been perfectly confirmed by two independent teams (IAU Circ 6251). This discovery however revealed a very strange planet with orbital parameters quite far of what was theoretically expected: a jovian planet but with a separation as small as 0.05 AU: A kind of Jupiter in hell!

If the planet hunting was successful at the end of 1995, it is also true that the first part of that year had started on an ill omen. Summarizing the results of the decade–long survey of the Canadian radial velocity programme, Walker et al. (1995) concluded with a rather pessimistic statement: "When our negative result is combined with other searches, one can say that, so far, no planets of the order of a Jupiter–mass or greater ($\geq 0.001\ M_{\odot}$) have been detected in short–period, circular orbits around some 45 nearby solar–type stars". As a matter of fact, no detection of planetary companions had been achieved by mid–1995 by



*C. Eiroa et al. (eds.),*
*Infrared Space Interferometry: Astrophysics & the Study of Earth-Like Planets*, 63–69.

any of the three other teams active in the field of precise radial velocity searches (G. Marcy, P. Butler at Lick, B. Cochrane, A.P. Hatzes at Texas University, R.S. McMillan at Arizona University).

At the very beginning of 1996 the discovery rate of planets was boosted by the efforts of the team at Lick's. In January 1996, G. Marcy and P. Butler annonced the discovery of two very low mass companions to the solar–type stars: 47 UMa and 70 Vir, with respectively 2.4 and 6.6 $M_J$ minimum masses and periods of 1090 and 117 days. The eccentric orbit of 70 Vir ($e \sim 0.4$) (Marcy & Butler 1996) as well as the rather large mass of its companion has inmediately revived the question about the limit between heavy planets and brown dwarfs. This point will be rediscussed in the section 3.

47 UMa (Butler & Marcy 1996) is also an interesting object. With its rather long period associated to a quasicircular orbit it looks like a normal Jovian planet. However, this is only partially true. In fact, its separation of 2.1 AU from the companion lies clearly inside the ice congelation point, probably closer to 4 or 5 AU (Boss 1995).

## 2. From March to July 1996

During the months following the Toledo meeting Marcy and Butler added three new planets to the list: 55 Cnc in April, $\tau$ Boo and $\upsilon$ And in June. All three planets have separations in the range of 0.04 to 0.11 AU. If at the time of its discovery 51 Peg could be considered to be a very unusual and exceptional object, this view can certainly not be sustained today. We definitely are in presence of a family of heavy planets, the "Hot Jupiters" or "Pegasian Planets", with masses from 0.5 to 4 Jupiters and ranging from 0.04 to 0.11 AU.

The sample of detected planets is still too small to draw a conclusion whether this is an accumulation of planets in a very limited range of separations or this is only the tail of a much broader distribution. We are still heavily affected by detection biases as the radial velocity amplitude is proportional to the mass $M_2$ and to $P^{-1/3}$. However, the very slow increase of the detection limit with the orbital period cannot alone explain the observed distribution of periods ($3.3^d$, $4.2^d$, $4.6^d$, $15^d$, $1090^d$)

Before the detection of 51 Peg the orbital decay of protoplanets by tidal interaction with the protoplanetary disk had already been considered by many authors (Goldreich & Tremaine 1980; Lin & Papaloizou 1986; Ward 1986; Takeuchi, Miyama and Lin 1996). We might, for example, mention the following statement by Goldreich & Tremaine (1980) "The angular momentum transfer is shown to be so rapid that substantial changes in both the structure of the disk and the orbit of Jupiter must have taken place on a time scale of a few thousand years". Apparently this statement had never been read by planet hunters!

Recently, Lin et al (1996) argued that 51 Peg might have been formed by core–collapse out of the congelation point at 5 AU and could then have suffered a strong orbital decay by tidal interaction with the disk. An outward torque is then necessary to stop the orbital decay when the planet arrives very close to the star (the planet is observed at 1% of its supposed formation position!). The tidal interaction with a (initially) fast rotating star or with a phenomena related to the inner-disk structure, could for example, induce such a "stop". The high number of planet observed in a very small range of separation could be an evidence for this scenario.

Prior to June 1996 five companions with circular orbits and minimum masses below 4 $M_J$ and two companions with eccentric orbits at 7 and 9 Jupiter–masses have been discovered. In the litterature very few detected spectroscopic binaries have companion with mass less than 40 $M_J$. In July 1996 we have detected 5 new substellar companions distributed over the range 10-40 $M_J$. All these very low mass companions show eccentric orbits. The three lighest ones are HD 110833 ($M_2 \sin i = 18M_J$), DM-4°782 ($M_2 \sin i = 23M_J$) and HD 112758 ($M_2 \sin i = 35M_J$) (Mayor et al. 1996)

## 3. From brown dwarfs to planets.

The distributions of orbital elements of spectroscopic binaries have been measured for a couple of unbiased samples of different primary masses ($M_1$) and different metallicities: the G dwarfs of the solar vicinity (Duquennoy & Mayor 1991), red giants members of open clusters ($M_1 \geq 2\ M_\odot$) (Mermilliod and Mayor 1992), deficient stars of the galactic halo (Latham et al. 1992). In all cases spectroscopic binaries with periods larger than the tidal circularisation period are never found with circular orbits. The only exception being the spectroscopic binaries affected by the evolution of the former primary component, presently a white dwarf.

Since the four heaviest planets of our own solar system have quasicircular orbits, the discontinuity of orbital eccentricities observed at about 10 $M_J$ was suggested by Duquennoy & Mayor (1991) to be a criterion to set the limit between heavy planets and brown dwarfs. Such an interpretation seemed adequate since the orbital eccentricities of protoplanets are damped by the dynamical interactions with the protoplanetary disk (Goldreich & Tremaine 1980) and the eccentric objects, probably formed by fragmentation, are heavier than 10 $M_J$ (Low & Lynden–Bell 1976). However, this interpretation is far to be certain. Artymowicz et al. (1991) and Artymowicz (1992) have shown that as a result of the dynamical interaction of a binary with its circumstellar/circumbinary disk, the orbital eccentricity of the system will increase if $M_2$ is larger than about 10 $M_J$ and damped if smaller than this limit. Thus the interpretation of the orbital eccentricity discontinuity at about 10 $M_J$ is still an open question. In consequence we do not have, to this day, an unambiguous criterion derived from observations to label a low mass substellar companion a "planet" or a very low mass "brown dwarf". The denomination being supposed to differentiate objects formed through different mechanisms. This ambiguity, in particular, concerns the two very low mass objects with eccentric orbits 70 Vir (Marcy, Butler 1996) and HD 114762 (Latham et al. 1989). Nevertheless, we can note that the change of the sign of $de/dt$ with the mass of the secondary is also a function of disk properties and orbital period. This dependence of $de/dt$ on parameters other than the secondary mass should introduce some smoothing of the discontinuity of orbital eccentricity distribution. With the existing detections we do not have any evidence of such a gradual change of $e$ with $M_2$.

Recently Black (1996) has questioned the fact that we have discovered any planets, suggesting for example that 51 Peg and 55 Cnc were brown dwarfs seen with quite small sin $i$. However, the present situation with five objects in the range of 0.45 to 4 $M_J$ is quite difficult to interprete as only due to brown dwarfs seen with small sin $i$. If we suppose that all these objects have masses larger than 10 $M_J$, the geometrical probability for the observed sin $i$ is respectively 0.1%, 0.3%, 0.4%, 2.6% and 8.3%. Such a distribution of small sin $i$ would imply a huge population of brown dwarfs with short periods –a population not observed–.

Fig. 1 illustrates the ($e$, log $M_2$) plane including substellar companions detected up to July 1996 (planets or brown dwarfs) as well as spectroscopic binaries in orbit around G dwarfs of the solar vicinity.

If the distribution of orbital eccentricities does not seem to provide an unambiguous criterion to establish a limit between planets and brown dwarfs, we still believe that distributions of orbital elements could offer that possibility in future. For example, the semi–major axis distributions should provide some hints on the difference of formation mechanisms for planets and brown dwarfs. We obviously need many more orbits of substellar companions to establish orbital distributions like $f(e, a, M_2)$. The vigorous efforts presently undertaken or planned for the very near future at Lick, KECK, OHP and ESO will undoubtedly provide the numerous discoveries needed.

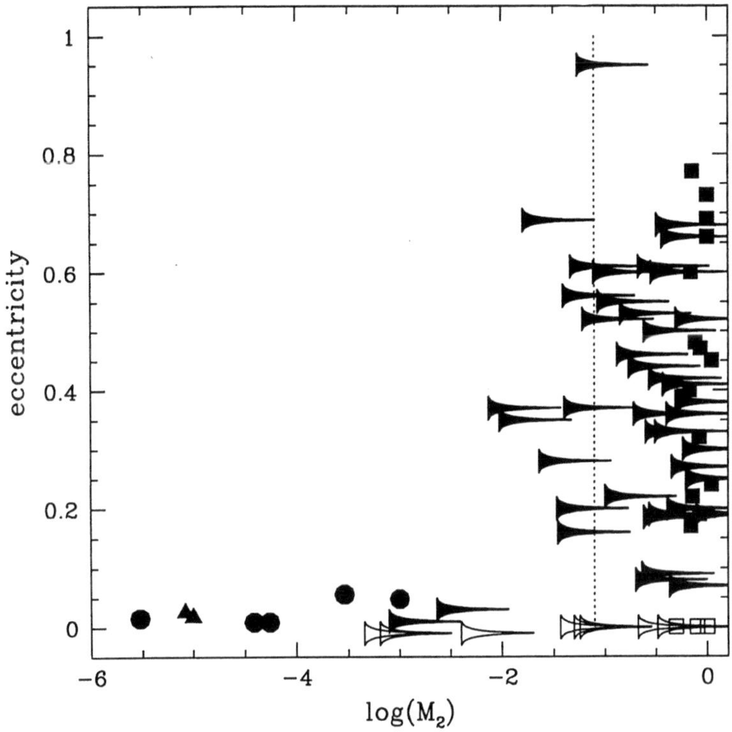


*Figure 1.* Orbital eccentricities of planets and stellar or sub-stellar companions of nearby G and K dwarfs as a function of their mass $M_2$. The planets of the solar system are indicated by filled circles and the planets around the pulsar PSR B 1257+12 by filled triangles. Filled squares indicate SB2 binaries. SB1 binaries are display with elongated symboles where the aera is proportional to the sin *i*-probability. Empty symboles are used for stellar and sub-stellar companions with periods less than 10 days as well as for planetary companions with periods less than 5 days. This limit corresponds roughly to a typical distance where the orbit of the companions undergoe a circularization by tidal effect from the primary star. The dotted line indicates the brown-dwarf limit

## References

Artymowicz, P., Clarke, C.J., Lubow, S.H., Pringle, J.E.: 1991, *ApJ* **370**, L35
Artymowicz, P.: 1992, *PASP* **104**, 769
Baranne, A., Mayor, M., Poncet, J.L.: 1979, *Vistas in Astronomy* **23**, 279
Baranne, A., Queloz, D., Mayor, M., Adrianzyk, G., Knispel, G., Lacroix, D., Meunier, J.-P., Rimbaud, G., Vin, A.: 1996, *A&AS* **119**, 1
Black, D.C.: 1996, *Sky and Telescope*, August p. 21
Boss, A.P.: 1995, *Science* **267**, 360
Butler, P., Marcy, G.: 1996, *ApJ* **464**, L153

Conche, M.: 1992, Lettres et Maximes d'Epicure, Presses Universitaires de France
Duquennoy, A., Mayor, M.: 1991, *A&A* **248**, 485
Goldreich, P., Tremaine, S.: 1980, *ApJ* **241**, 425
Latham, D.W., Mazeh, T., Stefanik, R.P., Mayor, M., Burki, G.: 1989, *Nature* **339**, 38
Latham, D.W., Mazeh, T., Torres, G., Carney, B.W., Stefanik, R.P., Davis, R.J.: 1992, in Binaries as tracers of stellar formation, A. Duquennoy, M. Mayor (Eds.), Cambridge Univ. Press, p. 139
Lin, D.N.C., Papaloizou, J.: 1986, *ApJ* **309**, 846
Lin, D.N.C., Bodenheimer, P., Richardson, D.C.: 1996, *Nature* **380**, 606
Low, C., Lynden-Bell, D.: 1976, *MNRAS* **176**, L367
Marcy, G., Butler, P.: 1996, *ApJ* **464**, L147
Mayor, M., Queloz, D.: 1995a, *Nature* **378**, 355
Mayor, M., Queloz, D.: 1995b, 9th Cambridge Workshop on Cool stars, stellar systems and the sun, Florence, R.Pallavicini (Ed.), ASP Conf. Ser., in press
Mayor, M., Queloz, D., Udry, S., Halbwachs, J.L.: 1996, in Astronomical and biochemical origins and the search for life in the universe, IAU coll 161, C.B. Cosmovici, S. Bowyer & D. Werthimer (Eds.), in press
Mermilliod, J.C., Mayor, M.: 1992, in Binaries as tracers of stellar formation, A. Duquennoy, M. Mayor (Eds.), Cambridge Univ. Press, p. 183
Takeuchi, T., Miyama, S.M., Lin, D.N.C.: 1996, *ApJ* **460**, 832
Walker, G.A.H., Walker, A.R., Irwin, A.W. et al.: 1995, *Icarus* **116**, 359
Ward, W.R.: 1986, *Icarus* **67**, 164
Wolszczan, A., Frail, D.A.: 1992, *Nature* **355**, 145
Wolszczan, A.: 1994, *Science* **264**, 538

## 4. Questions

*Question* : What is the status of the possibility of a second planet in the 51 Peg system?
*M. Mayor* : We have to be patient and wait the next observing season

*Question* : Over which range of spectral types do radial velocity searches work, and how does the accuracy depend on the spectral type of the stars?
*M. Mayor* : We are limited on the blue side by the dramatic increase of rotation line broadening at about F5. On the red side the photon noise is the main limitation factor. The optimum domain is from G to K spectral types.

*Question* : A way to disentangle giant planets from brown-dwarfs is that planets should be in planetary systems, hence there should be more than one planet in a given star. Complementary approaches are probably need to detect them!
*M. Mayor* : We fully agree

*Question* : The statistic that you presented (13% of SB and 1% of BD companion with period shorter than 3000d) for G-dwarf are also adequate for K-dwarf?

*M. Mayor* : Probably yes, but we have to wait the full analysis of our CORAVEL K dwarf survey.

*Comment* : I should have a word of caution about differentiating between brown dwarfs and planets based solely on their eccentricity. There are ways to pump up eccentricity of a planet (eg. by interaction with an accretion disk, as shown by Lin & Papaloizou). Besides, a $2M_\oplus$ planet with an eccentricity of 0.23 has been discovered at 7AU from a pulsar (Shabanova, ApJL, Nov 1995). The nature of 70 Vir B is at least uncertain. In fact we should ask, what is a brown-dwarf?

*Question* : How did you determine the sin $i$ for 51 Peg? How did you determine the spin period of 51 Peg?
*M. Mayor* : The spin period is derived from the level of chromospherical activity. The orbital inclination is derived from the line broadening of the spectrum ($v \sin i$) on the assumption that the spin axis and orbital axis are not far to be collinear.

*Question* : What steps are you taking to improve the velocity accuracy and how well do you expect to do?
*M. Mayor* : First, increase the sampling of spectra by using smaller pixels. Second, increase the resolution up to 50000. Finally, add a scrambler on the optical fibers.

# ARRAY CONFIGURATIONS TO DETECT AND CHARACTERIZE EXTRASOLAR PLANETS WITH A SPACE INFRARED INTERFEROMETER

B. MENNESSON
*Observatoire de Paris, DESPA, F-92195 Meudon principal, France*
*e-mail: bertrand@bluenote.obspm.fr*

**Abstract.** We list here some of the basic design requirements that have to be fulfilled by an infrared nulling interferometer, in order to properly detect and characterize extrasolar planets.Three main constrains can be identified: the interferometer must provide a very strong suppression of the starlight, a good spectral coverage (from 7 to 18 microns) with a fixed baseline, and be able to clearly distinguish planets from local dust disc emission.We review and compare here the possibilities of existing projects and present a new pupil geometry with five 1.5m telescopes deployed in an elliptical array.The telescope array, whose dimensions are about 50m×25m, has been optimized so that the exozodiacal emission is strongly extinguished, very weakly modulated by rotation about the line of sight, and concentrated at a few even frequencies.The planet's signal on the contrary is strongly modulated at many distinctive frequencies. A simple cross correlation method recovers a single image of a solar system twin at 10 pc distance, in about 30h. The spectroscopy of the planets can then be undertaken with the same baseline and would reveal absorption features of water, ozone and carbon dioxide of an Earth-like planet at 10 pc in less than one month.

**Key words:** extrasolar planetary systems, space infrared interferometry, zodiacal light, image processing.

## 1. Introduction

A general presentation of the DARWIN project and of its scientific objectives can be found in this Special Issue (see the parts by J.M. Mariotti and A. Léger). We concentrate here exclusively on the geometry of the entrance pupil and its optimization. We describe in section 2 the general design constrains which are critical in order to detect Earth-like exoplanets with a space infrared nulling interferometer and compare the capabilities of existing concepts formed of linear or two dimensionnal arrays.In section 3 we define a new design with 5 telescopes, discuss its properties and study its potential to recover images and spectra.



*C. Eiroa et al. (eds.),*
*Infrared Space Interferometry: Astrophysics & the Study of Earth-Like Planets,* 71–75.

## 2. Design Requirements

We focus here on the design and operating constrains set by the main sources of noise and spurious signal, namely the residual parent star flux, the solar and exo-zodiacal emissions.

The first requirement is to achieve a very strong and deep nulling on the line of sight in order to extinguish the starlight.To detect Earth-like exoplanets within reasonnable integration times, the rejection factor on the starlight has to be greater than $\sim 10^6$. This condition can only be realized if the transmission of the interferometer near the axis is proportional to $\theta^4$ or even better $\theta^6$, where $\theta$ is the angular separation from the optical axis, when pointing the star. This first condition sets some simple relations between telescopes locations in the array, and phase shifts applied to each of the incoming beams (Mennesson and Mariotti, in preparation).

To take benefit of this strong nulling of the starlight, the background due to the solar zodiacal emission in the IR must be kept lower than the residual emission from the star.As shown by Léger et al. (1993 and 1996) it is possible even with 4 or 5 1.5m class telescopes, when operated approximatively at 4 AU from the Sun.

Assuming that a rejection factor of $10^6$ is achieved, and that local zodiacal emission is strongly suppressed, the likely dominant source of spurious signal is the zodiacal emission from the exo-system under study, referred to as exo-zodiacal light ("exo-ZL" hereafter)

The fundamental drawback of a telescope array that exhibits a central symmetry, is that it makes no difference between the signals emitted by a point source or by two point sources located symmetrically with respect to the central star (Léger et al. 1996). If we are looking at a solar system twin from a 10pc distance, at $10\mu$m, under an incidence of 30 degres (average incidence), using a zodiacal light model developed by Good (1994), the dust disc modulation is already a few times stronger than the one of an Earth-like planet.

An attractive solution to this problem is to use an odd number of telescopes, regularly located on a 50m diameter circle (Léger et al. 1996). With this configuration, the central symmetry is now broken, the planets and the exo-ZL exhibit different frequencies.If the array is continuously rotated at a frequency $f_0$ ($\sim$ 1turn.h$^{-1}$) about the line of sight, the signal emitted by a planet located on the first set of bright fringes will exhibit a $5f_0$ frequency, whereas the exo-zodiacal light will exclusively concentrate at 10 $f_0$.The planet and the exo-ZL can now be distinguished without any ambiguity.

But this solution also presents a big drawback: for given wavelength and planet location, there are some holes of transmission and regions of very weak modulation that make it necessary to change the baseline in orbit, in order to get a more uniform spectral coverage, as in the case of linear fringes.

## 3. An optimized configuration with 5 telescopes

We propose here an alternative solution derived from the 5 telescope circular array, keeping the same central extinction in $\theta^4$, and same discriminative properties with respect to the exo zodiacal emission.We found a family of solutions derived by affinity, that is by simple contraction of the former circular network. The degree of contraction, i.e. the eccentricity, is a free parameter.Eccentricities ranging from 0.8 to 0.97 were found to provide the best results:

For any planet location and wavelength there are zones of high transmission within a rotation of the interferometer. The variance of the transmission, that measures the amplitude of the modulation is also better than in the circular case, as it never drops dramatically for any wavelength or planet location. *Then there is no more need for a variable baseline with this configuration.* The transmission map of an elliptical array, (corresponding to the case in which one dimension is contracted by a factor of two, eccentricity$\sim$ 0.87) shows a very asymmetric fringe pattern.Consequently the signal emitted by a very close planet still exhibits many distinctive frequencies.This yields to easy distinction with the signal of the exo zodiacal light (concentrate at a few even frequencies), and easy image recovery with *only one position per planet.*

We examined the case of a solar system "twin" as seen from a 10pc distance, with this 5 telescopes elliptical array (eccentricity of 0.87). The modulation of the exo zodiacal light is very weak, i.e. of the same order of magnitude as the Earth, and can be easily substracted without affecting the planet's image.

Fig. 1 shows the image recovered by a simple cross correlation method developed by Angel and Woolf (in preparation).The advantage of this method is to use the informations given independently by all the different wavelengths, with the same resolution as in the spectroscopic mode. A 30 hour integration time allows a 3 to 5 $\sigma$ detection of all three telluric planets.

In the spectroscopic mode, the interferometer is rotated about the line of sight exactly like in the detection phase.The correlation is made on the interferometer rotation angles, for each spectral element. Fig.2 shows the recovered spectrum of the Earth, as compared with the exact spectrum obtained with a resolution of 20.The continuous line corre-

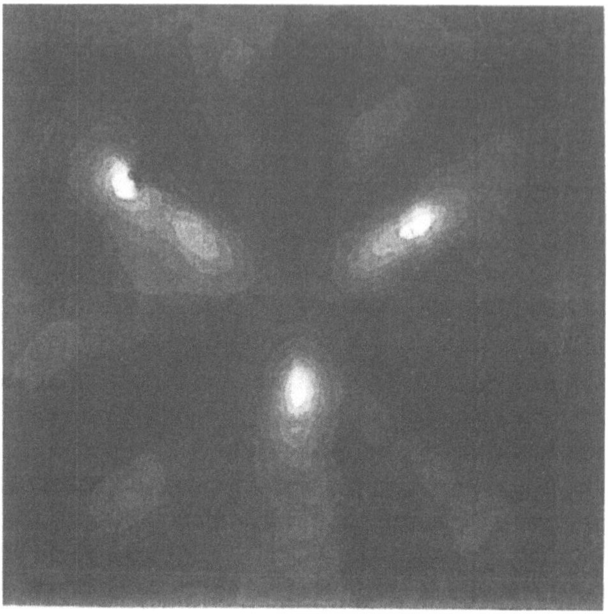

*Figure 1.* Image recovery of Venus, the Earth and Mars, as seen from a 10pc distance with a 5 telescopes elliptical array. The integration time is 30 hours .The exo-system is seen under a 30 degres inclination, the spectral range is 7 to 18 microns, with a resolution of 20. The flux of the Earth at 10 microns is $\lambda N_\lambda$=3.5ph.m$^{-2}$.s$^{-1}$. The ozone band is supposed to be 25% half deep. The collecting area corresponds to 5 telescopes of 1.5 m diameter. The overall detection efficiency is 0.15 The rejection factor on the starlight is $10^6$. The dust disc of the exo system exhibits a circular symmetry, inhomogeneities are at the level of photon noise.

sponds to the emission of a blackbody at 300K.The absorption features of water, ozone and carbone dioxide are visible.

## 4. Conclusion

We examined here some of the critical design constraints and proposed a very efficient interferometric configuration with five 1.5 meter class telescopes disposed on an elliptical array of about 50m×25m. With this new configuration there is no need for a variable baseline (like in the case of other two dimensionnal arrays), no problem of distinction with respect to the exo-zodiacal emission (case of most linear arrays), the image and spectrum recovery of the exoplanetary system are straight forward without any ambiguity on the planet location.

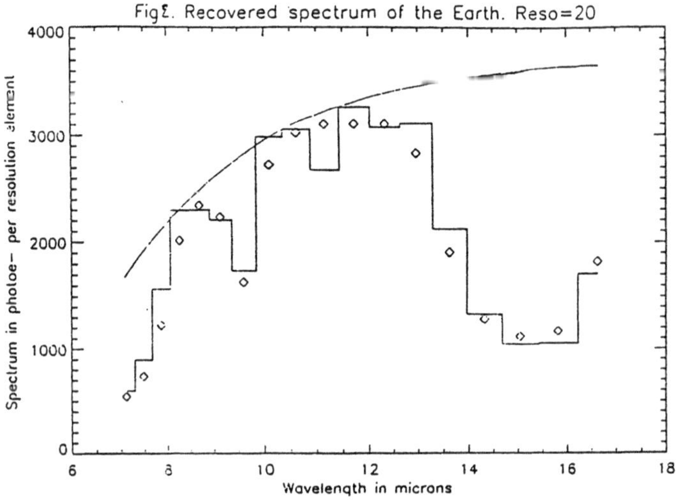


*Figure 2.* Recovered spectrum of the Earth (Resolution=20, Integration time=500h). Absorption features of water (< 8$\mu$m), ozone (9.6 $\mu$m) and carbone dioxide (15 $\mu$m) are visible. The lozenges show the exact spectrum at this resolution. The solid curve shows the emission from a blackbody at 300K.Same assumptions as for Fig.1 except that integration time is 500h.

## References


Angel, J.R.P., Woolf, N.: 1996, in preparation

Good, J.: 1994, *IRAS Sky Survey Atlas Explanatory Supplement. JPL publication 94-11*

Léger, A. et al.: 1993, *Proc. R. Soc. Land. (B)* **189**, 167

Léger, A.,Mariotti, J.M., Mennesson, B., Ollivier, M., Puget, J.L., Rouan, D., Schneider, J.: 1996, *Icarus*, in press

# RESOLVING DISKS IN YSOS

A. NATTA
*Osservatorio Astrofisico di Arcetri*
*Largo Fermi 5, I-50125 Firenze, Italy*

H. BUTNER
*Dept. of Terrestrial Magnetism, Carnegie Institution of Washington*
*5241 Broad Branch Road, N.W. Washington, DC 20015, USA*

**Abstract.** We discuss using high-spatial resolution mid-IR observations to detect disks in very embedded objects and consider how one might study their properties.

**Key words:** disks, mid-infrared, spatial resolution

## 1. Introduction

Our current picture of the star formation process requires the existence of circumstellar disks over a large fraction of the life of young stars, from the very earliest stages of the protostar's formation till the time the star settles onto the main sequence.

In this talk, we will address the impact that very high spatial resolution mid-infrared observations can have in our understanding of the very early phases of star formation. Evidence that disks exist even in the youngest objects detected to date (the so-called Class 0 objects; see, for a review of their properties, André 1995) is growing, as compact emission in the submm and mm range is detected in an increasing number of embedded sources by interferometric measurements (Terebey et al. 1993; André 1995). These measurements indicate that while most of the matter is still in the almost spherically symmetric infalling cloud, some mass has already accumulated in a disk. A determination of the disk properties, and, in particular of the disk accretion rate (as opposed to the spherical infall rate) is particularly important, and can only be obtained from the disk luminosity. In turn, this requires that the disk emission in the mid-infrared, where the disk is likely to be optically thick, is separated from the emission of the envelope.

A measure of the disk accretion rate in embedded objects is particularly interesting, among other reasons, because these objects often have powerful molecular outflows (Bontemps et al. 1996). The driving mechanism of such outflows is probably related to the accretion mechanism. In T Tauri stars, part of the accretion energy can be transformed into kinetic energy of the outflowing matter by the combined effect of rota-



*C. Eiroa et al. (eds.),*
*Infrared Space Interferometry: Astrophysics & the Study of Earth-Like Planets*, 77–84.

Table I. Disk Size as a Function of Environment

| Disk Type | Level | 5$\mu$m | 10$\mu$m | 20$\mu$m | 840$\mu$m |
|---|---|---|---|---|---|
| | | | ($10^{-3}$ arcsec) | | |
| Naked Disk | 1/2 | 0.2 | 0.23 | 0.25 | 0.3 |
| | 1/10 | 0.5 | 0.85 | 1.2 | 2.5 |
| | 1/100 | 1.2 | 2.4 | 4.6 | 42 |
| Irradiated Disk | 1/2 | 0.2 | 0.24 | 0.26 | 0.3 |
| | 1/10 | 0.5 | 0.85 | 1.3 | 2.9 |
| | 1/100 | 1.2 | 2.9 | 6.1 | 143 |
| L1551 Disk | 1/2 | 0.5 | 0.5 | 0.6 | 0.7 |
| | 1/10 | 1.4 | 2.2 | 3.0 | 5.5 |
| | 1/100 | 3.7 | 8.6 | 16 | 1000 |

tion and magnetic field (Shu et al. 1994). We do not know, at present, if similar mechanisms can account for the Class 0 outflows.

Finally, we will briefly comment on how mid-IR observations of high spatial resolution and sensitivity are likely to be the best way of studying binary systems in the very early stages of their formation through detection of the emission from the circumstellar disks.

## 2. Disks in T Tauri Stars

The expected properties of circumstellar disks can be better understood by considering first the properties of circumstellar disks associated with the optically visible T Tauri stars. The predicted spectral energy distribution (SED) for a typical accretion disk model for an isolated ("naked") star+disk system is shown in the inset of Fig. 1. The model parameters are specified in the figure caption.

The model-predicted brightness profile of this disk at different wavelengths is plotted in Fig. 1 as a function of the distance from the central star. The disk's apparent size is very small at all wavelengths, much smaller than its physical size (outer radius of 80 AU, or about 0.6 arcsec). Table 1 reports the full width of the brightness profile at different levels (1/2, 1/10 and 1/100 of the peak brightness) for 5, 10, 20 and 840 $\mu$m, respectively. The capability of spatially resolving such a disk in the mid-IR depends on the dynamical range of the observations. Typically, at 10 $\mu$m and with a dynamical range of 10, one needs a resolution of about few$\times 10^{-4}$ arcsec and a sensitivity of $\sim$1 mJy per resolution element.

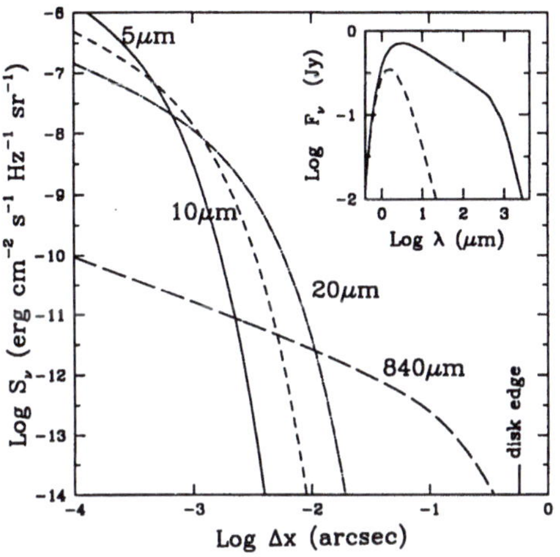


*Figure 1.* Disk surface brightness as function of the projected distance from the star in arcsec at 5 $\mu$m (solid line), 10 $\mu$m (short-dashed line), 20 $\mu$m (dot-dashed line) and 840 $\mu$m (long-dashed line). The inset shows the SED of the same disk (solid curve) and of the central star (dashed curve). Star and disk parameters are typical of T Tauri stars. The star has luminosity 0.75 $L_\odot$, effective temperature $T_\star$=3800 K; the disk has an intrinsic viscous luminosity of 0.25 $L_\odot$, mass 0.06 $M_\odot$, inner radius $R_{d,i}=R_\star$, outer radius $R_{D,out}$=80 AU, surface density $\Sigma \propto r^{-1.5}$. The distance is 140 pc, the inclination angle 0 deg.

These values do not change significantly if we consider the more realistic case of a disk embedded in a cloud of dust of moderate optical depth (Natta 1993, 1995; D'Alessio et al. 1996). The disk remains very small in the mid-IR, as shown by the values of the full width sizes in Table 1 ("irradiated disk") for the same disk as before, surrounded by a cloud of dust with optical depth at $\tau(1\mu m)$=0.4.

## 3. Disks in Embedded Objects

If the envelope surrounding the disk is very optically thick, as in deeply embedded objects, its effects on the disk temperature profile is much stronger, making the disk brightness profile significantly wider at all wavelengths. However, a thick envelope absorbs the radiation emitted by the disk and, much more important, emits at all infrared and mm

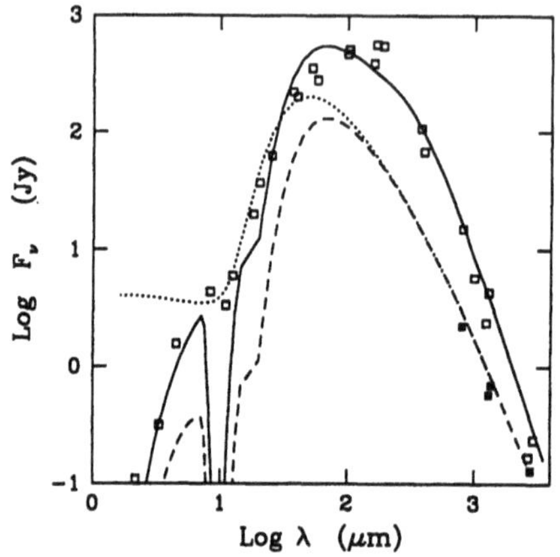


*Figure 2.* SED of L1551. Open squares are single dish observations, filled squares are interferometric measurements (see Butner et al. 1994 for details). The model predicted SED (as seen in a large beam) is shown by the solid curve. The dotted line shows the SED of the star+disk system, the dashed line the same spectrum attenuated by the envelope. The parameters of the model are as follows: the star has luminosity 14 $L_\odot$, effective temperature $T_\star$=5500 K; the disk has an intrinsic luminosity of 10 $L_\odot$, mass 0.1 $M_\odot$, inner radius $R_{D,i}=R_\star$, outer radius $R_{D,out}$=80 AU, surface density $\Sigma \propto r^{-1.5}$. The envelope density profile is described by two power-laws, an inner one (from 3 $R_\star$ to $R_{D,out}$) $\propto r^{-0.5}$,with radial optical depth at 1 $\mu$m of 0.2, and an outer one (from $R_{D,out}$ to 0.2 pc) with $n \propto r^{-1.5}$ and $\tau$=30 at 1 $\mu$m. The distance is 140 pc. These parameters have been chosen to approximately match those of L1551.

wavelengths. We illustrate the situation using as an example the well known object L1551. L1551 is more evolved than Class 0 objects, but it is very illustrative of the situation.

Fig. 2 shows the SED of a model for L1551 which involves a central star, a disk and an infalling envelope, computed following Butner et al. (1994) to include the effect of the envelope on the disk temperature ("backwarming"). The model parameters are specified in the figure caption. Note that this disk, which has a mass of 0.1 solar masses, is optically thick at 840 $\mu$m up to a radius of about 40 AU. The open squares in Fig. 2 are single-dish observations; filled squares are observations with interferometers, including the Lay et al. (1994) measurement at 840 $\mu$m. The dotted line shows the SED of the star+disk

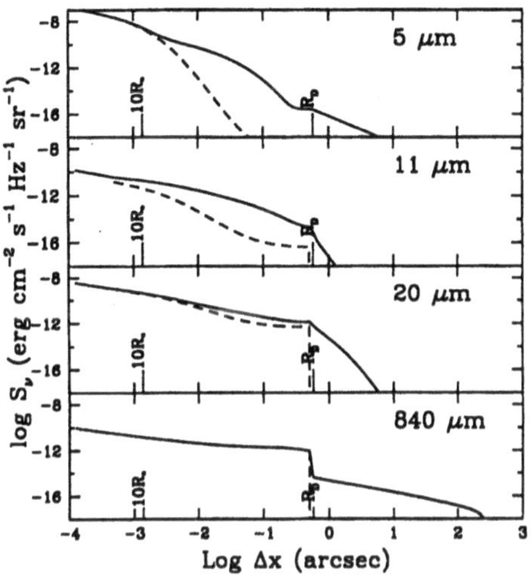


*Figure 3.* Predicted surface brightness profiles for L1551 at 5, 11, 20 and 840 $\mu$m(from top to bottom). The solid lines show the total surface brightness, the dashed lines the contribution of the disk. Model parameters as in Fig.2.

system, the dashed line the same spectrum as seen through the envelope. Note that the small-beam, interferometer observations are well fit by the predicted disk emission. The solid line is the resulting overall SED, which provides a reasonable fit to the single-beam observations.

Fig. 3 shows the brightness profile of the predicted L1551 emission at various wavelengths. The solid line is the total emission (disk +envelope), the dashed line the disk only. The physical size of the disk (80AU radius) is indicated by the vertical arrow. Full width sizes at 5, 10, 20 and 840 $\mu$m of the disk emission are given in Table 1. Fig. 3 shows that the envelope emission overwhelms the disk at scales larger than $\sim 3 \times 10^{-3}$ arcsec at 5 $\mu$m, $10^{-3}$ arcsec at 11 $\mu$m (where the disk emission is strongly absorbed by the silicate feature), $10^{-2}$ arcsec at 20 $\mu$m. At this wavelength, the envelope emission is at most twice as large as that of the disk. At sub-mm (and millimeter) wavelengths, the disk emission still dominates at scales comparable to the disk physical size; the envelope surface brightness is about two orders of magnitude lower, resulting in a sharp drop of the brightness profile in coincidence with the physical radius of the disk. This behaviour explains why Lay et al.

(1994) were able to resolve the L1551 disk at 840 $\mu$m with a resolution of about 0.3 arcsec.

In summary, in the case of L1551 we see that in order to well resolve the disk emission against the background of the envelope in the mid-IR, one needs a resolution of about $10^{-3}$ arcsec at 5 $\mu$m, and of $10^{-2}$ arcsec at 20 $\mu$m. The required sensitivity is of the order of 0.2 mJy per resolution element at 5 $\mu$m and 2 mJy at 20 $\mu$m. The situation in the 10 $\mu$m region depends strongly on the actual depth of the silicate feature.

The above results depend on the adopted model, and, in particular, on how the backwarming effect of the envelope is treated. In addition, the disk brightness will be very sensitive to the envelope dust properties and overall optical depth. However, they should be indicative of the actual situation in very embedded objects.

## 4. Binary Stars in Embedded Objects

The studies of binary stars in pre-main–sequence have shown that more than half T Tauri stars are binary (Simon et al. 1995). The formation mechanism is still under discussion. It would be very important to extend the statistical work to the very early phases of star formation, i.e., detect and study binaries in the embedded phase. This is at present a prohibitive task, because of the high extinction of the infalling envelope.

Millimeter observations have detected a few examples, including IRAS 16293-2422 (Mundy et al. 1992) and NGC 1333 IRS 4 (Lay et al. 1995). However, mm observations can succeed only if the binary has a separation large enough that each star keeps an extended circumstellar disk, which emits strongly in the mm. If, as is currently believed, the disk is disrupted by the companion on scales comparable to the binary separation, one can only detect at 1 mm binaries wider than 50 AU. Observations of binaries appear to confirm this hypothesis (Jensen et al. 1996). For closer binaries, which among the T Tauri stars in Taurus account for 50% of the binaries (Leinert et al. 1993; Simon et al. 1995), the circumstellar disks will be truncated at smaller radii and will emit strongly only at shorter wavelengths. The best way of studying them is to image in the mid-IR the inner parts of the embedded sources with high sensitivity (since the circumstellar disk emission is strongly attenuated by the envelope) and with milliarcsecond resolution, so that the envelope emission can be separated.

A special class of binaries, which can only be studied in the mid-IR, is that where an optical pre-main–sequence star has an infrared companion. From the ground, there are a handful known, including T Tau (Zinnecker & Wilking 1992; Herbst et al. 1996a,b). The possibility of detecting more when better spatial resolution will be available may have important consequences for our understanding of binary formation.

## References

André, P.: 1995, *Ap&SS* 224, 29

Bontemps, S., André, P., Terebey, S., Cabrit, S.: 1996, *A&A* in press

Butner, H. M., Natta, A., Evans, N. J.: 1994, *ApJ* **420**, 326

D'Alessio, P., Calvet, N., Hartmann, L.: 1996, *preprint.*

Herbst, T.M., Koresko, C.D., Leinert, Ch.:1996, Disks and Outflows around Young Stars, S.Beckwith, A. Natta and J. Staude (Eds.), Kluwer, in press

Herbst, T.M., Beckwith, S.V.W., Glindemann, A., Tacconi-Garman, L.E., Kroker, H., Krabbe, A.: 1996, *AJ*, in press

Jensen, E. L. N., Mathieu, R. D., Fuller, G. A.: 1996, *ApJ* **458**, 312

Lay, O. P., Carlstrom, J. E., Hills, R. E., : 1995, *ApJ* **452**, L73

Lay, O. P., Carlstrom, J. E., Hills, R. E., Phillips, T. G.: 1994, *ApJ* **434**, L75

Leinert, Ch., Zinnecker, H., Weitzel, N., Christou, J., Ridgway, S.T., Jameson, R., Haas, M., Lenzen, R.: 1994, *A&A* **278**, 129

Mundy, L.G., Wootten, A., Wilking, B.A., Blake, G.A., Sargent, A.I.: 1992, *ApJ* **385**, 386

Natta, A.: 1993, *ApJ* **412**, 761

Natta, A.: 1995, *RevMexAA Serie de Conferencias* **1**, 209

Shu, F., Najita, J., Ostriker, E., Wilkin, F., Ruden, S., Lizano, S.: 1995, *ApJ* **429**, 781

Simon, M., Ghez, A. M., Leinert, Ch., et al.: 1995, *ApJ* **443**, 625

Terebey, S., Chandler, C. J., André, P.: 1993, *ApJ* **414**, 759

Zinnecker, H., Wilking, B.A.: 1992, in Binaries as Tracers of Star Formation, A. Duquennoy, M. Mayor (Eds.), Cambridge University press, p. 269

## 5. Questions

*C. Beichman* : Can you comment on the size of the disks as seen at 5, 10 and 20 $\mu$m? Do your models show disks shrinking more quickly than $\propto \lambda$ ? It will be important to choose an optimum $\lambda$ for observing these systems with a given interferometer baseline.
*A. Natta* : From the values of the disk sizes given in Table 1, one can see that between 5 and 20 $\mu$m the FWHM is almost constant while the FW at 1/10 of the maximum increases less than $\lambda$ - in the case of L1551 disk, only by a factor of two. However, in this case the emission of the envelope also plays a role, as discussed in the text.

*S.K. Dunkin* : Will all the binary systems have both a circumbinary and individual circumstellar disk, or will some have only one or the other?

*A. Natta* : I do not think we know enough either theoretically or observationally to answer this question.

*E. Martín* : Brown dwarfs have been detected in the Pleiades cluster which is at similar distance than the nearest star forming regions. We should not be surprised if very young brown dwarfs are relatively bright and they might form through luminous and cool accretion disks. If infrared companions to T-Tauri stars are brown dwarfs in formation, could that account for the high frequency of companions found by infrared speckle surveys among T-Tauri stars?
*A. Natta* : I think that the frequency of binaries in T-Tauri stars (admittedly still quite uncertain) is significantly different from main sequence stars.

*F. Paresce* : Can you rule out the possibility that T Tau S is a young, bright brown dwarf?
*A. Natta* : I do not think we can estimate the mass of T Tau S as its location in the HR diagram is not well known.

# GROUND-BASED OPTICAL/IR LONG BASELINE INTERFEROMETRY

F. PARESCE
*European Southern Observatory*
*Karl Schwarzschild Str. 2*
*D-85748 Garching b. Munchen, Germany*

**Abstract.** The present status of ground-based efforts in very high spatial resolution astrophysics in the optical and IR range by means of long baseline interferometry is briefly reviewed with emphasis on the scientific objectives that can be expected to be reached in the near future with this technique. It is shown that, if everything goes according to plan, extensive capabilities in this field will be available to the astronomical community by ~ 2005. In particular, it will be possible, by this time, to image with high accuracy in the near and thermal IR the outer structure of accretion and debris disks around pre- and main sequence stars within a few hundred parsecs of the Sun, the circumstellar environment of AGB stars out to several kpcs and the inner regions of our galactic center and nearby AGNs. High precision narrow angle astrometry should also allow detection of low mass companions to nearby stars down to ~ 10 Earth masses and the use of the large telescopes planned for the end of the century should even permit direct detection of massive or young planets and brown dwarfs identified by radial velocity techniques. Thus, ground-based interferometers will represent powerful and versatile facilities whose scientific and technical achievements will critically impact the design of any space-based device.

**Key words:** interferometry: scientific and technical capabilities

## 1. Introduction

Since it is well known that the cost of an astronomical facility in space tends to be a factor of ~ 100 times higher than a similar one on the ground, understanding what scientific objectives the latter can achieve that the former cannot become of paramount importance. A first step in this endeavour is a sober and realistic assessment of the expected status of ground-based optical/IR long baseline interferometry at the time of the first expected launch of an interferometric space mission i.e. in the 2005–2010 time frame (NASA 1995). The bottom line is that much will have been accomplished by then since a large number of small to medium sized interferometry programs already exist or are expected to come very soon into operation. These will address many exciting and important astrophysical questions and even greater progress can be expected with the several large collecting area and long baseline projects that are presently under study and will most likely start routine operations in the 2000–2005 time frame including the Keck, LBT



*C. Eiroa et al. (eds.),*
*Infrared Space Interferometry: Astrophysics & the Study of Earth-Like Planets*, 85–95.

and VLT interferometers. If all these projects do indeed see the light of day as planned, competition for any space interferometer will be substantial.

In this paper, I will briefly summarize the main characteristics of the existing and planned facilities and concentrate on ESO's VLT Interferometer (VLTI) perhaps the most ambitious interferometry program presently under construction on Cerro Paranal in the Chilean Atacama desert. I will then show with these examples what the scientific capabilities and the most serious limitations of ground-based optical/IR interferometry are most likely to be and, therefore hopefully, shed light on where and how space interferometers will be expected to contribute to the rapid advancement of astronomy into the next century.

## 2. Existing and Planned Optical/IR Interferometers

In recent years, interferometric projects have begun to play a central role in ground-based high-resolution astronomy, and numerous instruments have been completed or are in the process of construction. A summary of their fundamental characteristics including the number of simultaneous baselines, the maximum available baselines and the telescope or siderostat collecting areas are listed in Table 1 for easy reference. These parameters critically affect the (u,v) plane coverage, the ultimate achievable spatial resolution and the overall sensitivity of the facilities, respectively. Already many technical and scientific breakthroughs have been achieved (Baldwin et al. 1996, Schwarzschild 1996) with the moderate to small collecting area first or second generation devices. These are mainly devoted to imaging of stellar surfaces and the precise determination of stellar positions, proper motions and diameters (Armstrong et al. 1995).

Interferometers of the next generation that exploit the large collecting area telescopes at very good sites such as Mauna Kea and Paranal will probably start limited operations by the turn of the century and, funding permitting, reach their design goals by $\sim$ 2005. These facilities will, of course, represent a quantum leap in sensitivity and precision. The simultaneous use of smaller outrigger telescopes will alleviate both the bigger telescopes' limited (u,v) plane coverage and the intense pressure for observing time. By the first years of the next century, a powerful array of optical/IR interferometers will be operational and available to the community covering the whole sky both in the northern and southern hemispheres with spatial resolutions approaching the milliarcsecond (mas) or better in the optical and, for the best sites, 10 mas in the thermal IR. In the first phases of operation, for example, the Keck

Table I. Current Ground-based Optical/IR long baseline interferometers Projects

| Program (Nation) | No. of simult. Baselines (ultimate) | Maximum Baseline [m] | Element Diameter [m] | Year of Operation |
|---|---|---|---|---|
| I2T (F) | 1 | 140 | 0.27 | operational |
| GI2T (F) | 1 | 65 | 1.52 | operational |
| ISI (USA)$ | 1 | 35 | 1.65 | operational |
| COAST (GB) | 3 (6) | 100 | 0.40 | operational |
| SUSI (AUS) | 1 | 640 | 0.14 | operational |
| IOTA (USA) | 1 (3) | 45 | 0.45 | operational |
| NPOI (USA) | 3 (6, 15) | 250 | 0.35 | operational |
| ASEPS-0 ITT(USA) | 1 | 100 | 0.45 | operational |
| CHARA (USA) | 10 | 350 | 1.00 | 1997 |
| KIIA (USA) | 1 / 6 / 15+ | 75/180* | 10/1.5 | 1998 |
| LBT (USA/I)@ | 1 | 20 | 8. | 1999 |
| VLTI (EUR) | 6 / 3 / 6+ | 128/200* | 8/1.8 | 2000 |
| MAGELLAN (USA) | 1 | 20 | 6.5 | ¿2000 |

+ beam combination main / auxiliary / hybrid
* between main / auxiliary telescopes
$ heterodyne, to be changed into a homodyne interferometer
@ monolithic array

interferometer will concentrate on measurements of exo-zodiacal dust clouds around nearby stars at $\sim$ 10 microns in order to characterize the background for the search for Earth-like planets and on developing the necessary technology to achieve the astrometric precisions required to detect the reflex motion of bright nearby stars due to orbiting planets.

## 3. The VLT Interferometer

As a prime example of the class of large interferometric arrays, ESO's VLTI has a number of interesting characteristics. For one thing, it is probably the only large facility that was designed from the beginning with interferometry in mind. The original concept called for coherent combination of at least seven telescopes in an optimal configuration covering both a fair fraction of the top of Cerro Paranal and a wide range of wavelengths from $\sim$ 0.4 to 25 microns (von der Lühe et al. 1994). Another unique aspect of the VLTI is its ability to combine four 8 m diameter telescopes yielding an unprecedented level of sensitivity. Although this very ambitious scheme was put into hibernation in 1993 due to severe budget constraints, the basic interferometric design specifications for the site and the telescopes have been maintained unaltered.

This means that all of the infrastructure that will be required when and if the interferometric capability is turned back on will be immediately available including,for example,the tracks for the moveable telescopes, the delay line tunnels and the interferometric laboratory.

A first reduced version of this powerful facility has been recently approved for immediate implementation by the ESO Council so that, barring unforeseen circumstances, we can expect to have a basic working system on the mountain by the beginning of the next century. A conceptual overview of the VLTI design is shown in Fig. 1. The outer dotted line represents the limit of the flat top of Cerro Paranal whose longest diagonal dimension is $\sim$ 200 m. The basic concept consists in the coherent combination of up to four 8 m telescopes (UTs) whose positions are indicated by three concentric rings in this figure. They will be supplemented by a number of smaller 1.8 m diameter telescopes (ATs) that can be moved to $\sim$ 30 stations whose positions are given by the small circles along the tracks indicated in the figure. The tunnel at the center of the picture has room to accomodate up to eight 60 m stroke delay lines while the interferometric lab at the center supports the image or pupil plane beam combination instrumentation.

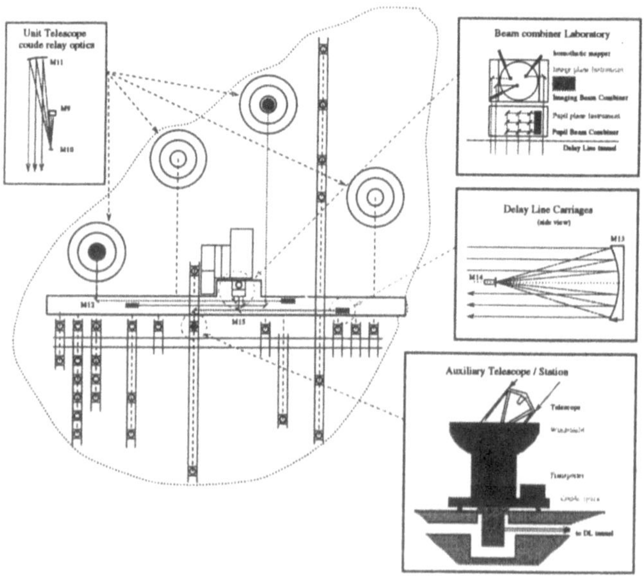


*Figure 1.* Schematic Configuration of the VLTI on Cerro Paranal

In the earliest phases of implementation of VLTI, two UTs will be equipped with coudé trains and two delay lines installed in the tunnel together with the control system, fringe sensor, image stabiliser and pupil plane beam combiner with the goal of detecting first fringes at $\sim$ 10 microns by the end of 1999. In parallel, two ATs will be developed and installed on Paranal by $\sim$ 2001 so that, by about that time, interferometry with any combination of 2 UTs and 2 ATs yielding reasonable (u,v) plane coverage will be available for routine operations at 2–10 microns. Since the present budget does not foresee sophisticated Coudé train adaptive optics (AO) systems, early operations will be restricted to those wavelength regions where the telescopes are essentially diffraction limited and a simple tip-tilt compensation device sufficient for efficient operations i.e. the 10–20 micron region for the UTs and the near IR for the ATs. Budgets permitting, the next phases would consist of bringing on line the other two UTs, a third AT, and two more delay lines in time for a $\sim$ 2003 commissioning. Simplified AO devices for the coudé trains or fiber optical links from Nasmyth AO systems, if available, are also being studied in order to allow operation of all five telescopes together at 2 microns where the advantage of ground- over space-based interferometers reaches a prominent maximum.

## 4. Science with Large Ground-Based Optical/IR Interferometers

It is expected that, with appropriate design and timely implementation, ground-based interferometers will contribute significantly and uniquely to a number of important areas of present and future astrophysical research (Paresce et al. 1996). Of these various worthy objectives, three are particularly crucial in the next decade i.e.: 1) detect and characterize objects of mass $\lesssim$ 0.2 $M_\odot$ in order to determine the luminosity and mass function of stellar and substellar populations of different ages and composition, 2) image the circumstellar environment in nearby star forming regions in order to understand the basic stellar and planetary formation mechanisms and 3) probe the center of our galaxy to establish the presence or absence of a central massive black hole and the central regions of nearby AGNs in order to test the unified model. In the following paragraphs, some of the planned ground-based capabilities in these areas that are most relevant to a direct comparison with the expected performance of future space-based devices are briefly summarized.

### 4.1. Low Mass Stars, Brown Dwarfs and Planets

Little is known presently about the stellar and substellar mass and luminosity functions of objects of mass $\lesssim 0.2\ M_\odot$ although it is expected that radial velocity and microlensing surveys will start filling in the gap in the near future. The accurate determination of the mass and luminosity of red dwarf stars near the Hydrogen burning limit, brown dwarfs, and planets in different environments and stages of stellar evolution will be a critical objective in the next decade. For this, it will be necessary to cover a range of several decades in mass and distances extending out to at least r $\sim$ 500 pcs. The latter range is such that it would include a large number of stars of different spectral types and luminosity classes and of young and old open clusters and star forming regions of different ages and metallicity.

Many of these objectives can be met, in principle, by accurate narrow angle astrometry (NAA) carried out over a period of time to determine the precise orbits of low mass companions to MS and PMS stars. The ultimate accuracy achievable with this technique with ground-based interferometers has been estimated to be $\sim$ 10 microarcseconds (Shao and Colavita 1992, Quirrenbach 1996). This number can be used to set boundaries in physical parameter space as shown in Fig. 2 where the mass of an object is plotted against the physical separation from its companion or its associated orbital period assuming zero orbital eccentricity for simplicity since most objects below $\sim 0.01\ M_\odot$ will most likely be found in circular orbits (Mayor and Queloz 1995). The sloping solid lines in Fig. 2 represent the limit below which objects at the distances indicated cannot be detected because their reflex motion is less than 10 microarcseconds. All the lines assume a solar mass primary except the one marked $0.1\ M_\odot$ for a system located at a distance of 1 parsec shown here for completeness. The dashed lines sloping in the opposite direction instead represent the reflex velocity (RV) lower limits taken here as 1 m $s^{-1}$ for the cases of 1 and 0.1 solar mass primaries and $\sin i = 1$ where $i$ is the orbit inclination angle with respect to the plane of the sky. This latter technique, of course, does not depend on distance.

The two techniques complement each other very nicely in the area of the plane located above the dashed lines and to the right of the appropriate solid line since both have inherent ambiguities that can be effectively eliminated by combining them. The RV method is especially sensitive to the $\sin i$ uncertainty while the NAA suffers from uncertainties in the proper motion of the reference stars. Except in the case of double-lined spectroscopic binaries, both methods are subject to errors in the mass of the primary. On average, we should expect to achieve

$\sim$ 4% relative precision in the mass of the companion of a large number of stars in the solar vicinity (Duquennoy et al. 1995). For the regions extending below the dashed lines, however, only the NAA technique can detect such low mass objects from the ground. Fig. 2 shows that gas giant cores (Black 1995) at 5 AU could be detected at this accuracy level provided they are located around 1 or 0.1 solar mass stars not much farther than $\sim$ 10 or $\sim$ 100 pc, respectively. Earth-mass planets need to be orbiting very nearby M dwarfs for them to impart a detectable wobble to the primary.

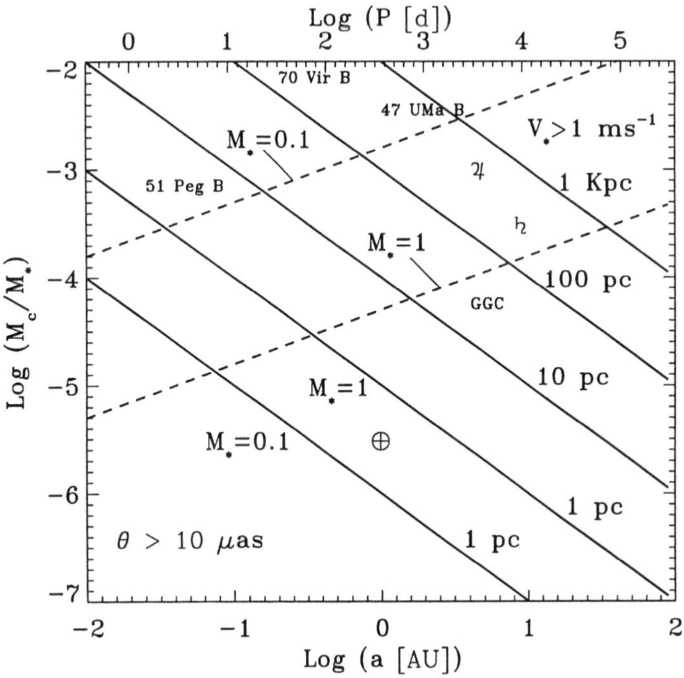


*Figure 2.* Mass-separation plane for substellar companion orbits

It should be emphasized that there is an exciting area waiting to be covered around the Hydrogen burning limit where the mass function is very poorly determined if at all and which can be readily accessed even by relatively modest position accuracies of 100–500 microarcseconds. One should also keep in mind that NAA techniques will allow coverage of objects that cannot be measured accurately by the RV method because the primary star's spectrum is unsuitable. These include many PMS and early type stars whose emission lines are scarce and broad.

Since only indirect information on the object's luminosity can be obtained by either of these methods, however, large collecting area interferometers will be essential in order to carry out a more detailed study of suitable candidates by directly detecting the low mass companion. Although this is an ambitious goal, the technology is definitely within reach with the VLTI and Keck, for example. The possibility exists to detect and measure the intrinsic broad-band thermal IR luminosity of a faint companion to a bright star corresponding to an intrinsic $\sim 10^4$ contrast ratio at 10–20 microns. Low to medium resolution spectroscopy to obtain the object's SED in the IR should also be feasible. This dynamic range is compatible with the expected contrast ratio of a Jupiter-like planet orbiting around a solar type star at separations of order 1–2 AU or with younger and hotter planets that may be orbiting young stars in the nearest star forming regions (Saumon et al. 1996) provided the observations are carried out with 8–10m size telescopes in the N or Q bands.

### 4.2. Circumstellar Environments of PMS Stars

The essential ingredient in both the stellar and planetary formation mechanisms is expected to be the accretion disk. The prototypical disk is probably only be viewed through severe extinction in the most favorable conditions. The size of the observable disk increases with wavelength much faster than the spatial resolution decreases, however, so that observations in the thermal IR are particularly useful (Malbet and Bertout 1995). As time progresses in its evolution towards the MS, the stellar photosphere and halo are slowly cleared by radiation pressure and P–R drag and the remnant of a disk should become easier to see in the near IR as is the case, for example, for the IR excess MS stars (Backman and Paresce 1993). At this point, the disk may extend out to several hundred AU but be detached from the star by $\sim$ 1–2 AU due to sublimation and/or the action of protoplanets.

Accurate imaging of these disks will clearly require the triple combination of high sensitivity, resolution, and dust penetrating ability of an IR optimized large collecting area interferometer. For the closest star forming regions located at $\sim$ 150 pcs, the accretion disk will be $\lesssim$ 100 mas in extent and require spatial resolutions in the range 1–20 mas for accurate mapping of its outer portions. These spatial resolutions will be sufficient, for example, in detecting gaps or flaring that are very sensitive indicators of the presence of low mass companions such as planets. Spectral resolutions from a few for high sensitivity to dusty featureless disks to several thousand for kinematic studies of the inner disk, wind and jet features should be straightforward.

The study of the IR excess MS stars that have large dusty disks surrounding them also require observations in the thermal IR where their emission peaks and their temperature structure determined most unambiguously. The latter is the critical observational parameter needed to constrain the accretion disk models presently available. At the same time, for an understanding of the later stages of PMS evolution and, in particular, the circumstellar envelope, the wind and the jets, high spatial and spectral resolution in the NIR will be of fundamental importance. Good coverage of the (u,v) plane, unvignetted fields of several arcseconds and good AO systems will be prerequisites for this field but nothing would seem to preclude their routine availability by $\sim$ 2005.

### 4.3. The Galactic Center and Active Galactic Nuclei

The IR sensitivity and angular resolution of the new generation ground-based interferometers presently being planned are sufficient, in principle, to resolve the central region of our galaxy since 10 mas would correspond to $\sim$ 100 AU at the galactic center. The targets that will certainly be attacked first with these devices will be the star cluster centered on IRS 16 and the infrared sources close to the position of SgrA*. It is expected that the 3D velocity field of the cluster could be determined all the way in to the core where single telescopes would be limited by crowding. This information, together with the continuum spectrum of every object therein, will shed considerable and, perhaps, definitive light on the nature of our galaxy's center including, in particular, whether or not it harbors a massive black hole and whether or not there is local and continuing star forming activity. The spectra will also be searched for signs of a possible collapse of the cluster core and the role of binary star formation due to an enhanced collision rate. Finally, good (u,v) plane coverage should also ensure the capability of imaging directly in the IR an accretion disk surrounding the black hole and to reveal the distribution of warm dust associated with SgrA*.

A key issue ground-based interferometers are sure to address by the next decade is the structure of the nuclear region of AGNs. The central one parsec or so will certainly be probed in about 20–30 nearby Seyfert galaxies that are bright enough to be used as references for fringe tracking. These measurements are quite likely to answer some of the basic and extant questions concerning AGNs like whether the observed differences are due to evolution, environment or viewing geometry. Detailed imaging of the obscuring torus of dust and gas is certainly the best way to directly investigate the unified model of AGNs.

## 5. Conclusions

The three key research areas discussed very briefly above serve perfectly to highlight, among other things, what performance specifications space-based interferometers will have to meet to fit seamlessly into the overall plan for very high angular resolution astrophysics into the next century. In the low mass companion area, it is quite clear that no instrument on the ground will, in the foreseeable future, have enough sensitivity and precision even in the near IR to detect and properly characterize Earth-like planets. Only a space-based interferometer with approximately ten times the astrometric accuracy achievable from the ground, with the ability to operate in the far IR where the inherent star-planet contrast ratio is particularly favorable and where many if not most of the unique spectral features reside and, finally, with the ability to operate from beyond the solar system's zodiacal cloud can have a prayer to carry out such an ambitious but essential undertaking. This then should be the basic and fundamental objective of such a complex and costly device.

Similar considerations can be made for the other areas outlined above. For the AGNs, for example, the highest possible spatial resolution imaging in the far IR to locate and study the heated dust commonly thought to dominate the IR emission from the central regions can best be accomplished from space where atmospheric absorption and emission effects can be eliminated. The ground will certainly cover the nearest objects but to get a better idea of the evolution of the phenomenon and to probe more distant and diverse objects space again is essential. An effective study of the star-disk transition region located at several stellar radii and the point of origin of the jets and bipolar outflows absolutely require what no ground-based device can possibly achieve for the moment i.e.: 0.1 mas resolution to image down to the stellar surface, simultaneous access to the optical range for the jets and the far IR for the disk.

**Acknowledgements:** I wish to thank all the members of the ESO Interferometry Science Advisory Committee (Paresce 1996) who have vigorously participated in defining the science drivers and the preliminary technical specifications for the first phase of VLTI and without whose assistance and advice this article could not have been written.

## References

Armstrong, J.T. et al.: 1995, *Physics Today* May, p. 42

Backman, D., Paresce, F.: 1993, in Protostars and Planets III, U. Arizona Press, p. 1253
Baldwin, J.E. et al.: 1996, *A&A* **306**, L13
Black, D.C.: 1995, *ARAA* **33**, 359
Duquennoy, A. et al.: 1995, in Science with the VLT, J.R. Walsh and I.J. Danziger (Eds.), Springer-Verlag, p. 150
Malbet, F., Bertout, C.: 1995, *A&AS* **113**, 369
Mayor, M., Queloz, D.: 1995, *Nature* **378**, 355
Exploration of Neighboring Planetary Systems, Mission and Technology Road Map, NASA Presentation, 1995
Paresce, F. et al.: 1996, *ESO Messenger* **83**, 14
Quirrenbach, A.: 1996, these proceedings
Saumon, P. et al.: 1996, *ApJ* **460**, 493
Schwarzschild, B.: 1996, *Physics Today* April, p. 17
Shao, M., Colavita, M.M.: 1992, *A&A* **262**, 353
von der Lühe, O. et al.: 1994, *Proc. SPIE* **2200**, 168

SER
TOLEDO

# INFRARED INTERFEROMETRY WITH THE VLTI

A. QUIRRENBACH
*Max-Planck-Institut für Extraterrestrische Physik*

## 1. Introduction

The VLT Interferometer, currently under construction on Paranal by the European Southern Observatory, will be the major ground-based interferometer open to the European community at the beginning of the next century. The baseline plan for the VLTI (see the reports by the ESO VLT Interferometer Panel, 1989 and 1992) includes combination of the four 8 m telescopes and a number of 2 m class auxiliary telescopes providing better coverage of the *uv* plane, and baselines of up to 200 m. The VLTI will operate at wavelengths ranging from about 0.45 $\mu$m to 20 $\mu$m. The scientific objectives of the VLTI have recently been summarized by Paresce et al. (1996) and by Paresce (these proceedings). It is expected that the VLTI will have a large impact on many fields of stellar research, including star formation and stellar evolution, stellar surface structure, low-mass stars and brown dwarfs, circumstellar matter, and mass loss from pre-main-sequence objects and from evolved stars. In extragalactic astronomy, the unprecedented combination of sensitivity and angular resolution will open new vistas for studies of the Galactic Center, of Seyfert nuclei, quasars, and high-redshift objects.

## 2. The VLTI at 10 $\mu$m and 20 $\mu$m

It will be relatively easy to implement the interferometric mode of the VLT in the 10 and 20 $\mu$m atmospheric windows ($N$ and $Q$ bands), since the instrumental tolerances are not as stringent as at shorter wavelengths. Most importantly, even the 8 m telescopes will be nearly diffraction-limited with simple tip-tilt guiding; no complicated adaptive optics system is required. The delay lines can be dithered at a rate of $\sim$ 20 Hz, providing for simple fringe detection and extremely effective rejection of sky noise. For background-limited observations in the thermal infrared, there is no loss in sensitivity for dispersed fringe detection, therefore interferometry with modest spectral resolution ($R \approx 200$) comes "free of charge". In this observing mode, the interferometric



*C. Eiroa et al. (eds.),*
***Infrared Space Interferometry: Astrophysics & the Study of Earth-Like Planets*, 97–99.**

field-of-view is quite large ($\gtrsim 1''$), and coherent operation of the interferometer can initially be achieved without active fringe tracking. The main challenge is efficient baffling of the focal plane instrument to keep the background at a tolerable level. The sensitivity will depend strongly on the baffling efficiency; it should be possible to track fringes at $10\,\mu$m on sources of a few 100 mJy under conservative assumptions. Interferometry in the thermal infrared will be an important component of most of the observing programs mentioned above. In particular, the VLTI working at $10\,\mu$m and $20\,\mu$m will be the ideal instrument for observations of disks around main-sequence stars, i.e., for gathering information on exo-zodiacal light. For these and other observations requiring very high contrast it may be possible to use two elements of the VLTI as a nulling interferometer. VLTI observations in the thermal IR will thus play an important role in the scientific and technological preparation of space missions aimed at the detection of Earth-like planets.

## 3. Observing Extrasolar Planets with the VLTI

Astrometric methods for planet detection are quite similar to radial velocity techniques, insofar as they look for the motion of the stellar photocenter around the barycenter of the planetary system. The VLTI has the potential of being an extremely powerful instrument for precise narrow-angle astrometry (see Shao and Colavita 1992). For instance, the atmospheric limit for determining the distance vector between two stars separated by $10''$ in a half-hour integration is about $10\,\mu$as (von der Lühe, Quirrenbach, Koehler 1995). The technical realization of this potential requires monitoring of the baseline vector with $\sim 50\,\mu$m precision, and measurement of the differential delay between the two stars with $\sim 5$ nm precision. This could be done with regular astrometric solutions and a laser metrology system. Implementation of an astrometric mode in the VLTI with these capabilities would enable us to detect Sun – Jupiter systems out to a distance of 1 kpc, Sun – Uranus out to 150 pc, and small planets (10 Earth masses) around the closest stars (Quirrenbach 1995). The following strategy could be adopted for the astrometry program: A list of $\sim 200$ target stars will be observed in the $K$ band with the auxiliary telescopes. These stars have to be bright enough for fringe tracking ($K \lesssim 12$), so that the astrometric references can be relatively faint ($K \lesssim 17$), ensuring that references can be found for almost any star. With an integration time of 30 min per star and night and 30 observations per target star, the total amount of observing time will be 300 nights, spread over a decade. The data for each star will be used to solve for relative parallax and proper motion; any residuals from this fit will indicate the presence of planets. The target list will include candidate planetary systems from radial velocity searches, for which the inclination and thus planetary mass can be determined with the VLTI, and potentially from a large-scale astrometric survey such as GAIA. There should also be a survey of other interesting objects, such as the closest stars – the VLTI can detect lower-mass planets than radial velocity searches for these –, IR-excess stars like $\beta$ Pic, and pre-main-sequence objects in low-mass star forming regions and in Orion.

While indirect methods will certainly yield a wealth of data about extrasolar planetary systems, the direct detection of photons from the planet itself would enable more detailed astrophysical studies: determination of chemical composition and temperature through spectroscopy, and study of surface structure and rotation through lightcurve analysis. It is extremely difficult to detect planets against the glare of the parent star in the visual and near-IR, but it may be possible with advanced adaptive optics systems (Angel 1994). The alternate approach is in the thermal IR, where the contrast is reduced by several orders of magnitude. A sensitivity calculation shows that Jupiter at a distance of 10 pc would not be detectable against the thermal background in a reasonable time with an Earth-based (and therefore uncooled) 8 m telescope (Saumon et al. 1996). It should be kept in mind, however, that other planetary systems are not necessarily similar to ours. In particular, quite warm giant planets may exist, either because of internal heating, which is stronger in younger and in more massive planets, or because of stronger external irradiation for planets with early-type parent stars or in close orbits (as in the case of 51 Peg, Mayor and Queloz 1995). There may be a realistic chance of detecting such warm giant planets in the solar neighborhood with the VLTI at 10 or 20 $\mu$m.

We can thus expect that a fair number of planetary systems can be studied astrometrically with the VLTI, and a few could even be observable directly. These systems would be excellent targets for follow-up studies with a space-based infrared interferometer.

## References

Angel, J.R.P.: 1994, *Nature* **368**, 203

ESO VLT Interferometry Panel: 1989, 1992, VLT Reports No. 59b and 65

Mayor, M., Queloz, D.: 1995, *Nature* **378**, 355

Paresce, F., et al.: 1996, *ESO Messenger* **83**, 14

Quirrenbach, A.: 1995, in Science with the VLT, Springer-Verlag, p. 33

Saumon, D., et al.: 1996, *ApJ* **460**, 993

Shao, M., Colavita, M.M.: 1992, *A&A* **262**, 353

von der Lühe, O., Quirrenbach, A., Koehler, B.: 1995, in Science with the VLT, Springer-Verlag, p. 445

# IMAGING WITH A SPACE-BASED INFRARED INTERFEROMETER

A. QUIRRENBACH and A. ECKART
*Max-Planck-Institut für Extraterrestrische Physik*

**Abstract.** Subarcsecond imaging throughout the infrared spectral range will be possible only with space-based interferometry. Major advances in virtually all fields of astronomy are expected from such observations. Examples are: characterization of dust around main-sequence stars, observing the earliest phases of star formation, unravelling the complex processes in the nuclei of active galaxies, and understanding the high-redshift universe. Imaging capabilities can be incorporated in a mission whose primary goal is the investigation of extrasolar planets with nulling techniques.

**Key words:** space interferometry, imaging, scientific case

## 1. Introduction

Space-based infrared interferometry with the specific goal of detecting and examining extrasolar planets, using nulling techniques as proposed by Bracewell (1978), is discussed extensively in many contributions to this volume. In this article we concentrate on complementary aspects of infrared interferometry from space: scientific goals and design considerations for an imaging instrument, which could address a wide range of astronomical questions. We will also consider the question how both nulling and imaging modes could be implemented in a single observatory.

## 2. Why Interferometric Imaging from Space?

The development of sub-arcsecond imaging at visual and near-IR wavelengths is one of the most significant recent advances in observational astronomy. Ground-based observations with modern telescopes in good seeing and the huge success of the Hubble Space Telescope (HST) show that improvements in angular resolution by factors of 2 to 10 can often give a dramatic increase in astronomical information. Sometimes such improvements open totally new fields. In the near-IR and mid-IR (at wavelengths between 1 and 20 $\mu$m) further progress is expected from diffraction-limited imaging with ground-based 10 m class telescopes using adaptive optics, and from the Next Generation Space Telescope (NGST), which might also have a diameter of up to 8 m. Even



*C. Eiroa et al. (eds.),*
*Infrared Space Interferometry: Astrophysics & the Study of Earth-Like Planets,* 101–107.

Table I. Resolution of an interferometer with a 30 m baseline.

| $\lambda$ [$\mu$m] | resolution | linear resolution at 10 pc | 1 kpc | 3 Mpc | $z = 0.5$ |
|---|---|---|---|---|---|
| 2 | 14 mas | 0.14 AU | 14 AU | 0.2 pc | 70 pc |
| 5 | 35 mas | 0.35 AU | 35 AU | 0.5 pc | 170 pc |
| 10 | 69 mas | 0.69 AU | 69 AU | 1 pc | 340 pc |
| 20 | 0.″14 | 1.4 AU | 140 AU | 2 pc | 700 pc |
| 50 | 0.″35 | 3.5 AU | 350 AU | 5 pc | 1.7 kpc |
| 100 | 0.″69 | 6.9 AU | 690 AU | 10 pc | 3.4 kpc |
| 200 | 1.″4 | 14 AU | 1400 AU | 20 pc | 7.0 kpc |

better angular resolution can be obtained with infrared interferometry; as an example, Table I lists numerical values through the infrared range for an instrument with a 30 m baseline. Note that such an interferometer operated at 10 $\mu$m has roughly the same resolution as the HST at 500 nm, and an 8 m telescope at 2 $\mu$m.

Ground-based optical and infrared interferometry has been developed during the past two decades. A number of instruments have produced significant astronomical results, and the technique will mature over the next few years with the development of multi-baseline and large-aperture arrays, which will be capable of executing a wealth of exciting observing programs in virtually all areas of astronomy (e.g., Paresce et al. 1996). However, ground-based arrays require a fairly bright unresolved source within the isoplanatic field for fringe tracking; the VLTI will have a fringe tracking sensitivity of a few hundred mJy in the N (10 $\mu$m) and Q (20 $\mu$m) windows (Quirrenbach 1996). In contrast, a cool ($\approx$ 40 K) space interferometer with modest aperture should easily reach a point source sensitivity of 100 $\mu$Jy at 10 to 20 $\mu$m with integration times of a few minutes. (It should be pointed out that ISO, which has a 60 cm aperture, has actually detected sources at that level.) Unrestricted access to the whole infrared range and data that are not corrupted by atmospheric turbulence are further important advantages of infrared interferometry in space.

## 3. Science Drivers

With good sensitivity in the near-IR and mid-IR ($\lesssim$ 30 $\mu$m) and unrestricted access to the far-IR (30 to 200 $\mu$m), combined with subarcsecond angular resolution, space interferometry will offer unique capabilities for detailed observations of cool objects. In the following sections, we try to identify a number of areas where these capabilities are needed to tackle open questions of fundamental importance. This list is by no means exhaustive, and it is quite likely that results from ground-based interferometry and from single-telescope infrared observatories (ISO, SIRTF, FIRST, NGST) will change our priorities significantly.

### 3.1. Dust Around Main-Sequence Stars

Prominent among the unexpected discoveries of IRAS is the detection of a large infrared excess around main-sequence stars, indicative of circumstellar solid material (for a review see e.g. Backman and Paresce 1993). After the first example of such an object found, this is called the "Vega phenomenon". While excess emission at $12\,\mu$m is rare, $\gtrsim 15\%$ of all main-sequence stars seem to display a far-IR (60 and $100\,\mu$m) excess. In many cases the far-IR emission is actually dominated by the dust emission, not by the stellar photosphere. Typical dust temperatures range from 20 to 150 K, and cloud or disk radii from tens to a few hundreds of AU. The Kuiper Belt in our Solar system has similar temperature and radius, so the Vega phenomenon may be a manifestation of dust in "exo-Kuiper Belts". Observations with ISO and SIRTF will certainly give much expanded statistics on IR-excess stars, and give a wealth of information on individual cases, allowing us to model cloud size and geometry, temperature, and grain size and distribution. With a space interferometer operating at $100\,\mu$m, one could go one step further and obtain real images of these objects; at a distance of 20 pc, the resolution will be of order 15 AU. It would thus be possible to check directly whether the geometry is cloudlike or disklike, and whether there are relatively empty zones of a few tens of AU around the central stars. An interferometer would also provide more sensitivity to detect weak dust emission than a single telescope, since it could easily distinguish between the pointlike photosphere and emission extended on the arcsecond scale.

### 3.2. Star Formation

The process of star formation as we understand it today is characterized by a number of complex phenomena: collapse of molecular cloud cores, infall, disk accretion, winds, outflows and jets. Observing the first phases of these processes is particularly challenging, because the protostar is still deeply embedded in the parent molecular cloud. Since this envelope is optically thick at wavelengths $\lambda \lesssim 30\,\mu$m, most of the luminosity emerges at wavelengths $50\,\mu\text{m} \lesssim \lambda \lesssim 100\,\mu$m from an extended dust photosphere, which has a typical size of 50 to a few 100 AU. At the distance of the nearest regions of low-mass star formation (140 pc), the resolution of a space interferometer with $\sim 30\,\mathrm{m}$ baselines should be just sufficient to measure the size and shape of this photosphere, and to separate the contributions of the accretion disk, the photosphere, and the extended emission from the surrounding cloud.

During the subsequent early evolution of the young pre-main-sequence object the envelope is dispelled, and the star becomes detectable at near-IR and then at visual wavelengths. Many of these pre-main-sequence stars still display a strong infrared excess at wavelengths up to $\sim 100\,\mu\mathrm{m}$, indicative of a circumstellar disk. Models of this disk are largely based on fits to the spectral energy distribution. It is known, however, that many of these stars are actually multiple systems (see e.g., Mathieu 1994); observations with subarcsecond resolution are therefore needed to locate the infrared excess. By clearly separating the contributions from disks associated with individual components of the multiple system, a large disk around the whole system, and an even more extended envelope, IR interferometry will contribute significantly to our understanding of the formation and early evolution of pre-main-sequence binaries. This in turn will provide essential input to the theory of cloud collapse and fragmentation, and thus to any detailed theory of star formation.

### 3.3. Nuclei of Active Galaxies

To understand the nuclei of active galaxies, we must consider a multitude of components and processes over a large range of spatial scales: an underlying population of old stars, molecular matter falling towards the center and forming young stars at a vigorous rate, ionized gas, a parsec-scale dusty molecular torus providing very large extinction ($A_V \gtrsim 50$) in edge-on objects, energetic jets, and accretion onto a central supermassive object. Infrared observations are indispensable for AGN studies, because of the high obscuration of the central region, and because many objects emit the bulk of their luminosity in the IR. Fine structure lines found throughout the whole IR range provide a superb diagnostic tool for measurements of the physical state of ionizied gas and, by inference, of the local radiation field. Molecular lines and dust continuum trace the cool material that is invoked by unified theories to explain the observed differences between objects with and without broad emission lines. The launch of ISO has opened new vistas in this field, and a rich harvest is expected from ISO as well as the next generation of IR satellites. However, it is already clear that lack of angular resolution will limit the capability of single-telescope missions to separate the nucleus from circumnuclear starburst rings and the emission of the underlying galaxy. Multi-line spectroscopic analysis is thus hampered by a mixture of lines originating in different regions with a variety of separations from the center and therefore with different physical conditions. Observations with an IR interferometer will give individual spectra of these regions. It should be possible to determine the size and geometry of the obscuring torus ($\approx 10\,$pc) in nearby objects through its dust emission, and to search for molecular lines from the same region. With sufficient spectral resolution ($R \sim 3000$), the kinematics of molecular and ionized gas can be studied on the pc scale. In quasars and BL Lac objects it will be possible to separate the synchrotron emission from the thermal dust continuum; in some cases, mid-IR images of the synchrotron jets and hotspots can be obtained.

### 3.4. Objects at Intermediate and High Redshift

The evolution of the Universe over cosmic time scales can be studied through observations of objects at intermediate ($z \gtrsim 0.5$) and high ($z \gtrsim 3$) redshift. The intrinsically brightest objects are naturally those most easily observed over these large distances. A space interferometer will be a powerful instrument to study starburst galaxies, AGN, quasars, and ultraluminous infrared galaxies (ULIRGs) at cosmological distances. In some cases, it is possible to take advantage of gravitational lensing, which amplifies and magnifies a distant object. The study of objects like F10214+4724 with a space interferometer will help us understand luminous objects in the early universe; observations of gravitational arcs detected in the visual wavelength range will give insight into more normal objects at high redshift.

Deep surveys with the aim of detecting the first generation of star-forming galaxies ("protogalaxies") at redshifts of 5 to 10 (?) are among the main scientific drivers of SIRTF, FIRST, and NGST; it is expected that such objects will be found at the $\approx 10\,\mu$Jy level at $10\,\mu$m. Since an interferometer with modest aperture will be less sensitive than NGST, and because of the limited field-of-view, an interferometric mission will probably not be a competitive instrument for a cosmological deep survey of "empty fields". Follow-up observations of fields previously observed with NGST may prove useful, however, to obtain high-resolution images of selected objects, or to resolve crowding problems.

## 4. Technical Requirements

To accomodate a broad range of scientific programs, a space interferometer should cover a large wavelength range (10 to 200 $\mu$m) and provide broad-band and spectroscopic ($R \lesssim 3000$) capabilities. The sensitivity at the longer wavelengths will depend critically on the operating temperature of the satellite; note, however, that even a passively cooled interferometer will be sensitive enough for many of the most interesting projects. Thronson et al. (1995) give an extensive discussion of the sensitivity of single-aperture space missions. Many of their considerations can also be applied to an interferometer with the same total collecting area, with the notable exception of source confusion. Because of the higher angular resolution, an interferometer is less affected by confusion than a single telescope, but the details depend in a complicated way on the instrument (baseline length, coverage of the *uv* plane, interferometric field-of-view) and on the small-scale clumpiness of the background. These parameters have to be taken into account for the optimization of a space interferometer, since the balance between detector noise, thermal background from the optics, and confusion is an important design criterion.

To achieve good sensitivity without the need for a bright reference star, the interferometer has to be very stable. The fringes must not move more than $\lambda/10$ or so during the detector integration time; their position must be known with the same accuracy for each coherent integration period. This requires a laser metrology system to monitor structural vibrations, but the tolerances are much less stringent than those needed in a nulling interferometer.

## 5. Imaging with a Nulling Interferometer

Imaging and spectroscopy of extrasolar planets is currently the main scientific driver for the development of infrared interferometry. This application requires the implementation of nulling techniques for the rejection of the light from the bright central star. Should and could such a nulling instrument be equipped with an imaging mode? There are a number of important arguments for the inclusion of imaging capabilities in a DARWIN type observatory:

- The scientific scope of the mission will be broadened considerably. Important results in many areas of astronomy can be expected, as detailed in Section 3 above.
- Observations in imaging mode can be used to verify the in-orbit performance of most subsystems of the interferometer and spacecraft.
- Many of the imaging projects will require only modest amounts of observing time. They can be performed during the first phase of

the mission, while the satellite is "in transit" from the Earth to the outer solar system. They will therefore not compete for observing time with the nulling observations.

- All targets in the extrasolar planet program can be observed in imaging mode to gather information on the strength and orientation of their associated zodiacal and far-IR emission. This can be done before the observations in nulling mode begin, so that a further pre-selection of the target stars is possible.

The technical realization of a dual-mode (nulling and imaging) interferometer is fairly straightforward: the interferometer will be equipped with two beam-combination instruments. One of these instruments is a "standard" image-plane or pupil-plane beam combiner; the other instrument contains the appropriate phase shifters needed for nulling. Slide-in mirrors are used to switch between the two instruments. The instruments are independent of one another; each one can therefore be optimized for its respective use, and be equipped with the necessary dispersing elements and detectors.

The quality of interferometric images depends critically on the coverage of the *uv* plane. In a dual-purpose interferometer the location of the individual telescopes will be mainly determined by the requirements of nulling, but some optimization of the imaging capabilities should still be possible. Among the configurations that have been proposed for a nulling interferometer, non-linear arrangements are clearly superior to linear arrays for imaging purposes. It is possible in all cases to enhance the imaging performance by adding one extra telescope to the nulling array. While a movable add-on telescope would be ideal to fill the *uv* plane, this solution may be prohibitively complex and expensive. However, even a linear four-element array with one additional telescope in a carefully chosen position, or a non-linear four-element array simultaneously optimized for nulling and imaging will provide a *uv* coverage sufficient for carrying out the scientific program described above.

## References

Backman, D.E., Paresce, F.: 1993, in Protostars and Planets III, E.H. Levy, J.I. Lunine (Eds.), University of Arizona Press, p. 1253

Bracewell, R.N.: 1978, *Nature* **274**, 780

Mathieu, R.D.: 1994, *ARAA* **32**, 465

Paresce, F., et al.: 1996, *ESO Messenger* **83**, 14

Quirrenbach, A.: 1996, in Science with the VLT Interferometer, F. Paresce (Ed.), Springer-Verlag, in press

Thronson, H.A., Rapp, D., Bailey, B., Hawarden, T.G.: 1995, *PASP* **107**, 1099

# INFRARED INTERFEROMETRY OF ACTIVE GALAXIES

G. M. VOIT
*Space Telescope Science Institute*
*3700 San Martin Drive, Baltimore, MD 21218*
*e-mail: voit@stsci.edu*

**Abstract.** IR interferometers operating at $2 - 20\,\mu$m over $\sim 100$ m baselines will be ideally suited to probing the interfaces between active galactic nuclei and their host galaxies. When these devices become operational, they will tell us how quasars generate infrared light and will clarify the relationships between the powerful black-hole engines at the centers of galaxies and the surrounding bursts of star formation.

**Key words:** active galaxies

## 1. Introduction

Interferometric instruments now planned for both terrestrial telescopes and space-based observatories in the coming decades will increase our potential resolving power in the optical and infrared bands by over an order of magnitude. Many articles in this volume document the burgeoning interest in interferometry as a way to discover and to investigate planetary systems around other stars. Although studies of exoplanets are perhaps the primary motivator for these kinds of instruments, interferometers will be useful in a multitude of other ways. This article outlines how infrared interferometers currently in the development or planning stages can be used to probe deeply into the hearts of active galaxies.

Why do we want to look at active galaxies? Most people find them interesting for one or more of the following reasons:

- They contain the most powerful engines in the universe.
- These engines are likely to be supermassive black holes that accrete matter from the surrounding galaxy.
- The most luminous active galaxies were much more common during the first few billion years of the universe's development, indicating that activity is linked somehow to the formation of galaxies.

Infrared interferometers are particularly well suited to studying the behavior of active galaxies at the interface between the black-hole driven engine and the interstellar medium of the host galaxy. Insights



*C. Eiroa et al. (eds.),*
*Infrared Space Interferometry: Astrophysics & the Study of Earth-Like Planets,* 109–118.

gleaned from the goings-on in this transitional region could help greatly in establishing how the central engines of active galaxies influence their hosts and vice versa. Such knowledge might eventually enable us to use the properties of distant quasars, the brightest high-redshift objects, to piece together the early history of the universe.

## 2. Active Galaxies: Basics

Active galactic nuclei (AGNs), including quasars, Seyfert 1 galaxies, Seyfert 2 galaxies, LINERS, and BL Lac objects, exhibit a wide variety of manifestations, but their underlying architecture is probably quite similar. The standard model for AGNs posits a supermassive black hole girdled by an accretion disk feeding it gas at rates ranging up to $10\,M_\odot\,\mathrm{yr}^{-1}$. These accretion disks, which radiate about 10% of the incoming gravitational potential energy, can emit luminosities up to $10^{47}\,\mathrm{erg\,s}^{-1}$, mostly in the form of UV and soft X-ray photons. A minority of AGNs also propel radio jets that move at relativistic speeds, presumably along the rotation axis of the black hole. For a good introduction to AGNs, see Blandford et al. (1990).

As UV and soft X-ray radiation flows out from an active nucleus, it ionizes interstellar gas at a wide range of distances and excites a rich emission-line spectrum. Strong forbidden lines with breadths of several hundred km $\mathrm{s}^{-1}$ embellish the spectra of almost all active galaxies. Collisional deexcitation suppresses these lines at densities $> 10^3 - 10^7\,\mathrm{cm}^{-3}$, so photoionization equilibrium constrains them to form at incident flux levels $< 10 - 10^6\,\mathrm{erg\,cm}^{-2}\,\mathrm{s}^{-1}$, corresponding to $10 - 1000\,\mathrm{pc}$ from the nucleus of a luminous quasar. Permitted lines like Ly $\alpha$ and C IV $\lambda1550$ are just as common, but in quasars and Seyfert 1 galaxies these lines are much broader, with widths of several thousand km $\mathrm{s}^{-1}$. The broad permitted lines come from a substantially denser region, less than a parsec from the center, that withstands UV/X-ray fluxes $> 10^8\,\mathrm{erg\,cm}^{-2}\,\mathrm{s}^{-1}$. These line-emitting regions are known as the *narrow-line region* and *broad-line region*, respectively. Even the most distant quasars exhibit strong lines from species like C IV and N V (Schneider, Schmidt, & Gunn 1991), indicating that their nuclear regions have undergone substantial star formation and chemical evolution in $< 1$ Gyr.

Evidence is mounting that all active galaxies have broad-line regions that sometimes go unseen. The nearest Seyfert galaxy, NGC 1068, has traditionally been classified as a Seyfert 2, since it lacks prominent broad-line emission. In polarized flux, however, its hydrogen Balmer lines clearly have broad wings (Antonucci & Miller 1985). Apparently, optically thick gas obscures our line of sight to the broad-line region, but a scattering medium redirects some of the broad-line emission around the obscuring material. NGC 1068 is not unique in this respect; spectropolarimetric observations continue to discover hidden broad-line regions in narrow-lined active galaxies (Miller & Goodrich 1990).

The prevailing opinion among investigators of AGNs is that much of their seeming diversity can be traced to orientation effects. These so-called "unified models" of active galaxies postulate that a thick torus of molecular clouds orbits the nucleus between the narrow-line and broad-line regions (Krolik & Begelman 1986; 1988). When seen on-axis, the torus reveals the broad-line region and accretion disk at its center. When seen edge-on, the torus hides the innermost part of the AGN, and only the narrow-line region can be seen directly. In some models, cloud-cloud collisions within the torus facilitate accretion of the host galaxy's interstellar medium into the nucleus.

The newest line of evidence favoring the standard picture of AGNs has come from interferometry of $H_2O$ masers in NGC 4258, a nearby spiral galaxy with a LINER nucleus (Greenhill et al. 1995; Miyoshi et al. 1995). A thin, warped disk of masing molecular clouds $0.13 - 0.25$ pc ($4 - 8$ mas) in radius orbits a $3.6 \times 10^7 M_\odot$ object at the center of NGC 4258 in perfect circular Keplerian rotation. The disk axis runs parallel to a large-scale H$\alpha$ jet directed almost perpendicular to the galaxy's rotation axis. Polarized broad lines and UV continuum emission from NGC 4258 indicate that the active nucleus itself is shielded from us (Wilkes et al. 1995). Infrared interferometry of this galaxy, and numerous others, at $\sim 10 - 100$ mas resolution will be ideal for studying the mysterious interface between the accretion-dominated AGN environment and the galactic disk of the host galaxy.

## 3. Infrared Emission from AGNs

To within about an order of magnitude, the broad-band spectra of active galaxies are flat in $\nu F_\nu$ from 100 $\mu$m through the X-ray band. These spectra typically peak in the UV, drop to a local minimum at around 1 $\mu$m, and rise again to a local maximum somewhere in the mid- to far-IR (Sanders et al. 1989). At wavelengths longer than 100 $\mu$m, AGN fluxes usually decline sharply, suggesting that heated dust grains emit the IR continuum from most active galaxies. The virtually universal minimum at around 1 $\mu$m adds weight to this hypothesis because dust grains hot enough to emit efficiently shortward of 2 $\mu$m quickly sublimate (Sanders et al. 1989; Phinney 1989). Across these two orders of magnitude in wavelength, AGN fluxes frequently remain constant in

$\nu F_\nu$ to within factors of a few, implying that the emitters, if thermal, must span a temperature range of 30 – 1500 K.

The limited spatial resolution currently afforded by IR instruments is insufficient for testing the various hypotheses for the origin of the IR emission: **(1) Warped Disk.** Illumination of a galaxy by an embedded AGN heats its interstellar dust. To reproduce the broad-band spectral characteristics of AGNs, dust at each order of magnitude in distance must intercept roughly equal amounts of UV flux, totaling $\sim$ 30% of the UV output from the central engine (Sanders et al. 1989). **(2) Torus + Galaxy.** A well-defined obscuring torus, as depicted in unified models, sits at a particular distance from the nucleus, so the temperature range of the dust it contains is limited. In between the broad- and narrow-line regions, dust temperatures are $\sim$ 60 – 600 K, leading to a peak flux in the mid-IR. If the torus shields the rest of the host galaxy from direct AGN illumination, the galaxy itself must supply the far-IR continuum (Pier & Krolik 1992; 1993). **(3) Starburst.** Certain heterodox models for active galaxies claim that there is no central engine (Terlevich & Melnick 1985). Instead, they argue that all the phenomena we associate with AGNs, except for jets, which are rare, can be explained as immense, centrally concentrated starbursts. In this case, the temperature of the IR continuum would not vary greatly with position, and the 2 $\mu$m continuum emission from quasars would spread over much more than a few parsecs. **(4) Synchrotron.** The IR continuua of some active galaxies, especially BL Lac objects, could be entirely non-thermal, with a far-IR/sub-mm decline owing to synchrotron self-absorption. Non-thermal far-IR synchrotron emission would not need to be extended at all, even on interferometric scales. Infrared interferometers, which will be able to resolve thermally emitting mid-IR regions in the brightest AGNs, will easily eliminate many of these hypotheses. If the mid-IR emission indeed comes from AGN-heated dust, these interferometers will be able to tell us much more.

## 4. Resolution & Brightness

Several articles in this volume describe ground-based interferometers currently under development and ideas for space-based interferometers optimized for studies of exoplanets. The Keck interferometer (Shao) and VLTI (Parsece) plan baselines of $\sim$ 100 m and OASES (Woolf) proposes a 75 m baseline, so we will adopt 100 m as a projected baseline for AGN studies. At 10 $\mu$m, the nominal resolving power of such an instrument is $\sim$ 20 mas.

Our ability to resolve a thermally emitting region using this beam-size depends primarily upon the optical/UV flux of the illuminating source, in this case an AGN. The equilibrium dust temperature, $T_d$, at some angular distance $\delta$ from an optical/UV source of flux $F_{\rm obs}$ is determined by the flux the dust sees, $F_d = \delta^{-2} F_{\rm obs}$. Since dust grains generally absorb UV radiation much more efficiently than they absorb infrared radiation, a sort of "dust greenhouse effect" elevates $T_d$ well above the effective temperature of the radiation field. Fig. 1 illustrates how $\delta$ varies with $F_{\rm obs}$, given in terms of $V$ magnitudes, for several interesting dust temperatures. Since emission from dust at $\sim$ 300 K peaks at 10 $\mu$m, instruments like those described above can resolve thermally emitting 10 $\mu$m regions around objects brighter than $V = 15$. Fainter 10 $\mu$m regions from somewhat cooler dust will be even more extended. Numerous nearby active galaxies satisfy this brightness criterion. The CfA survey identified tens of AGNs with $V < 14.5$ (Huchra & Burg 1992), including numerous Seyfert galaxies and LINERS, a BL Lac object, and one luminous quasar, 3C 273, at $z = 0.173$.

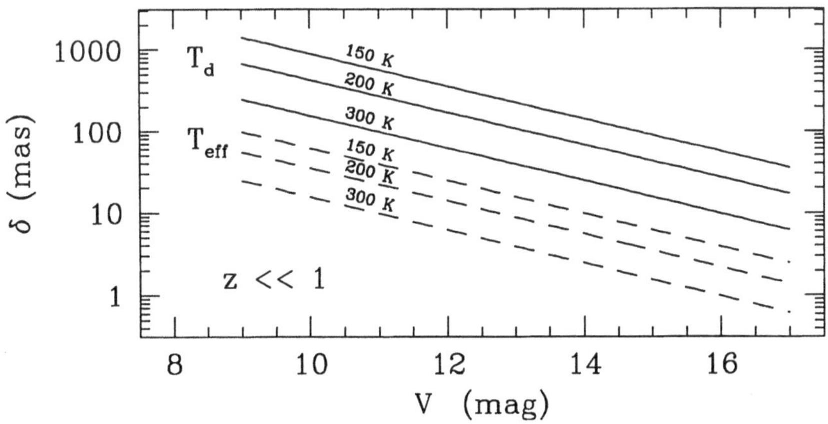


*Figure 1.* Angular sizes of thermally emitting regions. The solid lines give the angular sizes, $\delta$, of regions with typical dust temperature $T_d$ around AGNs with varying $V$ magnitudes. The dashed lines give the effective temperatures of the dilute UV/optical fields incident on these regions

Interferometric imaging of AGNs with $V \lesssim 15$ will enable the following kinds of studies: **(1) 2 $\mu$m Compactness.** If AGNs are just extreme starbursts, their 2 $\mu$m emission should be entirely resolved.

**(2) 10 $\mu$m Compactness.** If the IR continuum from AGNs is primarily non-thermal, their 10 $\mu$m emission will be unresolved. **(3) Warped Disk Morphology.** The signature of a thermally emitting warped disk will be an elongated IR image whose position angle changes systematically with wavelength. **(4) Accretion-Disk/Galaxy-Disk Relationships.** IR interferometers with 100 meter baselines will be tuned to spatial scales in between the VLBI/VLBA observations of masing disks in AGN and diffraction-limited IR imaging with single-element telescopes. **(5) UV Anisotropy.** Anisotropic emission of UV flux will systematically elevate $T_d$ along certain position angles. **(6) Torus Size/Morphology.** IR interferometers could conceiveably image the putative obscuring tori around AGNs, whose sizes are currently not well constrained. **(7) Starburst Knots.** Regions of intense star formation near AGNs will show up as bright spots at both 2 $\mu$m and 10 $\mu$m.

Many higher-redshift quasars are slightly fainter than $V = 15$. The Palomar-Green (PG) survey, which searched $\sim 25\%$ of the sky for UV-bright quasars, found over 60 quasars more distant than 3C 273 brighter than the survey's magnitude limit of $B = 16.16$, but only one with $B < 15$ (Schmidt & Green 1983). The IRAS 12 $\mu$m all-sky survey detected a number of these at the $\sim$ 100 mJy level (Sanders et al. 1989). At 10 $\mu$m, a 100 m interferometer can resolve the $T_d \approx 200$ K region around a $V \approx 15.5$ object when $z \ll 1$. Section 6 discusses how cosmological effects complicate matters at high redshifts.

## 5. Mapping Spectral Features

Ultraviolet radiation incident on clouds in the narrow-line regions of AGNs can excite a host of IR transitions. The fine-structure lines from ions of many abundant heavy elements populate the mid-IR regime and provide important diagnostics of density, metal abundance, and ionization level (Spinoglio & Malkan 1992; Voit 1992a). Some particularly informative lines are [Ne II] 12.8 $\mu$m, [Ne III] 15.6 $\mu$m, [Ne III] 36 $\mu$m, [Ne V] 14.3 $\mu$m, [Ne V] 24 $\mu$m, [O IV] 26 $\mu$m, [S III] 18 $\mu$m, [S III] 34 $\mu$m, [S IV] 10.5 $\mu$m, [Ar III] 9 $\mu$m, [Ar III] 22 $\mu$m. The brightest of these lines are quite luminous, outshining H$\beta$, but several can be seen only from space. The detectability of these lines with an interferometer depends on the currently unknown spatial distribution of the emitting clouds. Optical interferometry from the SIM mission (Shao, this volume) will tell us just how clumpy these clouds are, preparing the way for mid-IR emission-line imaging. Dynamical studies of narrow-line regions using these lines would require $\lambda/\Delta\lambda \gtrsim 1000$.

Mid-IR interferometers will also be well-suited to mapping the IR bands commonly ascribed to PAHs (Puget & Leger 1989). In starburst galaxies, these bands are quite prominent. In active galaxies, they are usually weak or absent, presumably because X-rays from the nucleus demolish the small grains/large molecules responsible for emitting them (Voit 1992b). Where these bands are seen in the vicinity of AGNs, they probably come from starbursting regions shielded from the AGN, perhaps by the obscuring torus. Maps of PAH emission in active galaxies and starbursts might therefore reveal much about the relationships between star formation and nuclear activity.

## 6. High-$z$ Active Galaxies

Quasars can be seen near the limits of the observable universe, some with $R < 18$ at $z > 4$. Such objects are immensely luminous ($\gtrsim 10^{47}\,\mathrm{erg\,s^{-1}}$). The fact that they appeared when the universe was still so young ($< 1\,\mathrm{Gyr}$) challenges our current ideas about how galaxies form and how black holes grow within them (Turner 1991). Space-based interferometers optimized for $2 - 30\,\mu\mathrm{m}$ observations measure the optical and near-IR light emitted by the most distant active galaxies, enabling studies complementary to those done at low redshift.

At high redshifts, the relationship between resolvability and brightness is less straightforward. The angular-size/distance relation in an expanding universe magnifies the circumnuclear regions, so that dust at a particular $T_d$ subtends a much larger angle than in low-redshift quasars of comparable brightness. However, the thermal radiation from this dust redshifts to longer, less easily resolvable wavelengths. Fig. 2 outlines the impact of redshift on resolvability for quasars of magnitude $R = 17-19$. The tradeoff between magnification and redshift turns out to be roughly even. An interferometer operating at $30\,\mu\mathrm{m}$ with a 100 m baseline marginally resolves thermally emitting regions around $R \approx 17$ quasars at all redshifts. Note, though, that at $z \approx 4$, this region corresponds to $T_d \approx 500$ K, potentially making it easier to resolve obscuring tori here than in all but the brightest low-redshift Seyfert galaxies.

Interferometry of bright high-$z$ quasars at $\sim 2\,\mu\mathrm{m}$ images them in their rest-frame optical band. While we might hope to see what quasar host galaxies look like in detail, reflection nebulosity will probably overwhelm the signal from starlight. Thermal emission models for AGN IR continuua require dust grains to intercept a sizeable fraction of the optical-UV continuum, and $\sim 10\%$ could easily be scattered. The reflected luminosity alone can exceed the host galaxy's total output and will be concentrated around the nucleus. Near-IR images of quasar

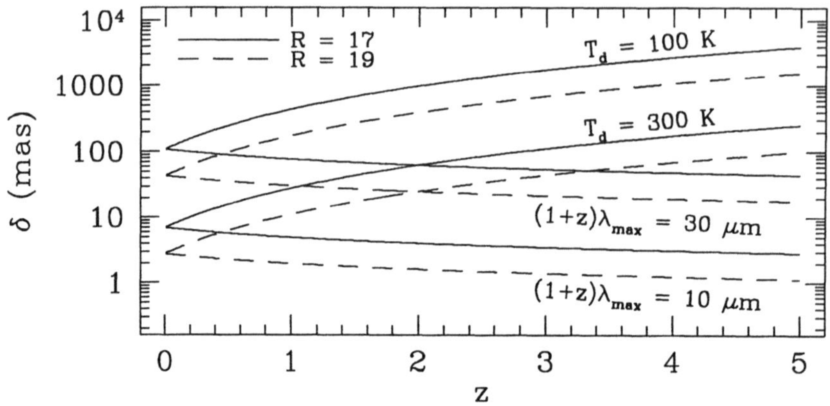


*Figure 2.* Angular sizes of thermally emitting regions at high redshift. The lines indicate the sizes of thermally emitting regions around quasars of $R = 17$ and 19. At constant $T_d$, the expansion of the universe magnifies apparent sizes but shifts the radiation to longer wavelengths. When the position of the observed peak in the thermal spectrum is held constant, the apparent size of the corresponding thermally emitting region drops gradually with increasing $z$.

reflection nebulae, in conjunction with Balmer-line mapping, might provide very useful information about the distribution of gas and ionizing radiation as close as $10-20\,\mathrm{pc}$ to the central engines of high-$z$ quasars. SIM observations of nearby AGNs will establish a baseline for this type of investigation.

## 7. Pushing the Limits

In keeping with the ambitious spirit of the Toledo meeting, I will conclude with two suggestions that push the limits of feasibility. The first is a scheme to measure $\Omega$, the cosmological closure parameter, by tracking quasar light echoes. The second proposes to exploit the properties of nulling interferometers to extract information about circumnuclear velocity fields at scales smaller than the nominal beamsize.

Reflection nebulae surrounding quasars carry information about their brightness histories. Around a variable quasar, light echoes should propagate across the plane of the sky at an angular rate depending upon $\Omega$, $H_0$, and the redshift of the quasar. Sparks (1996) shows that the angular speeds expected when $\Omega = 0$ and $\Omega = 1$, of order $0.03\,\mathrm{mas\,yr^{-1}}$, differ by 30% at $z = 1$ and by a factor of 2 at $z = 4$. A $2\,\mu$m interferometer operating over a several hundred meter baseline could measure $\Omega$ in a few decades. Perhaps much larger interferometers will be operating by then. Certainly we hope that $\Omega$ will already be known by other means when this kind of measurement is finally complete! In the meantime, we can content ourselves by confirming $H_0$ measurements through light-echo observations of nearer AGNs.

Extremely high resolutions are also needed to study the broad-line regions of high-$z$ quasars, thought to range up to $\sim 1$ mas in projected size. At $z \gtrsim 2$, the broad Balmer lines from quasars shift to $\sim 2\,\mu$m. The effective beamsize of a 100 meter nulling interferometer operating at $2\,\mu$m is $\sim 4$ mas, but the wings of the null will transmit more line emission from the outer parts of the broad-line region than from the inner parts. A single emission-line observation with the null pointed directly at the nucleus thus conveys information about the convolution of the null's shape with the projected velocity field. Scanning the null across the source or observing through nulls of different sizes and shapes, as in a multi-element nulling interferometer like OASES, can help to deconvolve the velocity field. These kinds of tomographic methods could potentially be used to weigh black holes in distant quasars, if the null characterization and pointing accuracy were sufficiently precise.

## References


Antonucci, R.R.J., Miller, J.S.: 1985, *ApJ* **297**, 621

Blandford, R.D., Netzer, H., Woltjer, L.: 1991, in Active Galactic Nuclei, T. Courvoisier, M. Mayor (Eds.), Springer-Verlag

Greenhill, L., Jiang, D.R., Moran, J.M., Reid, M.J., Lo, K.Y. and Claussen, M.J.: 1995, *ApJ* **440**, 619

Huchra, J., Burg, R.: 1992, *ApJ* **393**, 90

Krolik, J.H., Begelman, M.C.: 1986, *ApJ* **308**, L55

Krolik, J. H., Begelman, M. C.: 1988, *ApJ* **329**, 702

Miller, J.S., Goodrich, R.W.: 1990, *ApJ* **355**, 456

Miyoshi, M., Moran, J., Herrnstein, J., Greenhill, L., Nakai, N., Diamond, P., Inoue, M.: 1995, *Nature* **373**, 127

Phinney, E. S.: 1989, in Theory of Accretion Disks, F. Meyer, W. Duschl, J. Frank, & E. Meyer-Hofmeister (Eds.), Kluwer, p. 457

Pier, E. A., Krolik, J. H.: 1992, *ApJ* **401**, 99

Pier, E. A., Krolik, J. H.: 1993, *ApJ* **418**, 673

Puget, J. L., Léger, A.: 1989, *ARAA* **27**, 161

Sanders, D. B. et al. : 1989, *ApJ* **347**, 29

Schmidt, M. and Green, R. F.: 1983, *ApJ* **269**, 352

Schneider, D.P., Schmidt, M. and Gunn, J.E.: 1991, *AJ* **102**, 837

Sparks, W. B.: 1996, *ApJ* , in press

Spinoglio, L., Malkan, M.: 1992, *ApJ* **399**, 504

Terlevich, R., Melnick, J.: 1985, *MNRAS* **213**, 841
Turner, E. L.: 1991, *AJ* **101**, 5
Voit, G. M.: 1992a, *ApJ* **399**, 495
Voit, G. M.: 1992b, *MNRAS* ,**258**, 841
Wilkes, B. J. et al. : 1995, *ApJ* **455**, L13

## 8. Questions

*M. Rouan* : The mechanism to bring fresh matter from the 100 pc disk to the central engine is still unknown. Superdense stellar clusters, bars within bars, etc. have been suggested. Is there a way to make progress on this problem with IR interferometers?
*M. Voit* : Detection of stellar bars will be tricky. At 10 $\mu$m you are seeing mostly the distribution of warm dust. At 2 $\mu$m a stellar bar would produce a relatively minor modulation in brightness with position angle that could be hard for an interferometer to detect unambiguously.

*F. Paresce* : For the high-$z$ objects you mentioned, aren't you limited by the rapid drop in surface brightness with increasing $z$?
*M. Voit* : Surface brightness becomes a problem when the emitting area is much larger than a resolution element, but the thermal emission from high-$z$ objects will be difficult to resolve. As long as you are working near the resolution limit, you are limited only by flux.

# DUSTY DISKS AROUND MAIN-SEQUENCE STARS

C. WAELKENS
*Institute for Astronomy, Celestijnenlaan 200B, B-3001 Leuven*

L.B.F.M. WATERS
*Astronomical Institute University of Amsterdam*

**Abstract.** We review the evidence for dusty disks around main-sequence stars, with special emphasis on evolutionary aspects and on the relevance of the phenomenon in the context of the detection of exo-solar planets.

**Key words:** circumstellar matter, planetary systems, stars: evolution

## 1. The Vega phenomenon

The detection of an IR excess in normal main-sequence stars was one of the important surprises that came out of the IRAS mission. Aumann et al. (1984) detected an excess flux for Vega at 25, 60 and 100 $\mu$m, and soon Fomalhaut and $\beta$ Pictoris were added to the list. Surveys by various authors (e.g. Gillet 1986, Sadakane and Nishida 1986, Walker and Wolstencroft 1988, Stencel and Backman 1991 and Oudmaijer et al. 1992) have yielded a large number of other candidates. Various pieces of evidence, which we mention below, suggest that the circumstellar dust of these objects occurs in a disk and is related to the subject of this workshop, i.e. the formation of planets. The Vega phenomenon has been reviewed by several authors. For an extensive and comprehensive review, we refer to the contribution by Backman and Paresce (1993) in 'Protostars and Planets III'.

The relatively faint excesses in Vega, Beta Pictoris and Fomalhaut could only be detected by IRAS because of the high brightness of these nearby stars. The higher sensitivity of ISO is clearly required in order to determine accurately the fraction of stars which display such excesses. Preliminary IRAS-based studies suggest, however, that the phenomenon may be fairly common: Patten and Willson (1991) conclude that some 20% of the BAF main-sequence stars are affected, and Aumann and Good (1990) suggest that more than half of the G dwarfs have excesses. It then follows already now that the Vega phenomenon is not restricted to the early phases of evolution: in fact, Vega itself has already significantly evolved from the ZAMS.



*C. Eiroa et al. (eds.),*
*Infrared Space Interferometry: Astrophysics & the Study of Earth-Like Planets*, 119–128.

## 2. The 'Big Three'

It is customary to consider $\alpha$ Lyrae (Vega), $\beta$ Pictoris, $\alpha$ Piscis Austrini (Fomalhaut) and $\epsilon$ Eridani as the prototypes of the Vega phenomenon. So far, these nearby stars are the only ones for which spatially resolved information is available. In this section we discuss the former three, for which this evidence is most convincing.

### 2.1. Alpha Lyrae

It is well known that the IR excess of Vega was discovered during calibration observations of the IRAS satellite. This early detection allowed the scheduling of extremely-slow-speed scanning observations, from which the spatial extent of the extent can be measured with some confidence. Aumann (1991) determined an extent of 27" $\pm$ 4", but a more recent analysis by van der Bliek et al. (1994) yielded 35" $\pm$ 5". There is no evidence for non-circularity of the excess, which is in agreement with high-resolution optical spectroscopy, from which it follows that Vega is a rapidly rotating star that is observed pole-on (Gulliver et al. 1994).

From the combined study of the spatial extent of the excess region and the observed energy distribution, constraints can be found about the grain size: small grains reemit less efficiently the radiation they absorb and therefore need a larger area for a given excess. van der Blieck et al. concluded that the IR radiation of Vega is caused by grains that are smaller than 10 $\mu$m in size.

All ground-based coronographic observations have failed so far to detect the Vega disk in the optical. This non-detection seems at least partly related to the fact that the disk is viewed pole-on, and therefore has a low surface brightness.

The age of Vega is tightly constrained to some 400 Myr by the accurately known parallax of the star and by stellar evolutionary models. It certainly is a significant fact that a debris disk is observed for a star that has already spent about half of its expected main-sequence lifetime. In fact, the timescales of the Poynting-Robertson effect, that drags small grains onto the star, and of removal of larger grains by radiation pressure, is much shorter than the current age of the star. It therefore seems likely that the dust disk has been replenished recently, most probably by the collision of larger bodies, so that the debris disk indirectly points to the existence of such bodies orbiting Vega.

## 2.2. Beta Pictoris

Beta Pictoris has been observed only twice by IRAS, and its remarkable excess, that starts at 12 $\mu$m, was only detected after the mission. Therefore, the IRAS evidence for the spatial extent of the circumstellar disk is poor. However, this was largely compensated by the optical measurements, that started with the spectacular optical discovery by Smith and Terrile (1984) of a disk seen edge-on. Since, several other impressive coronographic pictures have confirmed this discovery (Paresce and Burrows 1987, Artymovicz et al. 1989, Golimowski et al. 1993, Lecavalier des Etangs et al. 1994, Burrows et al. 1995). On Fig. 1 we show the HST picture of Beta Pictoris by Burrows et al. (1995).

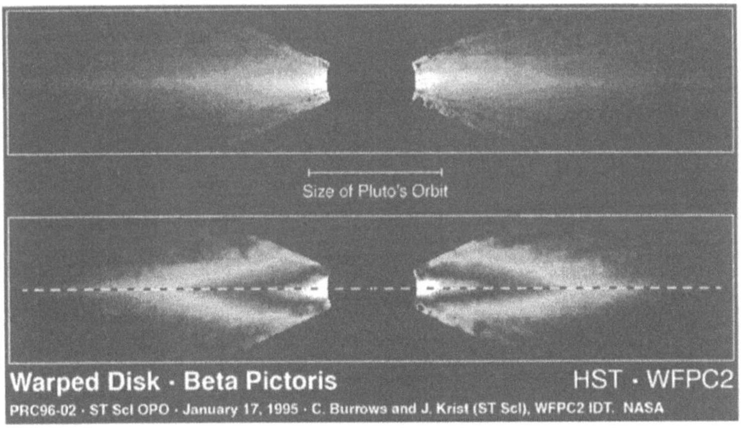


*Figure 1.* The HST image of the Beta Pictoris disk.

The vertical extent of the disk amounts to some 100 A.U., the outer radius being more than 1000 A.U. Combining the spatially resolved observations of Beta Pictoris with the spectral energy distribution reveals that the grain sizes are smaller than for Vega, being typically of the order of 1 $\mu$m (Backman and Paresce 1993).

Brightness profiles of the disk (e.g. Golimowski et al. 1993) suggest the occurrence of an inner hole. The presence of such a gap is best evidenced by the analysis of the IR image taken by Lagage and Pantin (1994). From the study of this image and the spectral energy distribution, Lagage and Pantin conclude that the disk is cleared inside the region that extends to some 40 A.U. from the central star.

Various indirect pieces of evidence have been put forward that the inner disk of Beta Pictoris may harbour at least one planet. Roques et al. (1994), Lazzaro et al. (1994) and Lagage and Pantin (1994) attribute the occurrence of an inner cleared region to tidal perturbations of a planet on the disk. According to Lagage and Pantin (1994) the presence of a planet may also account for the reported asymmetries of the disk.

The latter hypothesis was also put forward to explain the warp in the disk observed by Burrows et al. (1995).

Unlike Vega, Beta Pictoris clearly is a young star. Holweger and Rentzsch-Holm (1995) concluded from high-resolution optical observations that Beta Pictoris has the photospheric parameters of a ZAMS star with solar composition, contradicting previous evidence (Paresce 1991) that the star is metal-deficient. It is probable, then, that the disk of Beta Pictoris is the remnant of the proto-stellar disk in which it was formed.

The youth of the Beta Pictoris disk is also consistent with the remarkable presence of huge amounts of cometary material around the star. The frequent impact of comets onto the star is a natural explanation for the variable absorption features observed in the ultraviolet (e.g. Lagrange et al. 1989) and optical (e.g. Ferlet et al. 1993) spectra. The silicate feature between 8 and 12 $\mu$m observed by Telesco and Knacke (1991) and Knacke et al. (1993) is also very unlike the typical silicate feature observed around evolved stars, but matches quite convincingly that observed for comets and zodiacal dust in the solar system. Additional evidence that the disk of Beta Pictoris is continuously replenished by the desintegration of larger bodies is the observation that the iron content of the circumstellar gas is solar (Lagrange et al. 1995): indeed, in a stationary mixture of gas and dust, one expects that the iron atoms are mostly locked into the dust phase, as is observed in the interstellar medium.

### 2.3. Alpha Piscis Austrini

The two IRAS scans of Fomalhaut indicate that the image is extended and suggest that it is elliptic: in this sense this object is intermediate between the two others, probably showing a tilted disk. Attempts to resolve the disk at mm wavelengths have failed (Stern et al. 1994) as have all attempts to detect it in the optical. The gas content of the disk is low: the detection of some circumstellar gas was claimed by Cheng et al. (1994) but disputed by Ferlet et al. (1995). It appears clear that Fomalhaut is more alike Vega than alike Beta Pictoris. Also Fomalhaut has well evolved from the ZAMS; its age is estimated at some 200 Myr.

## 3. Evolutionary considerations

Statistical studies of the Vega phenomenon based on the IRAS data are hampered by the sensitivity limits of this satellite, since even for stars in the solar neighborhood the photospheric fluxes at 60 $\mu$m fall rapidly below the detection limit. From a study by Backman and Gillet (1987) of objects in the Gliese catalogue of stars within 25 pc from the sun, a lower limit of some 10 % was determined. This study will soon be supplemented by the ISO survey of Gliese stars. Plets et al. (1996) have scrutinized the ISO sample and found that from the 16 objects the photospheres of which could reliably be detected by IRAS, 9 show an excess. It is thus fairly probable that the fraction of stars showing the Vega phenomenon is already high in the old stellar population that characterizes our immediate neighborhood. The occurrence of the phenomenon among younger stars should also become clear from the ISO projects devoted to detect excesses in members of young clusters.

By all means, young Vega-type stars are not limited to Beta Pictoris. Jura et al. (1993) found that the M-type companion of HR 4796 (Jura 1990) is overluminous and therefore has not yet reached the main sequence. From lithium observations of this companion Stauffer et al. (1995) derived an age of 8 Myr for the HR 4796 system. The youth of the K-type excess star HD 98800, that shows an extreme IR-to-optical brightness ratio (Zuckerman and Becklin 1993), is attested by the cometary silicate feature (Skinner et al. 1992) it displays, by the strong lithium content and H$\alpha$ emission (Fekel and Bopp 1993) and by the relatively short rotational period of 14.7 days (Henry and Hall 1994).

The natural precursors to Beta Pictoris are the Herbig Ae/Be or Haebe stars, which are pre-MS stars in the mass range between 2 and 8 $M_{\odot}$. By definition, these stars primarily occur in star-forming regions, but several of the Vega candidates in the list by Walker and Wolstencroft (1988) can be considered as Haebe stars that recently left their star-forming region. Following Hillenbrand (1992) it is customary to classify the Haebe stars in Group I and Group II sources, where the far-IR excess in Group-II sources is the highest. It appears that Group II sources can often be resolved in the IR, while Group I sources appear pointlike at the current resolution (Li et al. 1994). The obvious interpretation is that the spectral energy distribution of the younger Group II sources is dominated by the lose circumstellar envelope, which can be considered as the remnant of the parent cloud, while the hotter IR excess of the Group I sources is confined to a region (disk?) much closer to the star. Marsh et al. (1995) were able to resolve the inner source of the relatively nearby Haebe star AB Aurigae, and found evidence for a disk-like structure with an inner cleared region.

From IRAS surveys (Walker and Wolstencroft 1988, Oudmaijer et al. 1992) several Haebe-like stars are found that occur near to but outside star-forming regions. It is natural to assume that these objects form the transition between the youngest Haebe stars and objects such as

Beta Pictoris. Their energy distributions typically peak in the mid-IR, suggesting that the parent cloud has dispersed completely. Whether the circumstellar matter in these objects occurs in a disk or is more spherically distributed, is presently controversial. However, the fact that for some objects with very similar energy distributions and spectral types such as HD 142666 and HD 144432 the former undergoes much circumstellar extinction and the latter not (Bogaert and Waelkens 1991), pleads for a non-spherically symmetric distribution of the circumstellar dust.

Waelkens et al. (1994) have studied the energy distributions of several isolated Haebe stars with the aim of determining an evolutionary sequence towards the Beta Pictoris stage, Haebe stars tending to have more hot dust. The observations do not suggest a sequence in which the IR excess gradually moves to the farther (60 $\mu$m) infrared as it fades, but rather a sequence in which the excess first disappears in the 10 $\mu$m region before the radiation at shorter wavelengths is removed. The appearance of a gap at 10 $\mu$m, i.e. at typical Jupiter distance from the central star, suggests the tantalizing possibility of the formation of a giant planet which then subsequently clears the inner part of the disk. However, it cannot be excluded that the near-IR radiation is produced by small grains out of thermal equilibrium, which then may occur at larger distances from the star. It can be hoped that the spectroscopic capability of ISO will partly resolve the ambiguity concerning the nature of the grains that cause the near-IR excesses of isolated Haebe stars.

The first ISO results nicely confirm the link between the isolated Haebe stars and Beta Pictoris. On Fig. 2 are shown the SWS spectra in the 10 $\mu$m region of HD 100546, HD 142527 and 51 Ophiuchi. Knacke et al. (1993) pointed out the similarity of the silicate feature of 51 Oph with that of Beta Pic.

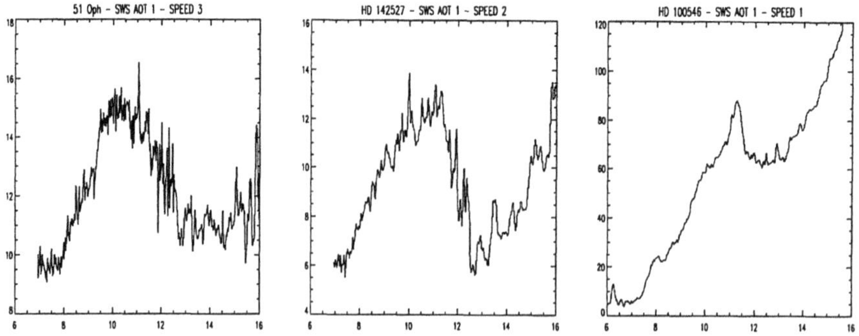


*Figure 2.* ISO-SWS observations of the circumstellar silicate features of the young objects 51 Oph, HD 142527 and HD 100546.

It is remarkable that at the high signal-to-noise ratio of the SWS spectra the features in HD 142527 and 51 Ophiuchi are virtually undistinguishible so far. The link should become clearer still when Beta Pic itself will become observable near the end of the ISO mission.

If circumstellar dust disks are common around main-sequence stars, it is to be expected that they will brighten again when the star increases in luminosity after the main sequence, since some of the remaining cometary bodies should fragment then and provide the necessary cores whereupon gas from the stellar wind can condense to dust grains (Matese and Whitmire 1989). Along these lines Jura (1990) has surveyed bright nearby giants for an infrared excess, but concluded that such excesses, if any, are rare. Zuckerman et al. (1995) and Plets et al. (1996) independently undertook a much larger survey of giants in the IRAS faint-source survey, and found that the small fraction already found by Jura is genuine. Further research on these G-K giants with IR excesses is needed in order to link more clearly these objects to the Vega phenomenon.

## 4. Interest for planet detection

The common occurrence of dusty disks around main-sequence stars is certainly a boost for theories in which the occurrence of planetary systems around stars is a general phenomenon. But it has to be pointed out that all evidence so far, such as the need for the replenishment of the disk of Vega and the various aspect related with the observations of the Beta Pictoris disk, is of an *indirect nature.* A more direct hint for a planet orbiting Beta Pictoris has been put forward by Lecavalier des Etangs et al. (1995), who discussed the peculiar photometric behaviour of this star in the Geneva system in 1981: in the midst of a slow rise in brightness by some 0.02 mag, attributed to a clearer zone in the gas surrounding the planet, a short occultation event may have been detected. The interpretation of this event is ambiguous, however. In fact, one of the present authors (CW) must present his apologies that this event was not followed up in more detail, since he contributed in 1982 to declassifying Beta Pictoris as a standard star in the Geneva photometric system!

In fact, the prospects of *direct* detection of planets of stars with dusty disks is not so good, since the IR radiation of the disk spoils the relatively favorable contrast effect of the planet with respect to the star in the infrared. Moreover, young stars such as Beta Pictoris tend to be broad-lined and are therefore not suitable for very-high-precision radial-velocity studies.

A more promising avenue for future research may be the possible anticorrelation between the presence of Jupiters and that of important amounts of zodiacal dust. Simulations of solar-system formation (Boss, this conference) suggest that (the early formation of) Jupiter has been instrumental in removing smaller bodies from the inner solar system. One could then expect that dust disks are rare in systems with Jupiters, while evolved Vega-type systems show a relative lack of giant planets. It can be hoped that, once the ambitious future censuses of both phenomena yield statistically reliable results, some pattern finally will emerge.

## 5. The need for high spatial resolution

The focus of this conference lies on high spatial resolution, which indeed is a primary requirement for significant advances in the field of the study of planetary systems and dusty disks around stars. A zeroth-order aspect is that in several, and maybe a majority of, cases, present interpretations are hampered by the unknown presence of a faint companion star. It has also become clear that the interpretation of infrared spectral energy distributions is not unique as long as no additional spatial information is available: only with such data will the controversies about disks versus envelopes and about grain sizes be settled. An additional dimension clearly is spectral resolution, on which ISO as well as ground-based instruments will soon deliver important contributions.

In the optical range, it remains a remarkable puzzle why only the Beta Pictoris disk has been reliably detected so far. The closeness of the system and the favorable inclination of the disk certainly contribute to the detectability of the Beta Pictoris disk, but it also seems unescapable that the albedo of the grains must be exceptionally high (Kalas and Jewitt 1996, Artymovicz 19965), a circumstance that may be related to the youth of the system. In order that the disks around other nearby stars are also detected in the optical, a much better contrast than that presently achieved is necessary (Kalas and Jewitt 1996).

In the infrared range, positive detections have been made for the 'Big Three' and also for SAO26804 (Skinner et al. 1995). As mentioned higher, a positive detection of the disk of the Haebe star AB Aurigae has been reported (Marsh et al. 1995). However, young stellar objects are rare in the vicinity of the Sun, and only a significant increase of the achievable spatial resolution will deliver results on a statistically relevant sample. Since also in the infrared a good contrast is needed, the case for a space mission is then obvious.

## References

Artymovicz, P.: 1996, in The Role of Dust in the Formation of Stars, R. Siebenmorgen, H.U. Käufl (Eds.), Springer-Verlag, in press
Artymovicz, P., Burrows, C., Paresce, F.: 1989, *ApJ* **337**, 494
Aumann, H.H.: 1991, in The infrared spectral region of stars, C. Jaschek, Y. Andrillat (Eds.), Cambridge University Press, p. 363
Aumann, H.H., Good, J.C.: 1990, *ApJ* **350**, 408
Aumann, H.H., et al.: 1984, *ApJ* **278**, L23
Backman, D.E., Gillet, F.C.: 1987, in Cool Stars, Stellar Systems and the Sun, J. Linsky, R.E. Stencel (Eds.), Springer-Verlag, p. 340
Backman, D.E., Paresce, F.: 1993, in Protostars and Planets III, E.H. Levy, J.I. Lunine (Eds.), University of Arizona, p. 1253
Bogaert, E., Waelkens, C.: 1991, in The infrared spectral region of stars, C. Jaschek, Y. Andrillat (Eds.), Cambridge University Press, p. 345
Burrows, C.J., Krist, J.E.; Stapelfeldt, K.R.: 1995, *BAAS* **187**, 3205
Cheng, K.P., Bruhweiler, F.C., Kondo, Y.: 1994, *ApJ* **425**, L33
Fekel, F.C., Bopp, B.W.: 1993, *ApJ* **419**, L89
Ferlet, R., et al.: 1993, *A&A* **267**, 137
Ferlet, R., et al.: 1995, *A&A* **297**, L5
Gillet, F.C.: 1986, in Light on Dark Matter, F.P. Israel (Ed.), Reidel, p. 61
Golimowski, D.A., Durrance, S.T., Clampin, M.: 1993, *ApJ* **411**, L41
Gulliver, A.F., Hill, G., Adelman, S.J.: 1994, *ApJ* **429**, L81
Henry, G.W., Hall, D.S.: 1994, *ApJ* **425**, L25
Hillenbrandt, L.: 1992, *ApJ* **397**, 613
Holweger, H., Rentzsch-Holm, I.: 1995, *A&A* **303**, 819
Jura, M.: 1990, *ApJ* **365**, 317
Jura, M., Ghez, A.M., White, R.J.: 1995, *ApJ* **445**, 451
Kalas, P., Jewitt, D.: 1996, *AJ* , submitted
Knacke, R.F., et al.: 1993, *ApJ* **418**, 440
Lagage, P.O., Pantin, E.: 1994, *Nature* **369**, 628
Lagrange, A.M., et al.: 1989, *A&A* **215**, L5
Lagrange, A.M., et al.: 1995, *A&A* **296**, 499
Lazzaro, D., et al.: 1994, *Icarus* **108**, 59
Lecavelier des Etangs, A., et al.: 1994, *A&A* **274**, 877
Lecavelier des Etangs, A., et al.: 1995, *A&A* **299**, 557
Li, W., et al.: 1994, *ApJ* **433**, 199
Marsh, K.A., et al.: 1995, *ApJ* **451**, 777
Matese, J., Whitmire, D.: 1989, *Icarus* **81**, 24
Oudmaijer, R.D., et al.: 1992, *A&ASS* **96**, 625
Paresce, F.: 1991, *A&A* **247**, L25
Patten, B.M., Willson, L.A.: 1991, *AJ* **102**, 323
Paresce, F., Burrows, C.: 1987, *ApJ* **319**, L23
Plets, H., et al.: 1996, in preparation
Roques, F., et al.: 1994, *Icarus* **108**, 37
Sakadane, K., Nishida, M.: 1986, *PASP* **98**, 689
Skinner, C.J., Barlow, M.J., Justtanont, K.: 1992, *MNRAS* **255**, 31P
Skinner, C.J., et al.: 1995, *ApJ* **444**, 861
Smith, B.A., Terrile, R.J.: 1984, *Science* **226**, 1421
Stauffer, J.R., Hartmann, L.W., Barrado y Navascues, D.: 1995, *ApJ* **454**, 910
Stencel, R.E., Backman, D.E.: 1991, *ApJS* **75**, 905
Stern, S.A., Festou, M., Weintraub, D.: 1994, *Nature* **369**, 766
Telesco, C.M., Knacke, R.F.: 1991, *ApJ* **372**, L29
van der Bliek, N.S., Prusti, T., Waters, L.B.F.M.: 1994, *A&A* **285**, 229

Waelkens, C., Bogaert, E., Waters, L.B.F.M.: 1994, in The nature and evolutionary status of Herbig Ae/Be Stars, P.S. Thé, M. Perez, E.P.J. van den Heuvel (Eds.), *PASP Conf. Series* **62**, 405
Walker, H., Wolstencroft, R.D.: 1988, *PASP* **100**, 1509
Zuckerman, B., Becklin, E.E.: 1993, *ApJ* **406**, L25
Zuckerman, B., Kim, S.S.., Liu, T.: 1995, *ApJ* **446**, L79

## 6. Questions

*J. Oró* : What would be the best method to detect the possibility of existence of planets within the inner zone of the $\beta$-Pic disk? How old is the disk? How many cometary bodies are falling into the central star? I realize the first question is difficult but the answer would be very important for exobiologists. Could you avoid the use of a coronograph by means of a interferometry detection?
*C. Waelkens* : As I said, direct detection of a planet there will be very difficult, because of the brightness of the disk. A coronograph or nulling interferometry are required to decrease the contrast with the central star. There are hundreds of comets falling on $\beta$-Pic every year, i.e. of the order of one every day.

*J. Mather* : The zodiacal disks around other stars may well have holes and gaps due to the presence of planets. These may actually be easier to measure than the planets themselves.
*C. Waelkens* : Yes, but we still need high contrast interferometry. The situation is like deducing the presence of the satellites of Saturn from the properties of the rings. The rings and gaps are much brighter than the controlling satellites. Even in the solar system, there are dust grain resonant orbit phenomena that do cause significant structure of the zodiacal cloud near Earth. The 10 $\mu$m luminosity of these resonances may be much larger than the 10 $\mu$m luminosity of the Earth itself. See the COBE DIRBE report of the resonant enhancement in Nature (1994, 1995, I think). The scale size of these resonances is of order 0.1 AU and they appear to share Earth's orbit, leaving or trailing by a short distance.

*M. Yanagisawa* : What is the magnitude of the $\beta$-Pic dust disk in optical wavelength?
*F. Paresce* : 16th mag per arcsecond.

# COMET-LIKE BODIES AROUND THE HERBIG AE STAR BF ORI

D. DE WINTER
*Dpto. Física Teórica, C–XI, UAM, Cantoblanco, 28049 Madrid, Spain*

C.A. GRADY
*Eureka Scientific, 2452 Delmer St., Suite 100, Oakland CA 96402, U.S.A.*

M.R. PÉREZ
*Applied Res. Corp., Suite 1120, 8201 Corporate Dr., Landover MD 20785, U.S.A.*

M.E. VAN DEN ANCKER and P.S. THÉ
*Astron. Inst., Univ. of Amsterdam, Kruisln 403, 1098 SJ A'dam, The Netherlands*

A.N. ROSTOPCHINA
*Crimean Astrophysical Observatory, Crimea, 334413 Nauchni, Ukraine*

**Abstract.** We have carried out a spectroscopic monitoring programme of several isolated Herbig Ae stars covering the He I 5876 Å, Na I$D_{2,1}$ and H$\alpha$ lines. In a first study we have demonstrated that for UX Ori a variable accretion activity similar to that seen towards $\beta$ Pic can be identified from redshifted components in these lines. Here we present, in part, the detection of similar events for BF Ori.

**Key words:** HAEBE stars, light profiles, solid bodies

## 1. Introduction

The uniqueness of $\beta$ Pictoris is that its disk is directly visible on images. However, to sample material close to the star high resolution spectroscopy is needed (Ferlet et al. 1993; Lagrange et al. 1988). Such studies indicated the presence of accreting material, which has been modelled in terms of the evaporation of infalling, comet-like bodies. Similar red shifted absorption components have now been routinely detected in both optical and UV spectra of Herbig Ae/Be stars, the likely progenitors of the $\beta$ Pic system. The first results were obtained for UX Ori. Prominent, red shifted absorption components (RACs) extending up to 200 km s$^{-1}$ were detected in the Na I $D_2$ 5890.0 and $D_1$ 5895.8 Å lines and weaker features in He I$\lambda$5876 (Grinin et al. 1994; de Winter 1996). Furthermore, monitoring of these lines and of H$\alpha$ revealed simultaneous variations from day to day (de Winter 1996).



*C. Eiroa et al. (eds.),*
*Infrared Space Interferometry: Astrophysics & the Study of Earth-Like Planets,* 129–132.

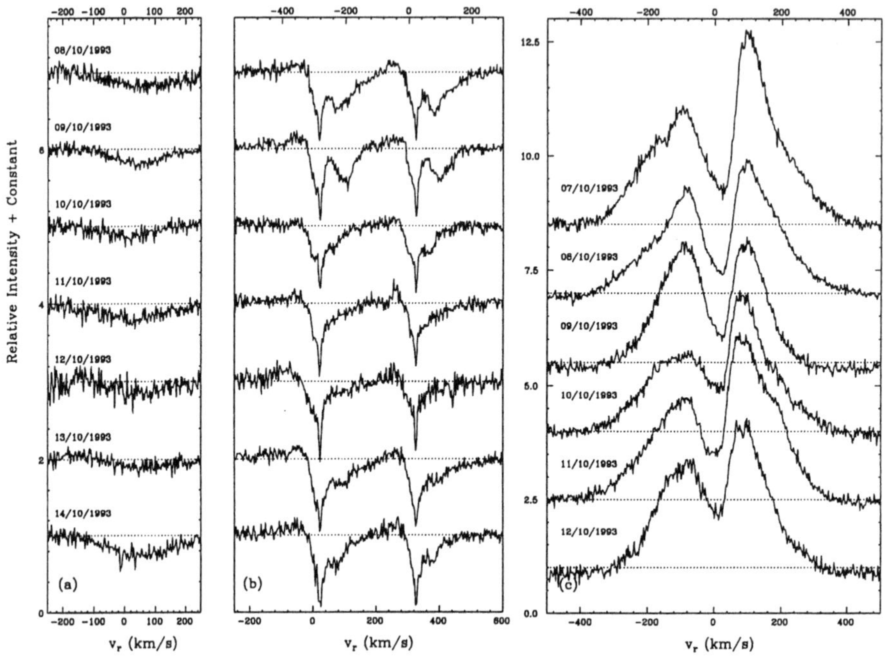


*Figure 1.* The observations: (a) He I and (b) Na I $D$ line profiles of BF Ori, $R = 55,000$, taken between the 8th & 14th of October, 1993. At this epoch the star was at intermediate brightness, $V = 11\overset{\mathrm{m}}{.}7$. (c) H$\alpha$ line profiles of BF Ori, $R = 50,000$, taken between the 7th & 12th of October, 1993. For all spectra the exposure time was 45 minutes. The normalised continuum levels are indicated by dotted lines.

## 2. Variations in the line profiles of BF Ori, UX Ori alike?

Similar to UX Ori, a high resolution monitoring programme for BF Ori was carried out covering the He I 5876 Å, Na I $D$ and H$\alpha$ lines for 8 consecutive nights, see Fig. 1. It was questioned whether BF Ori, an object equivalent to UX Ori in its astrophysical parameters, evolutionary phase and observational aspects, would have similar dynamics in the line profiles. As shown in Fig. 1, such evidence is clearly detected:
–) Well extended red wings of the Na I $D$ lines, which occasionally even developed into complete RACs.
–) The existence of He I$\lambda$5876 in absorption. Note the co-variations of this absorption profile and the redshifted Na I $D$ profiles.
–) The H$\alpha$ profile of BF Ori is double peaked, with the red peak the strongest. For UX Ori V/R > 1. However, the day to day variations of this profile are similar to those observed in UX Ori. Again the variations are in phase with those of the Na I $D$ lines. The strength of the red

H$\alpha$ peak clearly varies with the strength of the RACs and the velocities are cut-off to lower values when the RACs increase in strength.

Some differences with the UX Ori high resolution data are:

-) Some blueshifted absorption in the Na I $D$ lines.

-) Weak emission even further to the blue at these lines.

-) No double peaked absorption lines of Na I $D$ close to the rest velocity.

## 3. Interpretation

The model which explains the observed properties of UX and BF Ori, yields the infall of cool material from the outer circumstellar regions by comet-like bodies, which evaporate when approaching the star and cause the RACs of the sodium lines. When they interact with the material of the inner disk, collisionally ionized gas will be produced, which can be the origin of the observed He I absorption line being a volatile element and explains the co-variations with the lines of the refractory gas Na I (Grinin et al. 1994 and de Winter 1996). Furthermore, in this model the co-variations of the H$\alpha$ profiles are then well explained by the obscuration of certain parts of the H$\alpha$ line forming region by infalling material and the fact that hydrogen gas can be generated as volatile or refractory element as well.

The different V/R values of the H$\alpha$ profile of UX and BF Ori can be understood by the mechanism of a large rotating body in our line of sight that moves in opposite direction for UX and BF Ori and also obscures variable parts of the inner gaseous disk. For UX Ori this rotating body obscures the red part of the H$\alpha$ peak and we notice only "activities" in that part of the disk moving away from the observer. For BF Ori this body probably rotates towards the observer inducing activities, besides accreting bodies, in this part of the disk.

New and more intensive monitoring (de Winter 1996) on BF Ori supports this model and shows that the variations detected here are not variable on shorter time intervals. This suggests that from 8 until 10 October, we witnessed the evaporization of one comet-like body. Calculations (see de Winter 1996) show that this cometary had a possible mass of $5.73 \times 10^{17}$ g and a size of 5 km, comparable to comets in the solar system. The distance to the central star during evaporization was about 0.4 AU. However, this cometary could be a fragment of a larger orbiting body, of which the existence is suggested above.

The detection of comet-like objects in the disk of BF Ori, a 7–8 Myr old PMS A5e star, suggests that structures similar to those present in the solar system at the time of formation of the (proto-)planets are also present in other young stars and that planetary systems similar to our own my eventually evolve out of these.

## References

Ferlet, R., et al. 1993, *A&A* **267,** 137
Grinin, V.P., et al. 1994, *A&A* **292,** 165
Lagrange-Henri, et al. 1988, *A&A* **190,** 275
de Winter, D. 1996, PhD Thesis, University of Amsterdam.

# STELLAR DEATH: EJECTA AND CIRCUMSTELLAR MATTER

## *IR Interferometry of Dusty AGB Stars, Expanding Novae, & Explosive Supernovae*

D.H. WOODEN
*NASA Ames Research Center*
*MS 245-6 Moffett Field, CA 94035-1000, USA*
*e-mail: wooden@ssa1.arc.nasa.gov*

**Abstract.** IR 15 milli-arcsecond interferometry in the 5 – 15 $\mu$m wavelength region of the circumstellar environments of AGB stars probes the dust condensation zones close to the stellar photospheres. Interferometric measurements which spatially resolve Galactic novae will provide distance determinations, can reveal dense knots in the structure of the ejected shells, will probe nonspherical accretion onto the compact white dwarfs, and in many novae will witness dust grain growth despite grain destruction by photodesorption and sputtering. Interferometric imaging of nearby supernovae (d $\lesssim$ 50 kpc) will show the explosion structure frozen-in to the homologously expanding ejecta, the interaction of the ejecta with the surrounding medium, and possibly dust formation in an iron- and nickel-rich ejecta. Interferometric reconnaissance of the chemically anomalous clumpy outflows and ejecta of AGB stars, novae, and supernovae will scrutinize the theories for the microscopic processes of dust nucleation, grain cluster aggregation, grain sputtering, and photodesorption of grains, i.e., critical processes dictating the enrichment of the ISM.

**Key words:** dust condensation, C-rich and O-rich AGB stars, novae, supernovae

**Abbreviations:** AGB - asymptotic giant branch, ISM - interstellar medium

## 1. Introduction

IR interferometry is the next major technological step required to understand fundamental astrophysical processes. Current designs for a future space-based interferometer are for multiple 1.5m diameter apertures with a baseline of ~75m which permits an angular resolution of 15 milli-arcsecond (15 mas) at ~6 $\mu$m. If the interferometer orbits the Earth, then the expected IR interferometer spectroscopic sensitivities are scaled from previous estimates of space observatories (Thronson et al. 1995). For example, an observation of the continuum or of broad dust features at a resolving power of R=5 will yield a S/N=5 measurement of a 3 $\mu$Jy source at 5 $\mu$m in 100 sec. Observations of emission lines at a resolving power of R=100 will yield a S/N=5 on a $2\times10^{-19}$ W m$^{-2}$ line in 100 sec. The proposed spaceborne interferometer composed of 1.5m diameter apertures will have the sensitivity to observe the emission



*C. Eiroa et al. (eds.),*
*Infrared Space Interferometry: Astrophysics & the Study of Earth-Like Planets,* 133–145.

Table I. Galactic Sources of Star Dust

| Source | Carbon Dust $M_\odot$ kpc$^{-2}$ yr$^{-1}$ | Silicate Dust $M_\odot$ kpc$^{-2}$ yr$^{-1}$ | $< T_{rad} >$ K | $< v_{exp} >$ km s$^{-1}$ |
|---|---|---|---|---|
| C-rich AGB Stars | $2 \times 10^{-6}$ | | 2500 | 5-20 |
| O-rich AGB Stars | | $3 \times 10^{-6}$ | 2500 | 5-20 |
| Red Supergiants | | $2 \times 10^{-7}$ | 3000 | 10-30 |
| Planetary Neb. | $4 \times 10^{-8}$ | | 100,000 | 1000 |
| Novae | $3 \times 10^{-7}$ | $3 \times 10^{-8}$ | 250,000 | 500-3000 |
| WC Stars | $6 \times 10^{-8}$ | | 40,000 | 1500 |
| Type Ia SNe | $3 \times 10^{-7}$ | $2 \times 10^{-6}$ | 5000 | 3000 |
| Type II SNe | $2 \times 10^{-6}$ | $1 \times 10^{-5}$ | 5000 | 2000-5000 |

lines and continuum from the dust forming zones in cool stars, novae, and supernovae, at small spatial scales and at very high signal-to-noise.

Stars enrich the interstellar medium with their nucleosynthesized elements near the end of their lives. Cool red giant stars slowly ooze oxygen- or carbon-rich gas and dust into their circumstellar environments at velocities of tens of km per sec. Novae explosively burp off shells of enriched gas at velocities of a thousand km per sec. Some novae form dust within about a month of the event. Supernovae blow up, sending all but the innermost iron core of the precursor star into the surrounding circumstellar medium at velocities of many thousands of km per sec. Supernova SN1987A formed $\geq 10^{-4}$ $M_\odot$ of dust after ~2 years, so much dust that the dust blocked out most all of the visible radiation. Per object, supernovae (SNe) enrich the interstellar medium with at least 1 $M_\odot$ of nucleosynthesized heavy elements, far more than individual novae or cool giants. SNe numbers are, however, less than novae and much less than cool giants. Table 1 summarizes the sources of dust enrichment in the interstellar medium per unit area of the Galactic disk (Tielens 1995), along with the radiation field temperature of the central objects, and the expansion velocities of their outflows or ejecta. (An ejecta expansion velocity of 1500 km s$^{-1}$ corresponds to a resolvable size of 15 mas in 50 days at a distance of 6 kpc.) The AGB stars, novae, and supernovae dominate the dust enrichment process. The dust carries the chemical composition signature of its origin, including the enriched metals and isotopic anomalies, to new areas of star formation, i.e., to planetary systems in formation.

Dust nucleation is a current problem in theoretical and laboratory astrophysics. Cool O-rich AGB stars are observed to make copious amounts of dust, yet the gas densities in the dust condensation zones at many stellar radii are just enough to grow grains if growth is 100% efficient. Laboratory data and theoretical models predict, however, low

(~1%) grain nucleation efficiencies. Shells ejected in novae explosions are observed to make dust, but here grains must nucleate and grow in regions where the radiation fields are like that of planetary nebulae and the gas temperatures are like HII regions; where grains are expected to be destroyed by photodesorption (Johnson et al. 1993). Dust formation in supernovae has been theorized (Dwek 1988, Kozasa et al. 1991), but has been seen only in SN1987A. Instead of making silicates and carbon grains, SN1987A probably made Fe-, Ni-rich grains (Wooden et al. 1993, Wooden 1994). In all of these objects, it is proposed that grains form in clumpy media: clumpy media enhance densities of condensables in the gas phase and concur with the geometries deduced by other observables. These objects are IR bright but require observations at small spatial scales ($\leq$ 15 mas), and for some important diagnostics, at wavelengths inaccessible from ground-based observatories ($\lambda\lambda$ 5–8 $\mu$m, $\lambda \geq 14$ $\mu$m). IR spectroscopic interferometry of Galactic AGB stars, Galactic novae, and nearby supernovae ($d \leq 50$ kpc) will interrogate the regions of grain nucleation, and thereby disclose the physical conditions in which the major reservoirs of heavy elements in the interstellar medium – the dust grains – are formed.

## 2. Cool Mass-Losing Asymptotic Giant Branch Stars

Low- and intermediate-mass stars ($M_{star} \simeq 1 - 4.5\ M_{\odot}$) populate the Asymptotic Giant Branch (AGB) late in their lives, after having exhausted hydrogen and helium buring in their cores. In order to support their own weight by the paucity of photons which arise in hydrogen shell-burning, these stars puff themselves up, extending their low gravity tenuous atmospheres beyond Jupiter's orbit. The circumstances of this rather unstable point in their evolution yields conditions in which stellar pulsations, acoustic waves in their upper atmospheres, mass-loss by dense low-velocity stellar winds, and dust condensation coexist. Theoretical models for the winds of AGB stars *presume* mass-loss occurs and grains nucleate and thereby show that the wind momenta is a result of radiation pressure on the grains and grain-gas collisions. Mass-loss mechanisms including accoustic waves and first-overtone pulsations have been suggested, but these fundamental details are still absent from theoretical model atmospheres (Habing 1996). Grain growth on critical clusters or "seeds" can be modeled, but grain nucleation is still not understood. As described below, accoustic waves, clumpy outflows, and temperature-dependent grain chemistries contribute to solving the grain nucleation problem. The influence of all these processes at their relevant spatial scales can be investigated by IR interferometric observations of AGB stars.

Recent 11.15 $\mu$m Infrared Spatial Interferometer (ISI) observations by Danchi et al. (1994) of 9 bright O-rich AGB stars demonstrate that the inner radii of the circumstellar dust shells appear *either* 'close to the star' at radii $R_{inner\ dust} \simeq 2.4 - 6.8$ $R_{star}$ *or* 'far from the star' at $R_{inner\ dust} \simeq 8 - 19$ $R_{star}$. Constant mass-loss and grain formation is deduced for dust 'close to the star', while episodic grain formation ($\Delta t \approx 20$ years) is deduced for dust 'far from the star'.

The grain nucleation problem in AGB stars is easily posed. Consider the silicate grain condensation zone at $R_{con} = 6 - 12$ $R_{star}$ where $T_{con} =$ 1000 K grains are in thermal equilibrium with the stellar radiation field of $T_{star} = 2500 - 3500$ K. Mass-loss rates of $10^{-7} - 10^{-5}$ $M_{\odot}$ $yr^{-1}$ yield hydrogen gas densities of $n_H \simeq 10^8 cm^{-3}$ at $R_{con}$. Gas-grain collisions at this gas density are barely sufficient to grow grains, and insufficient to nucleate grains. For O-rich AGB stars, SiO molecules can form at these gas densities, but laboratory experiments which condense silicate grains from SiO molecules require SiO densities of $n_{SiO} \simeq 10^{12}$ $cm^{-3}$, a million times higher than the SiO densitites in the hydrogen-rich AGB star envelopes (Tielens 1989).

Based on the observed silicate condensation sequence for meteorites in the solar nebula, Tielens (1989) proposes that refractory Fe-poor grains such as sapphire or corundum ($Al_2O_3$, $T_{con} = 1760 - 1300$ K) condense close to the photosphere at $R \simeq 2R_{star}$. These Fe-poor grains are transparent in the visible and near-IR and couple poorly to the stellar photons, drifting slowly to the "condensation zone". At $R_{con} \simeq 3 -$ 7 $R_{star}$ the less refractory Fe-rich silicates such as enstatite ($MgSiO_3$, $T_{con} = 1350 - 1040$ K) and fayalite ($Fe_2SiO_4$, $T_{con} = 1100 - 950$ K) grow on the Fe-poor 'seeds'. The Fe-rich "dirty" silicate grains couple effectively to the stellar photons, collide with the gas and accelerate the wind. This grain condensation scenario predicts that the shape of the 9.7 $\mu$m silicate resonance ($\lambda\lambda$ 7.5 – 14 $\mu$m) depends on radius and the strength of the dust continuum shortward of the resonance ($\lambda\lambda$ 6.5 – 7.5 $\mu$m) is weaker (due to Fe-poor grains) close to the photosphere.

Single-dish observations of the silicate emission feature in O-rich AGB stars in fact show a variety of strengths and shapes. In the large sample of IRAS LRS spectra of AGB stars there are no correlations between silicate emission feature shape and variability class. That is, the large amplitude long period Mira variables, the semi-regular (SRa & SRb) variables, and the irregular (SRc) variables can have the same shape silicate emission features (Sloan and Price 1995). The *asymmetry* of the visual light curve and the dust emission feature shapes are, however, correlated (Vardya et al. 1986, Sloan et al. 1996). Miras with symmetric light curves show low-contrast broad silicate emission features, while Miras with increasingly asymmetric light curves show increasingly narrow silicate emission features more like interstellar silicates (Sloan et al. 1996). The asymmetric light curves are postulated to be due to accoustic waves in the tenuous atmosperes of AGB stars (Habing 1996). (The presence of these sound waves in the AGB stellar upper-atmospheres may even have been detected via circumstellar masers; the masers are thought to arise in the region below the dust condensation zone (Pijpers et al. 1994). While the different 9.7 $\mu$m silicate emission feature shapes are postulated to be due to evolutionary state (Stencel et al. 1990) or varying optical depth $\tau_{9.7}$ (Ivezic and Elitzur 1995), Tielens' scenario described above where grain chemistry depends on the physical conditions in the grain condensation zone is more consistent with shock waves affecting the resultant grain chemistry. The recent work by Sloan et al. (1996) on a new 13 $\mu$m feature, perhaps associated with corundum, shows that while few Miras show the 13 $\mu$m feature, most all (75%-90%) of the O-rich SRb AGB stars have the 13 $\mu$m feature. SRb AGB stars have typically thinner dust shells ($\tau_{9.7\mu m} < 1$) and weaker pulsations than Miras; the SRbs provide astrophysical 'dust laboratories' observationally neglected compared to the Miras. The accoustic waves probably act to enhance the gas densities in their dust nucleation zones, and therefore help the grain nucleation problem. IR interferometric imaging of SRb variable stars with asymmetric light curves over at least a full 'cycle' of their phase variations will probe the effect of accoustic waves on grain nucleation. IR interferometric imaging spectroscopy of SRbs in the 9.7 $\mu$m silicate emission feature (8.5 – 12 $\mu$m), in the 13 $\mu$m feature, in the SiO fundamental band (7.5 – 11 $\mu$m), and in the 6 $\mu$m and 14 $\mu$m continuum (shortward and longward of the resonances) will investigate the effect of accoustic waves and variable gas density on grain nucleation chemistry.

Clumpy outflows can also help the grain nucleation problem, increasing the density of condensables at several stellar radii. There are many indications that the outflows of AGB stars are asymmetric and clumpy, including studies of OH, SiO, and $H_2O$ masers, $J$, $H$, and $K$ band circumstellar polarization and speckle observations, and infrared maps of nonuniform dust distributions at very large radii (Habing 1996). The most recent visible interferometric imaging using nonredundant aperture masking (Wilson et al. 1992) and mid-IR 11.15 $\mu$m ISI interferometric visibility curves (Danchi et al. 1994) of the infamous O-rich AGB star $o$ Cet (Mira itself) show that the mass-loss and dust distribution is nonisotropic. The photosphere of $o$ Cet appears lop-sided at optical wavelengths, is 10% larger in the 710 nm TiO bands than in the adjacent 700 nm continuum, and maintains its shape for at least

4 pulsation periods (Wilson et al. 1992). The dust shell lies outside the photosphere and has the same odd shape (Danchi et al. 1994). The mass-loss may be in the geometry of a slow bipolar outflow where dense matter in the equatorial plane of the inner part of the molecular envelope is constraining the wind (as suggested by CO ($J$=1–0) interferometry maps (Planesas et al. 1990)). For *o* Cet, more dust is observed close to the stellar photosphere ($R_{con} \simeq 1.7\ R_{star}$) during the low luminosity phase of its light curves (largest physical extent of the atmosphere). The enhanced dust density at the inner edge of the dust shell results from *the acceleration of the wind by radiation pressure on the grains* (Danchi et al. 1994). The IR interferometric observations of Mira thus confirm this basic principle of wind models.

Although the detailed discussion of the grain nucleation problem has focused on the O-rich AGB stars, the grain nucleation problem also exists for the (far less numerous) C-rich AGB stars. C-rich AGB stars are observed to form carbon grains (Goebel et al. 1995). The chemical precursor to carbon soot formation is well known from the studies of industrial smoke stacks to be acetylene ($C_2H_2$). Despite the well-defined carbon chemical network, grain formation is still a problem in C-rich AGB envelopes because of the low gas densities at $R_{con}$ (Cherchneff and Tielens 1992).

AGB stars lose mass at a rate of $10^{-7} - 10^{-5}\ M_{\odot}\ yr^{-1}$ for a duration of $\sim 10^5\ yr$: yeilding a total loss of about $0.01 - 1\ M_{\odot}$ of gas to the interstellar medium with typical Galactic gas-to-dust ratios of $\sim 100$ (Habing 1996). *These dusty, cool AGB stars are the primary stardust birthsites.* High sensitivity, high-spatial resolution observations, especially across the 7.5 – 14 $\mu$m wavelength range of the strong 9.7 $\mu$m resonance *and* in the 5 – 7.5$\mu$m continuum shortward of the resonance will assess the abundances of Fe-poor and Fe-rich silicates as a function of radii and angular position. The greatest rewards may come from imaging the thinner dust shells around SRbs with asymmetric light curves (Habing 1996, Sloan et al. 1996).

## 3. Expanding Novae: Fast Polar Plumes and Slow Equatorial Ejecta

Novae occur when the older, compact white dwarf of a binary pair accretes enough gas from its companion to instigate a thermonuclear runaway (at time t=0) in a degenerate layer on the surface, and subsequently an outer layer of the white dwarf is expelled at high velocity. The ejection of a $10^{-6} - 10^{-4}$ $M_{\odot}$ shell at speeds of 500 – 3500 km $s^{-1}$ spews the heavy element enriched white dwarf gas and the hydrogen-rich gas accreted from the companion into the surrounding ISM. Slower expansion velocities are typically associated with low mass CO white dwarf novae, while higher expansion velocities are typically observed with intermediate mass ONoMg white dwarf novae.

At onset, novae appear to be 6000 – 10,000 K blackbodies. If both their Doppler expansion velocities and rate of brightening are measured during this short-lived "pseudophotospheric phase" of their light curves, then both t=0 and their distance can be determined by a modified Baade's Method for Cepheids, assuming that the novae are radiating an Eddington luminosity. This method is limited by the theoretical model of the expanding surface of the "fireball", the necessity that the novae be observed before peak brightness, and the assumption that the white dwarf is emitting an Eddington luminosity. This latter assumption is the most questionable, as some novae have been observed to have super-Eddington luminosities (Gehrz et al. 1995b, Harrison and Stringfellow 1994). IR interferometry of the hot continuum affords the opportunity to resolve even highly obscured novae later in their development, measure their expansion rates, extrapolate backwards to t=0, and thereby directly determine their distances from their apparent angular sizes.

The second phase in novae light curves begins at $t = 2 - 10$ days, after optical maxima. During this second "nebular phase", the ejected shell expands homologously and the luminosity drops as $t^{-2}$ (Gehrz et al. 1995b). The IR continuum is dominated by Brehmstrahlung (free-free) radiation and possible optically thick or optically thin thermal dust radiation. The transition between the "nebular phase" and the third "coronal line phase" is marked by the drop in hydrogen emission lines, and the appearance of strong IR forbidden lines, including [Al VI] 3.66 $\mu$m, [Ne VI] 7.64 $\mu$m, [Ar III] 7.99 $\mu$m, [Ne II] 12.8 $\mu$m, and [Ar II] 6.99 $\mu$m in the $\lambda\lambda$ 5 – 8 $\mu$m wavelength region obscured from ground-based telescopes by telluric water vapor. Theory predicts the detection of [Na VII] 4.67 $\mu$m, [Na VI] 8.61 $\mu$m and [Na VI] 14.2 $\mu$m, and a strong anti-correlation of the production of $^{26}$Al and $^{22}$Na with the mass and temperature history of the explosion for ONeMg or "neon" novae (Gehrz et al. 1994). The luminosity drops as $t^{-3}$ for the early part of this free-expansion phase, and at late times falls off exponentially: the exponential decay is of unknown origin (Gehrz et al. 1995b).

While other astronomical objects also make silicates, SiC, and carbon-rich dust, novae dust contains enhanced quantities of peculiar isotopes such as $^{13}$C, $^{15}$N, $^{22}$Na, and $^{26}$Al. Grains of SiC with radii between 0.3 $\mu$m and 1 $\mu$m are found in meteorites (Anders et al. 1989, Jura 1995), and the fraction of these inclusions with carbon abundance anomalies $^{13}\mathrm{C}/^{12}\mathrm{C} > 1$ must have originated in novae (Anders et al.

1989,Tielens 1990) and been incorporated into the solar system during the formation of the presolar nebula.

When dust condensation events are observed in novae, the grain sizes deduced are comparable to those needed as precursors to meteoritic inclusions; for example, the late 1970's IR photometric observations of Nova Ser 1978 (Gehrz et al. 1980a) and Nova Cyg 1978 (Gehrz et al. 1980b) yielded grain sizes of $a \approx 0.3 \mu m$. In order to explain the grain growth rates, the dust must form in clumpy structures: grains are sometimes observed to grow over a period of many weeks. If the gas which is converted to dust grains were distributed uniformly in a spherical shell, the thickness of the shell would be so thin that this shell would be subject to sonic disruption in about a day (Gehrz et al. 1980a). The regions within the expanding nova shell which form dust must represent density enhanced volumes within a thicker inhomogeneous shell. Kinetic models of carbon dust nucleation in novae ejecta show that photodesorption by UV photons is expected to be so strong that grains cannot form at all without a "sacrificial" hydrogen photodissociation layer (Johnson et al. 1993). The knots in which dust form must shield the small grains from the 100,000 – 250,000 K radiation field of the central source so that they can grow to sub-micron to micron sizes. If grains condense after 30 days in a nova ejecta expanding at a velocity of $v_{exp} = 1300$ km s$^{-1}$, the dust shell will have a radius of $\sim 3 \times 10^{14}$ cm (Gehrz et al. 1992) and a total apparent size of only 10 mas at a distance of 5 kpc. Dusty knots within the shell with a filling factor of 10% will, however, still be imagable by interferometers later on in the evolution of the ejecta: at 600 days, these dusty knots will be 20 mas is apparent size while the dust continuum is still bright (~1 Jy at 7.8 $\mu$m (Gehrz et al. 1992)).

Dust condensation in the three recent novae, (CO) QV Vul = Nova Vul 1987, (ONeMg) V838 Her = Nova Her 1991, and (CO) V705 Cas = Nova Cas 1993 reveal first the condensation of C-rich dust (graybody dust emission at $T_{con} = 1000$ K), followed by the condensation of O-rich dust (optically thin 9.7 $\mu$m silicate features). In Nova Vul 1987 C-rich dust formed within 3 months of outburst: carbon, SiC, and hydrocarbon or PAH (polycyclic aromatic hydrocarbon) dust appeared by ~100 days, long before silicates appeared at ~560 days (Gehrz et al. 1992). In Nova Cas 1993, the same condensation sequence was observed: PAH 3.29 $\mu$m emission was observed as early as 30 days, and the silicates appeared later at ~330 days (Gehrz et al. 1995a). In the very fast Nova Her 1991, a small amount of carbon dust was observed at 2000 K at only 8 days after outburst (Harrison and Stringfellow 1994). A large amount of 1000 K carbon dust was observed a month after shell ejection (Harrison and Stringfellow 1994, Woodward et al. 1992), and silicates were observed at about 3 months (Lynch et al. 1992). C-rich and O-rich dust are observed to form in the same novae! This recent data indicates that either there are considerable variations in the C/O abundances in a clumpy ejecta, or more likely, that the C-rich dust formed in fast-moving polar plumes (rich in products from the CNO runaway), while silicates (O-rich) formed in the slower (and most likely optically thick and predominantly hydrogen-rich nonprocessed) equatorial ring (Gehrz et al. 1992). These proposed fast polar plumes and slow equatorial ejecta have been imaged by the HST in the (dust free) Nova Cyg 1992 at ~460 days and ~700 days: the images show a dense layer of gas thrown off in the orbital plane of the binary system and an egg-shaped shell, with greater elongation perpendicular to the orbital plane (Paresce et al. 1995). Nova Her 1991's dust condensed in clumps filling only ~5% (Harrison and Stringfellow 1994, Woodward et al. 1992) of the volume of the ejecta (at 18 days, these dusty clumps covered 5% of a 12 mas diameter shell). The gas densities in these clumps were high enough for the gas to cool by Brehmstrahlung (free-free) radiation to temperatures of 1000 K in a few days (Harrison and Stringfellow 1994). The rapid cooling of the gas allowed grains to nucleate, and the clumps shielded the small grains from the ~250,000 K radiation field of the post-nova white dwarf and from the typical ~10,000 K gas which fills most of the nova shell.

Not only do the recent observations of carbon dust, PAHs, SiC, and silicates in the same novae challenge the previous consensus that slow CO novae form carbon grains and fast ONeMg novae form silicate grains, but the need for very high spatial resolution observations of novae is accentuated. IR spectral interferometry promises to reveal nucleation of grains in the dense knots within the anomalously-enriched ejecta and harsh radiation environment. Interferometric images of the thermal IR dust emission will reveal whether the ejecta has been compressed in random clumps or an axially symmetric flow during the explosion and subsequent expansion. The physical conditions (density, gas temperature, radiation temperature, expansion velocity) of the dust-forming regions will constrain both the nucleation kinetics and the hydrodynamic explosion/expansion models.

## 4. Explosive Supernovae

Supernovae are the most powerful and violent events in the cosmos, second only to the Big Bang. At about 6 hours after the core collapse of a massive star (type II SNe), when the explosion shock reaches the stellar surface, the fireball is expanding at one-tenth the speed of light and is as hot as the central star of a planetary nebula: evaporating all the dust in the circumstellar medium out to ~1000 AU, destroying the Kuiper belt and heating the Oort cloud (Wooden 1989). If an off-center explosion occurrs, then a piece of the precursor star can be expelled at a few thousand km $s^{-1}$ (Frail and Kulkarni 1991). In the first couple of months, the rapidly expanding outer atmosphere of the star is fully ionized, optically thick, increasing in luminosity as the radiating surface expands outwards, while effectively revealing lower and lower layers in the precursor star's increasingly tenuous atmosphere. During this "photospheric" phase of the light curve, an IR spectrum of the ~5000 K atmosphere is well-fit by a blackbody; combining the measured flux level with the expansion velocity inferred from the Balmer H$\alpha$ P-Cygni line profile, the distance can be determined (Kirschner et al. 1973). For a Galactic supernova, the ejecta could be interferometrically imaged in the IR forbidden line [Co II] 10.55 $\mu$m from radioactive cobalt, allowing a view of the explosively nucleosynthesized materials boosted from deep in the interior of the ejected star to the outer envelope of the star by instabilities in the explosion (Fryxell et al. 1991).

Once the supernova becomes optically thin after the peak in the light curve, all molecular-forming and dust-forming processes are revealed: due to the large velocity shear in the ejecta, the Sobolev approximation applies, and the IR observations peer right through the ejecta. In a velocity-resolved line profile, slices in velocity space are related to slices of the supernova ejecta perpendicular to the line of sight (McCray 1993). SN1987A's velocity-resolved near-IR and mid-IR forbidden emission lines of iron agree in shape, indicating that these lines originate in the same volume of gas, and have complex ripples, indicating that the ejecta is very nonuniform (Haas et al. 1990, Li et al. 1993). For a Galactic supernova, milli-arcsecond interferometry can expect to image rumpled shells of newly formed CO and SiO, and interior to that, a clumpy volume of iron-rich gas with a small amount of hydrogen mixed in (Li et al. 1993). For a nearby supernova ($d \leq 50$ kpc), at about 200 days during this "supernebula" phase of the light curve, the ejecta is large enough that it could be interferometrically imaged. The bolometric luminosity is falling as the e-folding time of $^{56}$Co, $\tau_{56Co} = 111\ days$, so the ejecta remains bright and can be imaged for a few years.

Supernovae have velocity-broadened lines ($\Delta v \simeq 2000 - 5000$ km $s^{-1}$), and a resolving power of $R = \lambda/\Delta\lambda \approx 50$ is sufficient to separate the species, while a resolving power of $R \geq 200$ is needed to measure the line profiles (Wooden et al. 1993). A handful of very important strong lines include [Ni II] 6.63 $\mu$m, [Ni I] 7.51 $\mu$m, and [Ar II] 6.99 $\mu$m which uniquely probe the ionization structure of the metal-rich volume of the ejecta. The ratio [Ni II] 6.63 $\mu$m/[Ni I] 7.51 $\mu$m yields a temperature and density insensitive measure of the ionization fraction (Wooden et al. 1993). Elemental abundances can be determined simply from the ground-state IR forbidden lines including [Ni II] 6.63 $\mu$m, [Ar II] 6.99 $\mu$m, [Co II] 10.55 $\mu$m, and [Ne II] 12.8 $\mu$m.

Although that theoretical models predict silicates should form in the oxygen-rich shell and carbon-rich grains should form in the carbon-rich shell of a supernova (Dwek 1988, Kozasa et al 1991), this has not yet been observed. SN1987A was observed to form dust at about 560 days after the event: optical light dropped and IR light increased, so that when the visible and IR luminosities are added, they sum to that expected from the radioactive decay. The dust emission from SN1987A never revealed any dust emission features. The dust formed in optically thick clumps (Wooden et al. 1993).

Detailed studies of the mid-IR iron lines ([Fe II] 26 $\mu$m,18 $\mu$m, and 24.5 $\mu$m) revealed a picture in which Rayleigh-Taylor fingers of radioactive material were shot to high velocities (Fryxell et al. 1991, Li et al. 1993, Wooden 1994). The radioactivity preferentially heated the iron-rich gas, causing the narrow fingers to do work on the surrounding medium, swelling, compressing the lighter metal shells of oxygen and carbon into smaller volumes. The iron-rich material which filled 1% of the volume at the time of the explosion expanded to fill 50% of the volume of the ejecta in the first week (Li et al. 1993). It is within these frothy fat iron fingers that dust formed (Wooden et al. 1993, Wooden 1994, Spyromilio and Graham 1992), covering 50% of the surface area of the ejecta. Within six months of the onset of dust condensation, over 90% of the visible light was being reprocessed into dust thermal emission (Bouchet et al. 1994). Of the light that did leak out, there was a $1/\lambda$ wavelength dependence to the color; the light was reddened by small grains (Lucy et al. 1989). Thus the graybody character of the dust emission at late times was the result of small grains forming in regions where the line-of-sight opacity was optically thick in the IR. The ejecta was inhomogeneous but did not necessarily having regions of higher density; the ejecta had regions with coherent, long-lines-of-sight which appeared optically thick out to far-IR wavelengths ($\lambda \geq 30$ $\mu$m) (Wooden et al. 1993, Wooden 1994).

The tale of SN1987A inspires visions of IR spectral interferometric images of the ejecta dominated by Rayleigh-Taylor fingers, dust destruction of the circumstellar medium, dust chemistry strongly dependent on geometry of ejecta, and macroscopic mixing of different regions of the ejected star. Observations of the continuum and of many of the important diagnostic lines in supernovae (Wooden et al. 1993, McCray 1993) require the wavelength range of an IR space interferometry platform.

**Acknowledgements:** Special thanks to Greg Sloan, Kay Justanont, Xander Tielens, to Chick Woodward, and to Hongwei Li for their discussions on AGB stars, novae, and supernovae. Thanks also to the SOC for travel support.

## References

Anders, E., Lewis, R., Tang, M., Zinner, E.: 1989, in Interstellar Dust, L.J. Allamandola, A.G.G.M. Tielens (Eds.), Kluwer, p. 389
Bouchet, P., Danziger, I.J., Gouiffes, C., Della Valle, M., Moneti, A.: 1994, in IAU Coll. 145: Supernovae and Supernovae Remnants, Cambridge University Press
Cherchneff, I., Tielens, A.G.G.M.: 1992, in Wolf Rayet Stars, K. van der Hucht (Ed.), Kluwer
Danchi, W.C., Bester, M., Degiacomi, C.G., Greenhill, L.J., Townes, C.H.: 1994, *AJ* **107**, 1469
Dwek, E.: 1988, *ApJ* **329**, 814
Frail, D.A., Kulkarni, S.R.: 1991, *Nature* **352**, 785
Fryxell, B., Miller, E., Arnett, D.: 1991, *ApJ* **367**, 619
Gehrz, R.D.: 1988, *ARA&A* **26**, 377
Gehrz, R.D., Grasdalen, G.L., Hackwell, J.A., Ney, E.P.: 1980a, *ApJ* **237**, 855
Gehrz, R.D., Greenhouse, M.A., Hayward, T.L., Houck, J.R., Mason, C.G., Woodward, C.E.: 1995a, *ApJ* **448**, L119
Gehrz, R.D., Hackwell, J.A., Grasdalen, G.L.: 1980b, *ApJ* **239**, 570
Gehrz, R.D., Jones, T.J., Matthews, K., Neugebauer, G., Woodward, C.E., Hayward, T.L., Greenhouse, M.A.: 1995b, *ApJ* **110**, 325
Gehrz, R.D., Jones, T.J., Woodward, C.E., Greenhouse, M.A., Wagner, R.M., Harrison, T.E., Hayward, T.L., Benson, J.: 1992, *ApJ* **400**, 671
Gehrz, R.D., Woodward, C.E., Greenhouse, M.A., Starrfield, S., Wooden, D.H., Witteborn, F.C., Sandford, S.A., Allamandola, L.J., Bregman, J.D.: 1994, *ApJ* **421**, 762
Goebel, J.H., Cheeseman, P., Gerbault, F.: 1995, *ApJ* **449**, 246
Haas, M.R., Colgan, S.W.J., Erickson, E.F., lord, S.D., Burton, M.G., Hollenbach, D.J.: 1990, *ApJ* **360**, 257
Habing, H.J.: 1996, *A&A* **330**, 999
Harrison, T.E., Stringfellow, G.S.: 1994, *ApJ* **437**, 827
Ivezic, Z., Elitzur, M.: 1995, *ApJ* **445**, 415
Johnson, D.J., Friedlander, M.W., Katz, J.I.: 1993, *ApJ* **407**, 714
Jura, M.: 1995, in Airborne Astronomy Symposium on the Galactic Ecosystem: From Gas to Stars to Dust, *ASP Conf. Ser.* **73**, 359
Kirschner, R.P., Oke, J.B., Penston, M.V., Searle, L.: 1973, *ApJ* **185**, 303
Kozasa, T., Hasegawa, H., Nomoto, K.: 1991, *A&A* **249**, 474
Li, H., McCray, R., Sunyaev, R.A.: 1993, *ApJ* **419**, 824
Lucy, L.B., Danziger, I.J., Gouiffes, C., Bouchet, P.: 1989, in Supernovae: The Tenth Santa Cruz Workshop in Astronomy and Astrophysics, S.E. Woosley (Ed.), Springer-Verlag, p. 82
Lynch, D.K., Hackwell, J.A., Russel, R.W.: 1992, *ApJ* **398**, 632
McCray, R.: 1993, *ARAA* **31**, 175
Paresce, F., Livio, M., Pack, W., Korista, K.: 1995, *A&A* **229**, 823
Pijpers, F.P., Pardo, J.R., Bujarrabal, V.: 1994, *A&A* **286**, 501
Planesas, P., Kenney, J.D.P., Bachiller, R.: 1990, *ApJ* **364**, L9

Sloan, G.C., Price, S.D.: 1995, *ApJ* **451**, 758
Sloan, G.C., LeVan, P.D., Little-Marenin, I.R.: 1996, *ApJ* **463**, 310
Spyromilio, J., Graham, J.R.: 1992, *MNRAS* **255**, 671
Stencel, R.E., Nuth, J.A., Little-Marenin, I.R.: 1990, *ApJ* **350**, L45
Thronson, H.A., Jr., Rapp, D., Bailey, B., Hawarden, T.G.: 1995, *PASP* **107**, 1099
Tielens, A.G.G.M.: 1989, in From Miras to Planetary Nebulae: Which Path for Stellar Evolution?, M. O. Mennessier, A. Omont (Eds.), Editions Frontieres, p. 186
Tielens, A.G.G.M., 1990, in IAU Coll. 126: Origin and Evolution of Interplanetary Dust, A.C. Levasseur-Regourd (ed.), Kluwer
Tielens, A.G.G.M.: 1995, in Airborne Astronomy Symposium on the Galactic Ecosystem: From Gas to Stars to Dust, *ASP Conf. Ser.* **73**, 3
Vardya, M., de Jong, T, Willems, F.: 1986, *ApJ* **304**, L29
Wilson, R.W., Baldwin, J.E., Buschner, D.F., Warner, P.J.: 1992, *MNRAS* **257**, 369
Wooden, D.H.: 1989, Ph.D. Thesis, University of California, Santa Cruz
Wooden, D.H., Rank, D.M., Bregman, J.D., Witteborn, F.C., Tielens, A.G.G.M., Cohen, M., Pinto, P.A., Axelrod, T.S.: 1993, *ApJS* **88**, 77
Wooden, D.H.: 1994, in IAU Coll. 145: Supernovae and Supernovae Remnants, Cambridge University Press
Woodward, C.E., Gehrz, R.D., Jones, T.J., Lawrence, G.F.: 1992, *ApJ* **384**, L41

# PART II

# INSTRUMENTAL AND TECHNICAL CASES

# KILOMETRIC BASELINE SPACE INTERFEROMETRY

P.Y. BELY
*Astrophysics Division, European Space Agency*
*Space Telescope Science Institute, 3700 San Martin Drive, Baltimore, MD 21218, USA*

**Abstract.** Two versions of a kilometric baseline interferometer with equivalent science capabilities have been studied, one located on the Moon and the other operating as a free-flyer. It has been found that the Moon is not an ideal site for interferometry because of the need for long delay lines and the large temperature swings from day to night. Both versions could attain the required scientific performances and each one needs the same type of metrology control but the free-flyer is intrinsically advantageous because of its reconfiguration flexibility, quasi-unlimited baseline length and higher observation efficiency.

**Key words:** space interferometers, Moon-based astronomy

## 1. Introduction

With an angular resolution 1000 times that of the Hubble Space Telescope, kilometric baseline interferometry from space at visible and near infrared wavelengths has the potential to revolutionize astrophysics. In our own galaxy, interacting binaries of all types could be examined, the pulsation of Cepheids could be observed, and surface features on stars could be mapped directly. In external galaxies, beamed synchrotron jets, gas motions in broad emission line regions of quasars and active galactic nuclei in nearby galaxies could be imaged. The impact of the resulting discoveries on our understanding of physical processes in the universe could truly be extraordinary.

Going into space provides the same advantages for interferometry as for conventional telescopes, i.e. the elimination of the turbulent and partially opaque atmosphere and provides a significant increase in resolution and sensitivity and an unlimited isoplanatic patch size. A number of connected-element space interferometers have been proposed (Traub & Gursky 1980, Noordam et al. 1987, Bély et al. 1989, Nein & Morgan 1989), but their practical baseline limit is probably around 100 meters. Free-flying (separated) spacecraft have also been proposed (Labeyrie et al. 1984, Stachnik et al. 1984, Kulkarni et al. 1994), but are widely expected to present daunting difficulties in station keeping at the wavelength scale. The Moon, on the other hand, has long been considered a suitable platform for a space interferometer because it combines the



*C. Eiroa et al. (eds.),*
*Infrared Space Interferometry: Astrophysics & the Study of Earth-Like Planets,* 149–155.

advantages of space with the baseline stability conferred by firm ground (Burns 1989, Labeyrie 1992).

In order to assess the relative merits of these two approaches, the European Space Agency (ESA) sponsored a comparative engineering of a free-flying and a lunar-based interferometer, which was performed by Matra-Marconi Space under the guidance of a team of scientists and interferometry specialists. We present here the principal results of this study, the details of which can be found in an ESA report (SCI 96-7 1996).

Most of the astronomical objects of interest at very high angular resolution have relatively complex morphologies which require imaging capability, a large dynamic range, and adequate sensitivity. We thus determined that the unit telescope size should be at least 1 meter for sufficient sensitivity and the number of telescopes should be at least 6 to allow for imaging with adequate phase closure and rapid coverage of the $uv$-plane.

We found that passive stabilization of the interferometer elements was extremely difficult both in space or on the Moon, and that active control of the optical pathlengths using laser metrology and a reference star was required. By monitoring the fringes from a reference star, the positioning tolerances on the interferometer elements are relaxed by several orders of magnitude. However, stars used as reference need to display a high apparent fringe visibility which means that their diameter must not be resolved. For a 5 km interferometer operating at visible wavelengths, this implies that stars should be at about 400 parsecs on average, i.e. with a magnitude of about 18.5. Based on traditional star count statistics, a field of 2.5 arcminutes in radius is necessary to find at least one such suitable reference star at the galactic poles with 95% probability.

## 2. Moon-based version

The general arrangement of the six-telescope Moon-based interferometer is shown in Fig. 1. The observatory is arranged in a 'Y' configuration tilted a few degrees away from the North/South direction with two telescopes in each arm. The observatory would be located near the Moon limb to avoid Earth straylight and at about 45 degree latitude to provide a good $uv$-plane coverage as the Moon rotates.

If the Moon offers a relatively stable platform for interferometry and variable baseline thanks to its rotation, a major drawback is that delay lines have to be used to compensate for the changing optical pathlengths. These delay lines are of a length comparable to the baseline

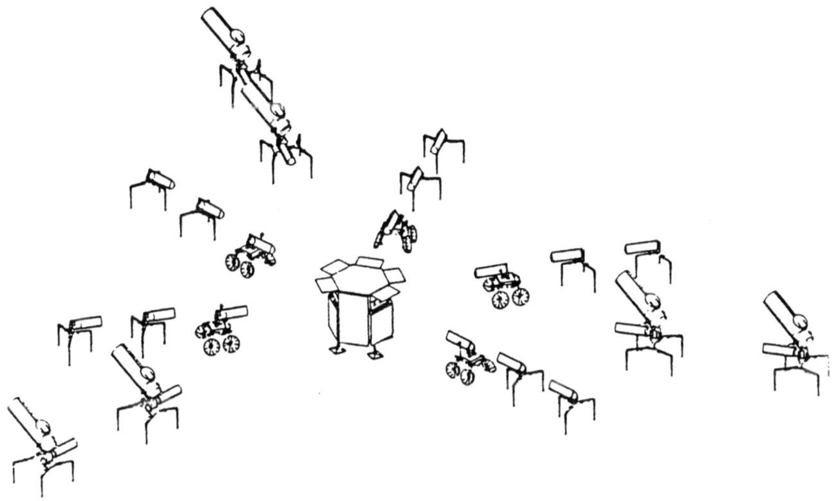

*Figure 1.* General arrangement of the 6 telescope array (not to scale). The telescopes are the larger elements arranged in a Y configuration and shown with their 2 meter long baffles deployed. The smaller mirrors, 5 of which are on mobile carts, constitute the delay lines systems.

in order to permit the observation of targets close to the horizon. We selected to use mobile carts with independent traction systems because of their much higher optical efficiency compared to banks of fixed mirrors and lighter mass compared to railed systems. These carts do not move during single observations, but must be repositioned from one observation to the next. A delay line inside the hub allows for optical pathlength changes during a given observation.

Another drawback of the Moon as an observatory platform is that heat and straylight during the lunar day prevent observations, and that a non-solar power system has to be used for supplying electrical power during the lunar night. For power, we baselined regenerative fuel cells which make use of the propulsion hydrogen and oxygen tanks of the landers and which are regenerated during lunar day with solar arrays.

The entire observatory can be delivered to the Moon surface in two upgraded Ariane 5 launches.

## 3. Free-flyer version

A view of the free flyer version is shown in Fig. 2. A powerful feature of the free-flying interferometer is its ability to be reconfigured to best match the mapping requirements of the observed source. It is indeed

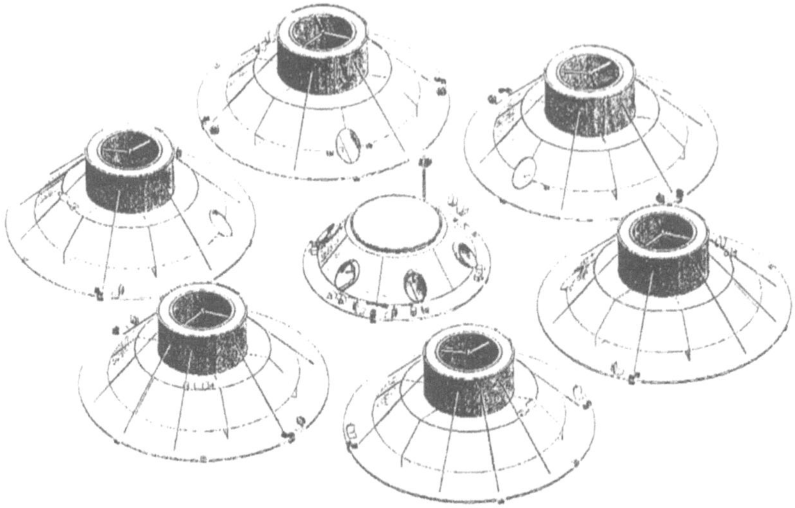

*Figure 2.* Free-flying interferometer shown in close formation for clarity. Under normal conditions the distance between the recombining hub and the individual telescopes varies from 50 to 2500 meters.

possible to adjust both the maximum baseline and the number of intermediary measuring points to optimize resolution and distribution of the visibility measurement points in the Fourier plane. This optimization should be done on a case by case basis but, as a rule, the majority of targets requires uniform mapping which can be obtained with one of the configurations worked out by Cornwell (1988). To cover the *uv*-plane more or less uniformly, the radial expansion or contraction should be accomplished with a velocity varying as the inverse distance from the hub.

The six 1-meter telescopes in the array simultaneously observe the science target, a reference star to cophase the array and a pointing star for attitude control. The reference star and pointing star are collected in the telescope's field of view with movable off-axis pick-off mirrors. The three beams are separated spectrally and collimated into a common 40 cm diameter beam for transfer to the recombining hub.

The beams coming from each of the 6 telescopes are collected inside the recombining hub with a 40 cm diameter cassegrain unit and are demagnified so as to be more manageable. Each composite beam first goes through a common delay line which compensates for small optical pathlength differences that may exist between each of the telescopes in the array and the recombining hub. The three beams are then separated spectrally, with the pointing star beam being sent to a pointing analyzer and the reference star and target beams sent to a focal plane instrument. The reference star beam goes through an additional delay line which corrects for the differential optical pathlength between target and reference star. A laser metrology is used to determine the respective position of the telescopes and hub so as to ensure that they stay within the common reference-star/target cophasing range.

An extremely low disturbances orbit is required and the observed sky field must be continuously visible during the full observing time which typically lasts several hours. These two conditions exclude low Earth orbits. Among higher orbits a halo orbit close to the second Lagrange point of the Sun-Earth system offers the most advantage.

Due to continuous presence of the sun in the selected orbit, it is indispensable that the telescopes and recombining hub be well protected. This is accomplished by keeping in shadow the portions of the spacecraft in the view of all other spacecraft, and by extensive baffling of the output and input transfer telescopes.

Because of this self shadowing, the portion of the sky that can be observed at any one time is about 14%. Over a one year period, however, the band between ±45 degrees of ecliptic latitude is observable, which represents about 72% of the entire sky.

Station keeping and attitude control is performed by a Field Emission Electric Propulsion system (FEEP) (Gonzales et al. 1993) which is very compact and provides extremely smooth actuation over a wide range of thrust, from a few $\mu$N to a few mN. The estimated total fuel requirement for a nominal 5 year mission is 8 kg per telescope.

The six telescope units, hub and transfer stage can be accommodated in a single Ariane 5 launch.

## 4. Comparison of the two versions

The study of the two versions of the interferometer indicate that they are both feasible, do not exhibit major show stoppers and can essentially meet the scientific requirements we had established.

Perhaps the main point is that locating an interferometer on the Moon does not bring any obvious simplification in construction or operation. One could argue, on the contrary, that the Moon-based system has to cope with a much more severe environment than its space-based counterpart: large temperature swings (day/night), ground disturbances, the impossibility of observing during the lunar day, dust, and lack of solar-generated power during the lunar night.

From the observational point of view, too, the space-based system has the advantage. There is no real limit to its baseline length. Its efficiency is about twice as much because of the impossibility of the

Moon-based counterpart to observe during the lunar day. Its array can always be configured optimally, while the lunar based has to contend with whatever source declination and geographic latitude provide. And for some scientific programs, the Moon simply rotates too slowly: super synthesis mapping requires half a lunar day or about 14 Earth days, which is too long compared to the rotation or time scale of physical processes of many sources.

On the logistical side, automatic deployment of the interferometer on the Moon would be very difficult and astronaut supervision and site preparation would be highly desirable. On the other hand, availability of man at a lunar base means that maintenance and upgrading would be possible, resulting in a longer lifetime.

We conclude that the free-flyer is better suited for implementation in the near or mid-term future, but that the Moon-based version should be considered in the long term in conjunction with a manned lunar base.

## References


Bély, P-Y., Burrows, C.J., Roddier, F., Weigelt, G.: 1989, in New technologies for Astronomy, ECO Proc 1130 (*HARDI*)

Burns J.: 1989, in NASA Conf Publ. 3066 (*LOUISA*)

Cornwell, T.J.: 1988, *IEEE Tr.* **AP-8**, 1165

Gonzales, J., Saccoccia, G., von Rohden, H.: 1993, in AIAA 23nd International Electric Propulsion Conf, IEPC 93-157

Kilometric baseline space interferometry, ESA Report SCI-96-7, 1996

Kulkarni S.R., Bély, P-Y, Colavita, M., DeCou, A.B., Jones, D.L., Krolik, J.H., Martin, C.D., McAlister, H.A., McGuire, J.P.: 1994, in Separated Spacecraft Interferometry, Proposal to NASA Res. Ann. 94-15

Labeyrie A.: 1992, in ESA Colloquium SP-354

Labeyrie, A., Authier, B., Boit, J.L., de Graauw, T., Kibblewhite, E., Koechlin, L., Rabout, P., Weigelt, G.: 1984, *BAAS* **16**, 3

Nein, M.E., Morgan, S.H.: 1989, in Next Generation Space Telescope, Proc. Space Tel. Sc. Inst.

Noordam, J.E., Greenaway, A.H., lePoole, R.S., Bregman, J.D.: 1987, in ESA Colloquium SP-273 (*OASIS*)

Stachnik, R.V., Ashlin. K., Hamilton, S.: 1984, *BAAS* **16**, 3 (*SAMSI*)

Traub W.A. and Gursky, H.: 1980, *Proc. SPIE* **228**


## 5. Questions

*P. Claes* : Where are the MOFFIT results published?
*P. Bély* : They will be published in April-May 1996.

*R. Simon* : Considering the current availability of large, low cost apertures on the ground (e.g., the ~10 m aperture of the Hobby-Eberly

telescope cost <$15M), is the proposed free-flyer mission truly competitive with what could be accomplished from the ground with a similar amount of funding?
*A. Greenaway* : Ground and space are probably better seen as complementary rather than competitive. Terrestrial interferometry looks likely to give superior performance in the near IR where thermal background is not a problem, but a (relatively) modest space mission should give sensitivity gains of ~4 magnitudes at least in the thermal IR at 10 $\mu$m and beyond. At short wavelengths, sensitivity gains of useful size are also achievable. However, for bright targets a terrestrial interferometer should give good performance accross the wavelength range. One needs to be sure that the extra sensitivity from space adds to the science programme in an essential way to justify a space mission. If useful science can be done within the sensitivity reach of a terrestrial instrument it should be done there. This would then give results to be extended in scope and sensitivity by a subsequent space mission - if the extra sensitivity is needed to complete the objectives of the science programme. We do not believe that it would be practical to complete the strawman programme from a terrestrial instrument.

*D. Tytler* : What factors determine maximum baseline for this free flyer design?
*P. Bély* : The main factors are: 1) gravity gradient (drifts during reference star integration need to be minimized); 2) beam transfer (the larger the distance, the more diffraction there is requiring larger transfer telescopes).

*D. Tytler* : For ExNPs the IR interferometer should be out at about 5 AU. Can you get enough power, from solar arrays or other ways, for the electric propulsion used by the free flyer?
*P. Bély* : This is an obvious problem and needs more study, but might be solved with the use of deployable large solar arrays.

*M. Yanagisawa* : What is the lifetime of the free flyer mission ? Does the lifetime depend on the amount of the propellant of the Field Emission Propulsion ?
*P. Bély* : The lifetime of the mission was set at 5 years. The amount of propellant required is small (kilograms) and not driving the lifetime.

*M. Faucherre* : Did you consider focal instrumentation in your study? Is it different for a Moon-based and a space-based mission?
*P. Bély* : Yes, we did. I did not have the time to describe it in this presentation but you will find more details in the report.

# THE MEASUREMENT OF DIRECTIONAL RADIATIVE PROPERTIES WITH APPLICATIONS TO PASSIVELY COOLED SPACE TELESCOPES

R. P. BLAKE and B. W. JONES
*Physics Department, The Open University, Milton Keynes, MK7 6AA, UK*
*(email: r.p.blake@open.ac.uk , b.w.jones@open.ac.uk)*

**Abstract.** With passively cooled infrared space telescopes being proposed in many missions for the next century by both ESA and NASA, accurate models are needed to predict their behaviour. However, commercially available packages do not model radiative heat transfer correctly. We tackle this problem with our own Monte-Carlo model of a space telescope with simplified geometry but sophisticated handling of radiation heat transfer.

However, for such models to produce accurate results, they need reliable data on surface properties for candidate materials. We therefore outline our ongoing experimental project to collect such data.

**Key words:** space telescopes, passive cooling, directional properties, experimental measurement, monte carlo method

## 1. Introduction

Space-based infrared astronomy will be one of the most exciting areas of research in the next century and will play a key role in observing extra-solar planets. Current designs for passive cooling (Hawarden et al. 1995) will produce telescopes more sensitive and economic than ever before. Even cryogenically cooled telescopes will benefit greatly from passive cooling.

In order to predict the effectiveness of passive cooling designs, the radiative exchange of heat between surfaces must be modelled very accurately. Until now, this has not been done since commercially available modelling packages make grossly simplified assumptions about radiative exchange, casting doubt on their results.

## 2. Modelling

We have developed a Monte Carlo model (Blake & Jones 1995) of a simplified telescope (Fig. 1) to investigate the importance of realistic modelling of radiative heat exchange. All the surfaces in the model were assumed to be gold, apart from the outer sunshield (Teflon coated silver) and the aperture (modelled as a disc with emittance = 0.8



*C. Eiroa et al. (eds.),*
***Infrared Space Interferometry: Astrophysics & the Study of Earth-Like Planets,*** **157–161.**

(Hawarden et al. 1992)). Note that, while the geometry was fairly primitive, the treatment of radiative transfer was very sophisticated.

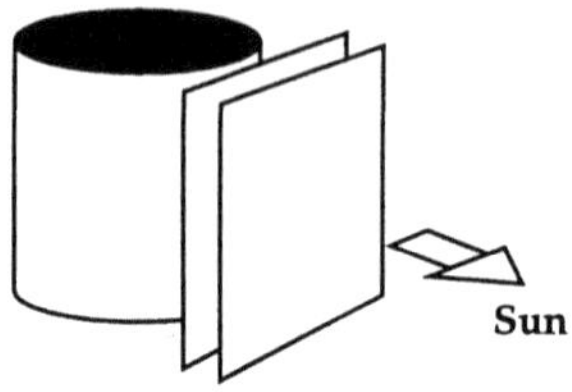


*Figure 1.* A simplified telescope model.

A key omission in other models is the directionality of radiative emission, which is assumed to be diffuse. Fig. 2 shows many values for directional properties of gold (a candidate material for passively cooled telescopes) and demonstrates that it is far from diffuse.

In Fig. 2, Case A shows the simplified assumption of diffuse emission ($\epsilon = 0.03$). Case B shows a more realistic emittance derived (Siegel & Howell 1992) from quoted values of the optical constants of gold at $5\mu m$ and scaled up so that the hemispherical average emittance is 0.03. Case C is the same as case B but with no scaling, while case D is derived from quoted optical constants at $100\mu m$ (being a more typical wavelength of emission at cryogenic temperatures), again with no scaling.

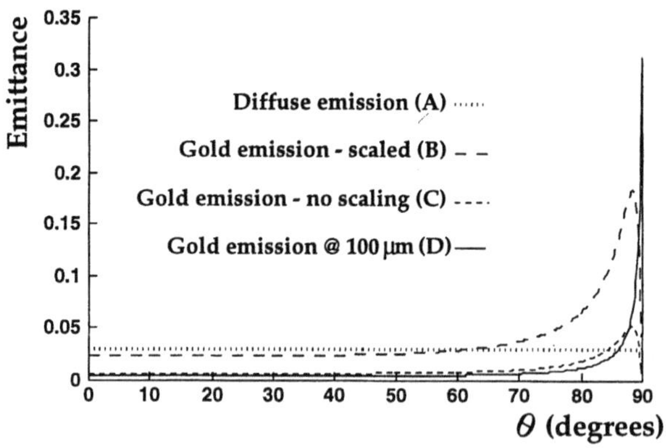


*Figure 2.* Directional values of the emittance of gold.

We modelled many such cases to test the sensitivity of the final telescope tube temperature on the emissive properties of the surfaces. Other properties varied included $H$ (the thermal conductance between

Table I. Summary of results from Monte-Carlo simulation.

| | $N$ | $H\mathrm{W/K}^{-1}$ | radiative properties of gold | $T_1$/K | $T_2$/K | $T_3$/K |
|---|---|---|---|---|---|---|
| 1 | 10 | 0 | high emittance, diffuse (case A) | 211.2 | 97.0 | 17.7 |
| 2 | 10 | 0 | high emittance (case B) | 211.2 | 90.0 | 15.7 |
| 3 | 10 | 0 | opt. con. @ 5$\mu$m, 300K (case C) | 213.5 | 68.3 | 6.6 |
| 4 | 10 | 0 | opt. con. @ 100$\mu$m, 300K (case D) | 214.1 | 61.5 | 4.9 |
| 5 | 10 | 0 | opt. con. @ 10,30,100$\mu$m, 300K | 212.6 | 66.7 | 6.0 |
| 6 | 10 | 0 | as line 5, with polarisation | 213.8 | 68.6 | 6.4 |
| 7 | 5 | 0 | as line 5, with polarisation | 213.8 | 57.6 | 4.8 |
| 8 | 2 | 0 | as line 5, with polarisation | 213.4 | 44.7 | 3.3 |
| 9 | 1 | 0 | as line 5, with polarisation | 213.2 | 37.3 | 2.9 |
| 10 | 10 | 0.1 | as line 5, with polarisation | 213.7 | 109.8 | 41.5 |
| 11 | 10 | 0.01 | as line 5, with polarisation | 213.8 | 78.6 | 21.5 |

The temperatures $T_1$, $T_2$ and $T_3$ correspond to the outer sunshield, the inner sunshield, and the telescope tube respectively. 'Opt. con.' indicates values derived from the optical constants at the given wavelengths (Ordal et al. 1983).

the parts of the telescope) and $N$ ($= \frac{\textit{length of side of sunshield}}{\textit{separation between sunshields}}$). The sunshield separation was equal to the minimum distance between the inner sunshield and the tube. Our results are summarised in Table I.

One of the major problems in modelling this kind of heat exchange is the lack of reliable data on the surfaces involved. We began by using the oft-quoted value of the 'emittance of gold in the infrared' of 0.03, but even this raised questions when directionality was being modelled. Ultimately we modelled all the cases in Fig. 2, and more (Blake & Jones 1996).

Our search for data also turned up a serious lack of information about properties at low temperatures. Clearly this will be vital in a case study of a passively cooled telescope and needs to be investigated.

Another omission of other models is the polarisation of the radiation. Fig. 3 shows how the emittance of gold varies for each polarisation. Clearly, even unpolarised radiation will become polarised after one reflection. This effect is, again, summarised in Table I.

This Monte Carlo model is not trying to be a fully accurate prediction of how such a telescope will perform, but it does highlight the importance of modelling radiative heat exchange accurately.

## 3. Measurements

Faced with such a serious lack of useful data, we have embarked on a program of measurements of surface properties of candidate materials.

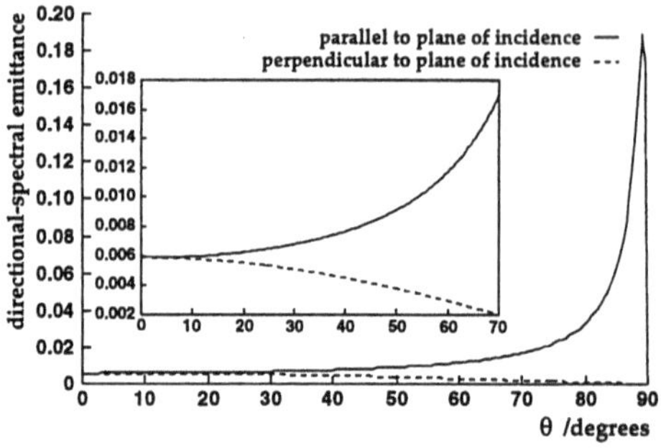


*Figure 3.* The variation of emittance with polarisation.

We are specifically interested in the directional-reflectivity $\rho'(\theta, \phi, \lambda, \mathrm{T})$ since it is relatively easy to measure and can be used to determine the emittance $\epsilon'(\theta, \phi, \lambda, \mathrm{T})$ and absorptance $\alpha'(\theta, \phi, \lambda, \mathrm{T})$.

Our equipment consists of a powerful monochromatic source (a laser) to drown out any background noise, a sample mounted on a rotating stage, and a detector also on a rotating stage able to sweep around the sample (Fig. 4). The sample and detector are simply moved through many positions and a reading taken at each one.

As the samples are highly specular, the bi-directional reflectivity in the specular direction (i.e. when $\gamma = 0$ and $\alpha = \beta$) is equal to the directional-reflectivity so only one reading is required at each angle of incidence. A second detector monitors the laser drift.

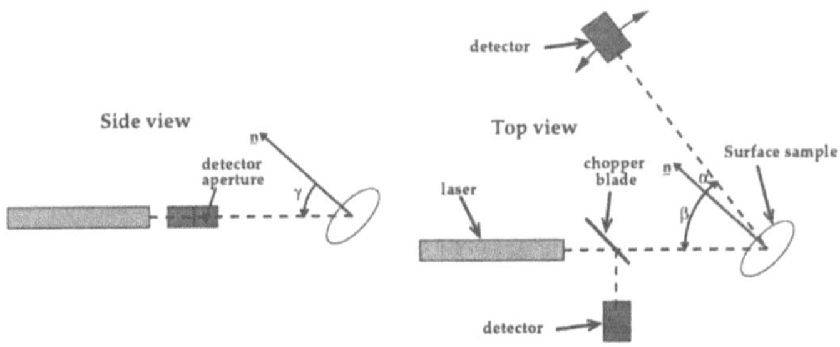


*Figure 4.* Design of a simple reflectometer.

This is a fast and simple way to build up a picture of directional surface properties, but has difficulty producing accurate results when studying the very high reflectivities involved in passive cooling design.

We have also built a cryogenic chamber for measuring reflective properties at low temperatures (as noted above, an area in need of study). At low temperatures surfaces become even more reflective and the equipment described in Fig. 4 would not be sensitive enough to collect accurate data. Thus, we have designed a multiple reflection cell (Fig. 5) which will enable us to improve accuracy.

In this case, the laser beam is bounced off the sample surface many times as it 'walks' down the two parallel mirrors, thus magnifying the reflectivity. It is then reflected back along its path by the smallest mirror and deflected into the detector by the beam splitter.

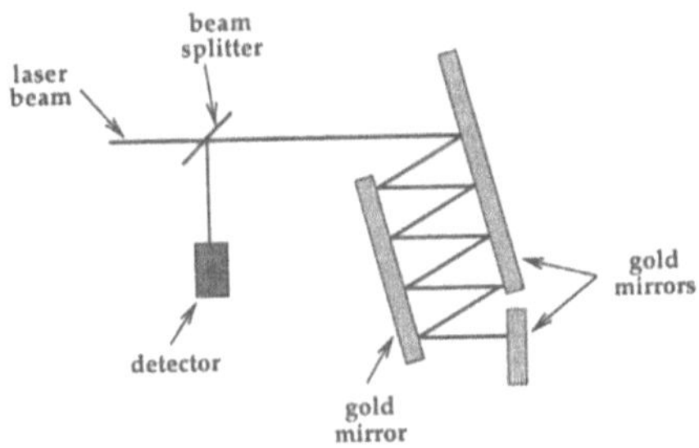


*Figure 5.* Design of a multiple reflection cell.

We will initially use it with 3 identical gold samples. One gold sample can then be replaced with any surface of interest. Although this equipment cannot be used for directional measurements, it is an order of magnitude more sensitive than the previous apparatus. It is hoped, in the near future, to develop a similar design to take a reading at a single large-$\theta$ angle, so giving some idea of directionality.

While early results have been achieved with the simple set-up, we are not yet at the stage of publishing, but were interested to see directional behaviour even from heavily weathered samples.

## References

Blake, R. P., Jones, B. W. : 1995, *Space Science Reviews* **74**, 175

Blake, R. P., Jones, B. W. : 1996, *Journal of Geophysical Research - Planets* **101**, 9303

Hawarden et al. : 1992, *Space Science Reviews* **61**, 113

Hawarden et al. : 1995, *Space Science Reviews* **74**, 45

Ordal et al. : 1983, *Applied Optics* **22**, 1099

Siegel, R., Howell, J. R. : 1992, in Thermal Radiation Heat Transfer (3rd ed.), Ch. 4-6.3, Hemisphere: London.

# CONCEPTS FOR A PRE-CURSOR SPACE INTERFEROMETRY MISSION WITH A MICROSATELLITE

K. BRIESS, C.J. SCHALINSKI, H.P. RÖSER and I. WALTER
*Institute of Space Sensor Technology (ISST), D.L.R.,*
*Rudower Chaussee 5, D-12489 Berlin, Germany*
*e-mail: schalinski@dlr.de*

**Key words:** space interferometry, small satellites, separate spacecraft interferometers, precursor missions, exoplanets

## 1. Introduction

Space Interferometry represents the key tool to eventially detect and characterize terrestrial planets outside our solar system, as identified by missions both by NASA (ORIGINS: SIM, Planet Finder) and ESA (HORIZON 2000 Plus interferometry cornerstone mission candidate DARWIN). In order to achieve theses goals, ambitious programs have to be undertaken to provide "21st century" key technologies for space interferometry, such as active controled structures, space qualified sub-nm laser metrology and optical delay lines, passive cooling techniques, and sensitive IR-detectors. Microsatellites (mass $\leq$ 100 kg, dimensions $\leq$ 1 m) bear a great potential to buy down risks and costs of the "big missions" by testing and validating critical technologies both on the component and system level under space conditions.

The "faster, cheaper, better"-approach by the use of microsatellites may provide fast innovation and substantial shortening of development times – with the fringe benefit of demonstrating that interferometry works in space! Furthermore, the extension of free flyers to separate spacecraft interferometers with its immense potential for astronomy (e.g. ESA's MOFFIT study; NASA's New Millenium 3 program), remote sensing applications/laser communications and gravitational wave physics (e.g. Johann et al., these proceedings) can be demonstrated with comparatively low cost micro-satellite precursors (Briess et al., in preparation).

DLR-ISST has recently established a "small satellite" program line, allowing the launch of a mission every 2-3 years. In 1997 a microsatellite targetting attitude control on the arcsec level (project in collaboration with the Technical University, Berlin) will be launched; Phase B stud-



*C. Eiroa et al. (eds.),*
*Infrared Space Interferometry: Astrophysics & the Study of Earth-Like Planets,* 163–168.

ies have commenced for a remote sensing satellite ("BIRD") with a multiple payload including a mid- infrared detector to be launched in 1999. Technological goals are the demonstration of "intelligent sensor"-technology, including on-board processing and advanced data compression techniques for thematic data reduction. In collaboration with German and Danish universities a concept for cDIVAR, a post-Hipparcos mission on a minisatellite, has been developed and awaits feasibility studies in 1997.

In this paper the concept of a small space interferometer on a microsatellite (50kg) platform with a base length of $\sim$ 1m and 2 telescopes (D $\sim$ 12cm) is described. For the technological experiments the mission design was carried out for a preferred Low Earth Orbit (LEO) (GPS receiver on board), although a High Elliptical Orbit (HEO) or a Geo-Transfer Orbit (GTO) may also be considered.

## 2. Critical Technological Areas

Some main problems of optical space interferometry are connected with the substitution of heavy optical structures on ground and the manual ground operations by light-weight and active optical structure elements and subsystems, and by remote and partly autonomous operations in space. Four of the most critical technological areas and some components to test in space are:

- **Static Alignment of Telescopes**
    - lightweight mirrors and optical calibration elements
    - remote control of alignment
    - adaptive alignment control system
- **Platform Stability**
    - suppression of vibration sources from telescopes
    - passive damping
    - active damping (smart structures)
- **Dynamic Pointing Control**
    - basic pointing (arcmin to few arcsec) by means of the spacecraft attitude control system with star sensors, gyros, wheels
    - fine pointing of telescopes via pointing star, stabilization using the interferometric system, accelerometer, gyros

– **Active OPD Control**

- measurement of low-frequency distortions (drift, deformation)
- measurement of high-frequency distortions (vibrations) by means of laser metrology, micro-accelerometer
- OPD compensation by means of delay lines or other optical elements

## 3. Mission Objectives

The objectives of a technological pre-cursor mission for space interferometry may be summarized as follows:

– to investigate active pointing control and platform stability of a space interferometer under space conditions

– to study the OPD control under microgravity and space conditions

– to study the influence of internal and external disturbances on the OPD control

– to investigate the accuracy of a laser metrology system

## 4. Launch System

The mission should be characterized by a piggy-back launch because of the low costs on the order of 6000US/$kg$. For this reason the microsatellite must be compatible in mass and start dimensions to the auxiliary payload adapter of several launchers: Ariane (A.S.A.P.), Zenit, Zyklon, Cosmos, Delta II and others.

## 5. Orbit Definition

Baseline for the mission design is Low Earth orbit (LEO) with a maximum coverage of the European region for maximum contact times with the existing DLR-ground stations in Germany. The required range for the inclination should be $\geq 53°$ to sun-synchronous. The required altitude range for a LEO is 450km to 900km. It is clearly not an optimum orbit for space interferometry because of the distortions and impacts on the instrument. But it is a range with a high probability for a launch opportunity. For the technological experiments fulfilling the primary objectives this orbit has no fundamental drawbacks. Other orbits, e.g. a highly elliptical orbit, are also possible. The microsatellite could be released for instance into the Geo-Transfer-Orbit with the launch of a geostationary satellite. But without own propulsion system the orbit has typically an inclination of about 7°. This requires additional collaborations and costs for non-domestic ground stations.

## 6. Space Segment

### 6.1. Payload

The following payload (a Michelson type interferometer) matches the mission objectives (see above):

- two telescopes with an aperture of about 12cm to receive the beams in the visible wavelength domain
- a beam combiner and a fringe detector (CCD matrix or photodiode)
- an active pointing system
- an OPD-control unit
- a GPS receiver for on-board time and orbit position determination.

### 6.2. Spacecraft Bus

The microsatellite mission is characterized by a three axis stabilized satellite with a size of 450mm x 450mm x 550mm and a mass of 50kg for a circular Low Earth Orbit ($i \geq 53°$). The satellite has a duty cycle of about 80% and is able to store 0.5Gbit on board. A data compression is foreseen to collect data from few orbits and send it down to a ground station. The special microsatellite design should be compatible to other LEO orbits with slight modifications of the bus. The availability of a piggy-back launch is an important design driver for the spacecraft. The proposed spacecraft design should be a first approach to these requirements. The main body is a cubic box compatible to most of the launch adapters for secondary payloads. It is a three-axis stabilized spacecraft. 3 sides of the spacecraft are completely covered with solar arrays and 1 side bears the telescopes, beam combiners and antenna platform. The remaining 2 sides are occupied by solar arrays and radiators.

### 6.3. Spacecraft Configuration

The spacecraft represents a microsatellite design with a mass of 50 kg within a volume of 130 $dm^3$. Due to the operation constraints one of two opposite sides of the rectangular solid carries the main payload and the other the solar panel system oriented to the Sun. To reach the required solar radiant area two side walls will be deployable from start to flight position (see Fig. 1 and 2), so that a solar cell area of about 0.65 $m^2$ is available after deployment. Fig. 1 shows the integration scheme of the micro-satellite. The eject adapter is fixed on the bottom front of this part to spread the launch loads to the structure elements. The solar panels are based on self-sustaining structures. The completed bonnet-shaped system will be integrated on the back surrounding with multi-layer-insulation (not shown in the figures). The S/C- structure mainly consists of the front and the back instrument plates (reinforced carbon fibre construction) connected by 6 cylindrical supports. The instrument units mounted between are attached to each of the instrument plates. Thus a rigid core compartment of the satellite can be built. Fig. 2 shows the flight configuration of the space interferometry satellite.

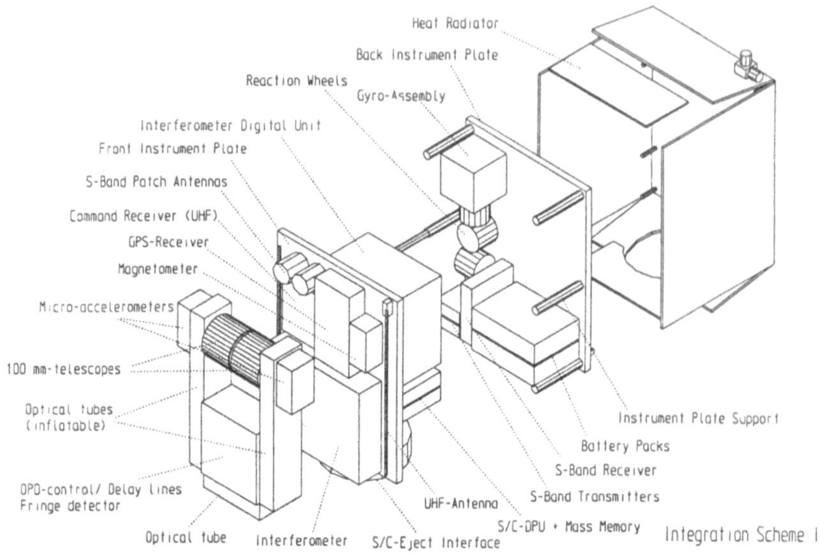


*Figure 1.* Integration scheme in start configuration.

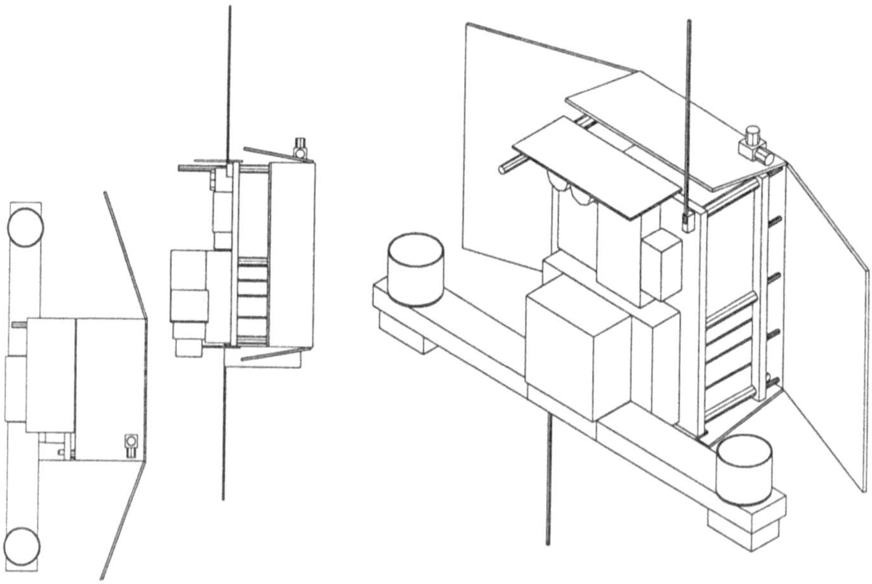

*Figure 2.* Flight configuration.

# LIGHT WEIGHT SIC FOAMED MIRROR FOR TELESCOPE TO BE OPERATED IN SPACE

O. CITTERIO
*Osservatorio Astronomico di Brera - Via E.Bianchi, 46; 22055 Merate - Italy*

G. PARODI
*BCV progetti Srl - Via S.Orsola 1, 1; 20123 Milano - Italy*

**Abstract.** By combining the excellent intrinsic thermo-mechanical properties of the SiC (Silicon Carbide) with a structural design based on a sandwich structure composed of two SiC face sheets CVD (Chemical Vapor Deposition) deposited on a foam core of the same material, it is possible to manufacture very light and stiff primary mirrors for telescopes to be operated in space. The paper presents an analysis of the mechanical properties of this structure and it points out the important saving in weight wich is possible with respect to other approaches used to manufacture light weight mirrors.

**Key words:** silicon carbide, light weight mirror, space telescope

**Abbreviations:** D = mirror diamater; q = unit surface mirror weight; Q = mirror weight; t = overall mirror thickness; $t_f$ = upper and lower faceplate thicknesss; $t_c$ = core thickness = $t - 2t_f$; $\eta$ = mirror lightening ratio; $\gamma$ = weight density of the material; $\rho$ = core (foam-rib) lightening ratio = *core weight* / (*core volume* $\times \gamma$)

## 1. Conventional Light Weight Mirror vs. Foamed SiC Mirror

Many mirror lightening techniques have been developped. Their aim is to create cavities in the substrate removing the material giving smaller contribution to the global stiffness, obtaining ribbed geometries between two faceplates. Lightened mirrors have been produced for example by casting, fusion, frit-bonding, machining, water jet machining, hot isostatic pressing. A characteristic parameter is the lightening ratio $\eta$ defined as the ratio between the mirror weight and the weight of a solid mirror having the same volume. Conventional lightening techniques permit to obtain $\eta$ in the range 0.2 - 0.3. On the contrary it is possible to reduce $\eta$ when the ribbed core is replaced by a foam, provided that the foam density is sufficiently low and its stiffness sufficiently high (Fig. 1). As we will show at the end of the paper, the design proposed for foamed SiC mirror has $\eta$=0.082.

Fig. 2 show that SiC foamed sandwich has bending stiffness considerably greater as regards to conventional light weight Zerodur mirrors having the same weight. The two technologies are comparable in terms



*C. Eiroa et al. (eds.),*
*Infrared Space Interferometry: Astrophysics & the Study of Earth-Like Planets*, 169–172.

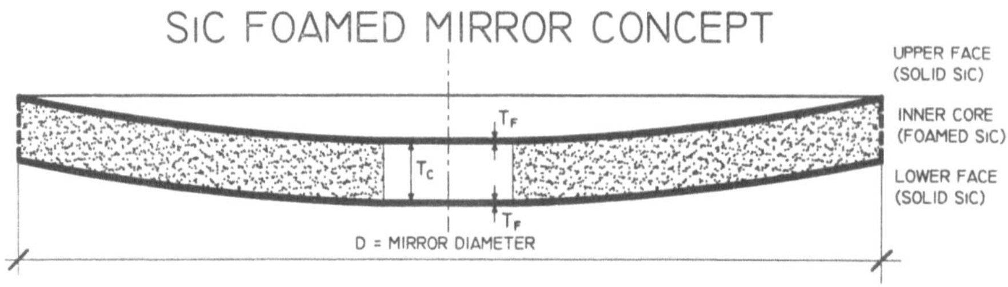


*Figure 1.* SiC foamed mirror concept: two thin CVD SiC faceplates are deposited on a SiC foam. By assign 3/4 of the available SiC mass to the foamed core and the rest to the two faceplates minimizes the global bending deflections (Gibson et al.). The foam weight density used is equal to 2 $kN/m^3$.

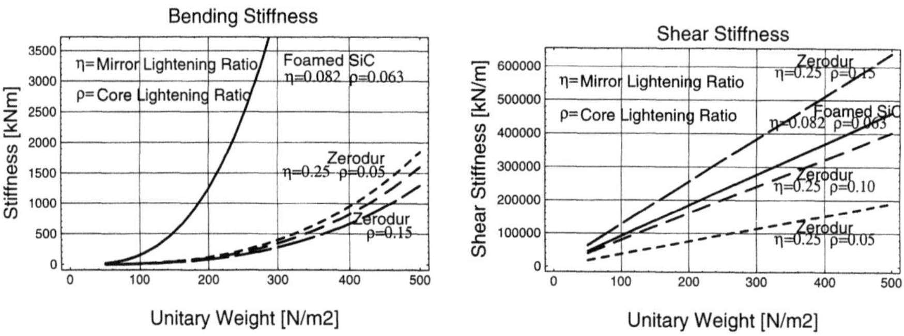


*Figure 2.* The bending and the shear stiffness of light weight Zerodur and foamed SiC mirrors are compared as function of the unitary weight q. For the Zerodur mirrors a lightening ratios $\eta$=0.25 has been used and the range 0.05 - 0.15 has been considered for the core density ratio $\rho$

of shear stiffness. From Fig. 3 it appears that foamed SiC sandwich shows a considerable gain in the global stiffness as regards to Zerodur light weight mirrors.

To check the sensitivity of SiC foamed sandwich, in terms of deflections of the optical surface, as regards to concentrated loads applied to the back plate we analysed the mirror zenith pointing, on three vertical supports placed at 2/3 of the mirror radius, 120 degrees spaced, loaded by the self weight.

These checks are particularly meaningful since the local deflections are mainly driven by the core stiffness that, in cellular solid as a foam, decreases with the square of the relative density.

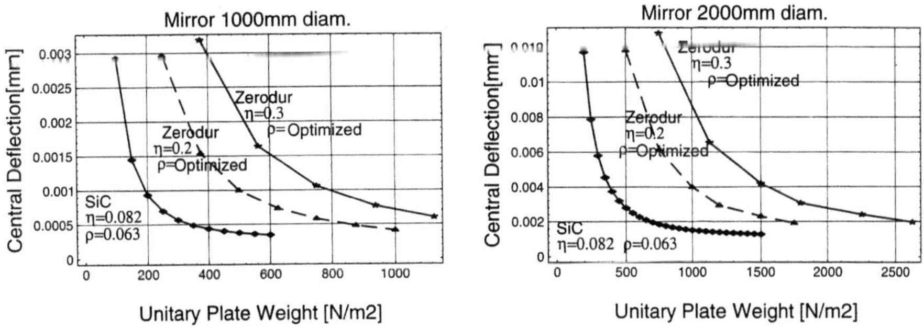


*Figure 3.* For two mirror sizes 1 and 2 m diameter, we compare the global stiffness of foamed SiC and Zerodur light weight mirrors in terms of the central deflection produced by self-weight when the mirror is simply supported at the outer edge. The deflection are plotted as function of the unitary mirror weight. Zerodur mirrors have an optimised design, i.e. for each value of the unitary weight "q", for each value of the lightening ratio $\eta$, the value of the relative core density ratio $\rho$ is stated by means of the deflection minimization.

By means of Finite Element Analyses (FEA) we computed the RMS, the Peak to Valley (PtoV) and the fundamental frequency of foamed SiC mirror. Two sizes 1 and 2 m diameter have been considered. For 2m size two mirror designs (thickness) have been considered. FEA results and some comparisons with other light weight mirrors are reported in the Table I.

Table I.

| a<br>mm | material | Tf<br>mm | Tc<br>mm | Q<br>kN | $\eta$ | PtoV<br>$\mu m$ | RMS<br>$\mu m$ | Freq.<br>Hz |
|---|---|---|---|---|---|---|---|---|
| 1000 | SiC | 0.5 | 49 | 0.102 | 0.082 | 1.05 | 0.23 | 352 |
| 2000 | SiC | 1. | 95. | 0.797 | 0.082 | 4.33 | 0.96 | 174 |
| 2000 | SiC | 1.5 | 147. | 1.232 | 0.082 | 2.59 | 0.58 | 220 |
| 1000 | Be (HIP) | 4.8 | 39.4 | 0.18 | 0.25 | | | 600 [1]) |
| 1828 [2]) | OHARA E6 | 26 | 200-315 | 5.2 | 0.31 | 1.68 | 0.38 | 201 |
| 2000 [3]) | OHARA E6 | 26 | 200-315 | 6.3 | 0.31 | 2.25 | 0.51 | 183 |

1) nine supports instead of three (Paquin & Gardopee 1991).
2) VATT primary mirror (Parodi 1987).
3) VATT primary mirror expanded from 1828 to 2000 mm diameter.

## 2. Final Remarks

The results reported in the table confirm that SiC foamed mirror are competitive, also for the local stiffness, with traditional glass light weight mirror. Moreover they give performances comparable to Beryllium mirrors with some saving in weight.

Depending on the magnitude of the shear contribution to the deflections with respect to the bending one, improvements of foamed SiC mirror performances could be possible by an optimisation of the foam density.

We have pointed out the save in weight obtainable by the proposed technology with respect to other approaches, but other advantages are offered i.e.:

- the manufacturing is simpler and faster with respect to other solutions;
- it is possible to increase locally the inner foam density in order to increase locally the stiffness;
- the solid material has good thermal properties;
- in the case of on ground applications, it should be possible to perform the thermal control of the mirror by an air flow crossing the foam and/or cavities drilled in the foam.

## References

Gibson, L.J. and Ashby, M.F.: in Cellular Solids Structure and Properties, Pergamon Press, p. 242

Paquin, R.A., Gardopee,G.J.: 1991, *SPIE Proc. Large Optics II* **1618**, 61

Parodi, G.: 1987, *BCV Technical Note n° 4 for Columbus Project (Steward Observatory)*

# RECENT ADVANCES IN CRYOGENIC OPTICS TECHNOLOGY FOR SPACE INFRARED TELESCOPE AND INTERFEROMETER SYSTEMS

D.R. COULTER and S.A. MACENKA
*Jet Propulsion Laboratory,*
*California Institute of Technology Pasadena, CA, USA.*

**Abstract.** In this paper we will describe recent advances in the development of optical systems for future space infrared telescope and interferometer applications which will operate at very low or cryogenic temperatures (T≤77K) with emphasis on beryllium and silicon carbide optics. New material formulations and advanced processing and manufacturing techniques are enabling the development of large, very low mass, high performance cryogenic optics. The design, manufacturing and cryogenic testing of several recently developed mirrors and optical assemblies will be discussed.

**Key words:** cryogenic optics technology, space infrared interferometry

## 1. Introduction

A number of future astrophysics and planetary science missions are currently being proposed and/or studied which will require the implementation of large space based telescopes and interferometers. Some of these include the Space Infrared Telescope Facility (SIRTF), the Space Interferometry Mission (SIM), the Next Generation Space Telescope (NGST), the Terrestrial Planet Finder Array (TPFA) and the Terrestrial Planet Mapper Array (TPMA). Schematic representations of concepts for two of these missions, NGST and TPFA, are shown in Fig. 1.

The NGST is currently envisioned to be a large deployable telescope with an ~8m aperture, passively cooled to 30-70K and performing imaging and spectroscopic studies between 0.5$\mu$m and 20$\mu$m (diffraction limited at 1-2$\mu$m). The TPFA concept shown in Fig. 1, is a multi-baseline interferometer composed of four 1.5m aperture telescopes, passively cooled to 35K, observing in the 7-17$\mu$m band and capable of achieving sufficient starlight nulling to enable imaging of Earth-like planets around distant stars.

NGST and TPFA, as well as many of the other mission studies, have highlighted the infrared (~1-20$\mu$m) as the key spectral region for observing programs aimed at the study of the early universe, the observation of extra-solar terrestrial planets, and the characterization of atmospheres of extra-solar planets in a search for markers point-



*C. Eiroa et al. (eds.),*
***Infrared Space Interferometry: Astrophysics & the Study of Earth-Like Planets*, 173–185.**

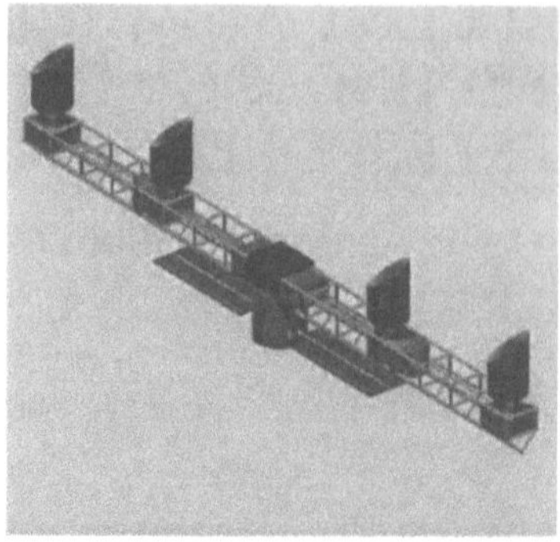

*Figure 1.* Future Cryogenic Space Optical System Concepts: Next Generation Space Telescope(1), Terrestrial Planet Finder Array(2)

ing to life. At these wavelengths, in order to not be limited by thermal emission from the observatory, it is necessary to cool the optics to cryogenic temperatures. The exact operational temperature depends on the longest observing wavelength but is typically in the 5-70K range. This requirement, to operate at very low temperature, has a major impact on the mission architectures. At the same time, there is a strong desire on the part of the international space agencies to reduce the cost of future missions which drives the mission architects to simplify designs, accelerate development, reduce development costs, reduce the mass of the flight system and utilize smaller launch vehicles. The combination of science goals, engineering considerations and programmatic limitations places some unique requirements on the optical system designs and materials.

## 2. Previous Cryogenic Space Telescopes

To date, there are two well known examples of major civilian cryogenic telescopes which have been developed, launched and have successfully performed scientific investigations in space. In 1983, NASA launched the Infrared Astronomical Satellite (IRAS). The 70kg, IRAS telescope (optics and structure) was manufactured from vacuum hot pressed beryllium and operated at ~4K, cooled with liquid helium carried on-board in a large dewar. The telescope was a Richey-Chretien type, with a 57cm aperture and was diffraction limited at ~20$\mu$m. Twelve years later, in 1995, ESA launched the Infrared Space Observatory (ISO) which is still in operation. The 50kg ISO telescope has an invar/aluminum structure and lightweighted fused silica mirrors and is also cooled to ~4K with on-board liquid helium. It too is a Richey-Chretien design, with a 60cm aperture and achieves diffraction limited performance at ~5μm, much better than IRAS.

## 3. Future Cryogenic Space Optical System Design

IRAS and ISO have been very successful forerunners to future cryogenic space optical systems and have provided an invaluable base upon which the future systems can be built. However, as discussed above, a number of new and different factors will drive the development of the future cryogenic space telescopes and interferometers. Among other things, future systems will have to be larger, lower cost and lighter weight than their predecessors and most will be passively cooled as opposed to carrying large quantities of cryogen into space.

Key design considerations for these future systems will include cost, mass, manufacturability, complexity of the optics and structure, wavefront error, control methodology, thermal performance and athermalization, mode of cooling, launch survivability, on-orbit durability and system level testing. The design considerations and ultimate performance are directly linked to the materials that will be utilized to manufacture the optics and the structure. Candidate materials for cryogenic mirror manufacturing include fused silica, silicon carbide, beryllium, aluminum, composites (carbon fiber reinforced polymers, metal matrix and ceramic type), and various hybrids incorporating multiple materials. Candidates for precision cryogenic structures include aluminum, silicon carbide, beryllium, invar and composites, as well as hybrids. There is no perfect material or combination of materials that are suitable for all applications. Each candidate material has both positive and negative aspects with respect to cryogenic space optical system applications. The key to successful design of future systems is to understand the science goals of the particular mission, their implication in terms of engineering requirements, and the available cost and schedule to find the optimum set of materials to achieve the desired performance. A summary of selected material considerations for cryogenic optics is given in Table 1.

## 4. Cryogenics Optics Technology Development for SIRTF

The next major cryogenic space optical system to be developed by NASA is planned to be SIRTF, an 85cm clear aperture telescope, cooled

Table I. Selected Mirror Materials Considerations for Cryogenic Space Optical Systems.

| Fused Silica | Silicon Carbide | Beryllium | Aluminum | Composites |
|---|---|---|---|---|
| **PROS** | | | | |
| Large Experience Base | High Stiffness | Very Lightweight | Very Low Cost | Low Cost |
| Low Surface Scatter | Low Surface Scatter | High Stiffness | Easy to fabricate | Very Low Mass |
| Good Figure Quality | Good Figure Quality | Good Figure Quality | High Thermal Conductivity | Tailorable Properties |
| Good Dimensional Stability | Good Dimensional Stability | High Thermal Conductivity | High Strength | High Stiffness |
| Low Specific Heat | Low Specific Heat | Easy to Mount | Easy to Mount | High Strength |
| Good Homogeneity | High Thermal Conductivity | Athermalized Systems | Athermalized Systems | Athermalized Systems |
| | Athermalized Systems | Durable | High Durability | High Durability |
| | Near Net Shape with RB | Near Net Shape by HIP | | Replication |
| **CONS** | | | | |
| Low Thermal Conductivity | Immmature Technology | Low Microyield | Very High Thermal Contraction | Poor Figure Quality |
| Difficult to Mount | Limited Availability | High Thermal Contraction | Heavy | High Surface Scatter |
| Difficult to Athermalize | Brittle | Null Figuring Required | Low Stiffness | Material Variability |
| Heavy or ... | Difficult to Mount | Limited Availability | | Anisotropic |
| Fragile if Lightweighted | Heavy | Limited Size | | Moisture Absorbing |
| | Extent of Possible Light-weighting Unknown | Surface Scattering | | Outgassing |
| | | Expensive | | |
| | Reaction Bonded | Hot Isostatic Pressed | | Carbon Fiber Reinforced Polymer |

to <5K and performing imaging and spectroscopy in the 3.5$\mu$m to ~160$\mu$m region of the spectrum.

In early 1993, in preparation for SIRTF, the Jet Propulsion Laboratory (JPL) embarked on a technology development program to demonstrate the viability of a lightweight, cryogenic telescope which had twice the collecting area of IRAS, half the mass and was diffraction limited at a substantially shorter wavelength (6.5$\mu$m vs. 20$\mu$m). The initial effort focused on the manufacturing and cryogenic testing of two subscale (0.5m diameter) test mirrors fabricated from the two leading candidate mirror materials - beryllium and silicon carbide. This was followed by development and testing of a full scale (85cm clear aperture), lightweight IR telescope called the Infrared Telescope Technology Testbed (ITTT). Hughes Danbury Optical Systems (HDOS) was selected, via a competitive proposal process, to develop the ITTT based on an a beryllium design.

### 4.1. Beryllium and Silicon Carbide Test Mirror Development

The 0.5m diameter beryllium test mirror was fabricated from a blank manufactured from specially processed I-70H powder at Brush-Wellman, Inc. using the HIP (hot isostatic pressing) process. The special processing of the beryllium powder was aimed at achieving a very homogeneous starting material for the blank. It involved additional steps (beyond the standard I-70H specification) designed to remove impurities and carefully control the particle size distribution. Following manufacture of the blank, precision machining, but no lightweighting, was done by Loral American Beryllium and optical finishing, to a spherical surface, was performed by Tinsley Laboratories, Inc. A key element in the manufacturing plan was repeated acid etching and thermal cycling of the mirror, following each major processing step, to relieve any built up internal stress. The finished optic, shown in Fig. 2, had a 2m radius of curvature, a room temperature rms wavefront error of 0.072$\lambda$ ($\lambda$= 633nm) and an rms surface roughness of 13Å.

Cryogenic optical testing at 77K and 4.4K was performed using optical interferometry at the NASA Ames Research Center in their Optical Test Facility which has been described previously (Young et al. 1990). The mirror proved to be the most stable large beryllium mirror ever measured (Gordon et al. 1995). The rms wavefront error at 77K was found to be 0.150$\lambda$. Further cooling to liquid helium temperature produced essentially no change and yielded a measured rms error of 0.140$\lambda$ at 4.4K. The thermal distortion from room temperature to 4.4K of <0.1$\lambda$ was comparable to that observed in most large fused silica mir-

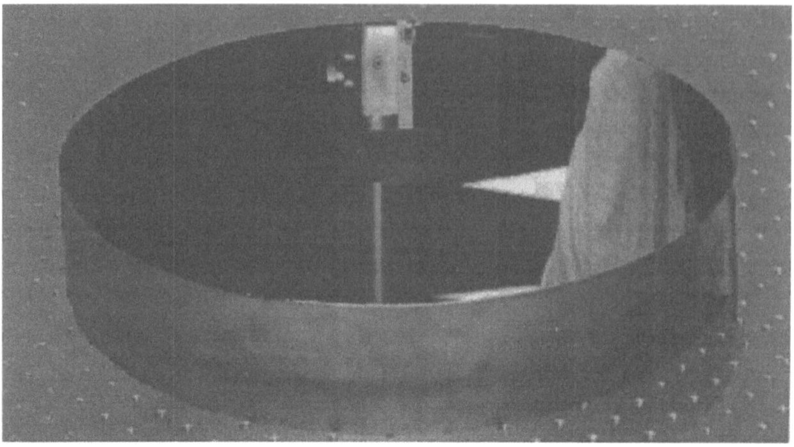

*Figure 2.* 50cm diameter beryllium test mirror.

rors. Even more importantly, there was no indication of measurable hysteresis in the figure of the mirror as a result of thermal cycling, a problem which was well known in previous large beryllium mirrors (Melugin et al. 1988).

The detailed causes of the observed hysteresis in previous beryllium mirrors has never been fully investigated. It is believed to be related to poor quality beryllium powder, poor consolidation prior to the HIP process and internal stresses built up in the optic during machining, grinding and polishing (Paquin et al. 1996). The manufacturing plan for the 50cm beryllium test mirror addressed all of these concerns. Ultimately, the combination of very clean beryllium powder with a uniform particle size distribution, care in the consolidation process and extensive stress relieving during machining, grinding and polishing, resulted in an excellent optic. Ultimately, the plan was to return the mirror to Loral American Beryllium for further machining and lightweighting followed by another round of cryogenic optical tests, however, this work has never been completed.

The 0.5m diameter silicon carbide test mirror was a closed back, lightweighted structure fabricated from reaction bonded optical (RBO) grade silicon carbide and rough ground to a sphere by United Technologies Optical Systems (UTOS). UTOS is no longer supporting this technology. However, Xinetics, Inc. in Littleton, MA, USA is a current supplier. Litton Itek Optical Systems (since acquired by HDOS in Danbury, CT, USA) was responsible for the optical fabrication and polishing. The mirror, shown in Fig. 3, had a 2m radius of curvature and an rms wavefront error of 0.053$\lambda$ ($\lambda$=633nm) at room temperature. Surface microroughness was never measured on this particular optic. However, surfaces of 30Å-40Å rms are achievable on RBO silicon carbide optics (Arnold & Laughlin 1991).

*Figure 3.* **50cm diameter RBO silicon carbide test mirror.**

Cryogenic optical testing of the silicon carbide mirror was performed at Lockheed-Martin again using optical interferometry (Cox et al. 1994). When cooled to 8K, the mirror showed an rms wavefront error of $0.126\lambda$. Upon return to room temperature, a slight hysteresis of $\sim 0.03\lambda$ was noted. A second cycle produced similar results. Lockheed believes that some of the error, and possibly some of the observed hysteresis was due to thermal effects on the test chamber window during the test which were difficult to control or quantify. There were plans to take the mirror to the Ames facility for further tests. However, when UTOS stopped producing the RBO silicon carbide optics, and this specific product was no longer available, those plans were dropped. The conclusion was that the silicon carbide mirror performance was comparable to that of the beryllium mirror and that on the basis of the subscale mirror evaluation program, both materials remained viable candidates for cryogenic optical system applications.

### 4.2. The Infrared Telescope Technology Testbed Design

In June of 1994, JPL issued the Infrared Telescope Technology Testbed RFP inviting industry and academia to propose to design and build a prototype telescope meeting the needs of the SIRTF mission. The principal requirements levied on the proposers were that the ITTT should achieve diffraction limited performance at 6.5$\mu$m, at 5.5K with an 85cm clear aperture and a total mass of <50kg. The primary mirror and system focal ratios were specified as F/1.2 and F/12 respectively. HDOS was selected to build the ITTT based on their concept for a (nearly) all beryllium telescope. The telescope is fabricated from hot isostatic pressed I-70H (special) beryllium identical to that used in the test mirror except for six titanium bipod flexures and several pins used to mount the primary and secondary mirrors. The design is based on a single arch primary mirror attached to a lightweight bulkhead via three of the flexures. The secondary mirror is mounted in a similar fashion to the secondary mirror assembly. The secondary mirror assembly is attached to a lightweight metering tower which incorporates the primary and secondary cone baffles and three longitudinal struts into a single machined piece. Copper cooling straps are used to facilitate cooling of the ITTT in the test chamber. The secondary mirror assembly is designed to accommodate a one degree of freedom focus mechanism but this element has not been incorporated into the current hardware. The total mass of the ITTT at completion is estimated to be 29kg.

### 4.3. The SIRTF Telescope Test Facility

The Ames facility is too small to accommodate the ITTT for cryogenic testing. Consequently, while HDOS was manufacturing the ITTT, JPL developed the SIRTF Telescope Test facility (STTF) shown in Fig. 4.a. Briefly, the STTF which has been described in detail elsewhere (Chave et al. 1995, Luchik et al. 1995) consists of three concentric shells. The outer shell maintains the vacuum, the intermediate shell is at liquid nitrogen temperature cooled by a single tank at the base, and the inner shell is at liquid helium temperature cooled by dual tanks at the top and bottom. These tanks also supply cryogen for cooling a vibration isolated precision gimbal mount and the experimental hardware which can be mounted either on the upper or lower tank. The upper tank is movable within the helium shroud thus accommodating optics of differing focal ratio. Each of the tanks has a cylindrical hole through its center to allow light to pass. The interior diameter of the helium shroud is 1.4m. In the testing of the ITTT Primary Mirror Assembly (PMA) which is described in detail in section 4.4, the hardware was mounted to the gimbal via three titanium flexures and the gimbal/PMA assembly was attached to the base of the upper tank with the mirror facing down. Copper straps to the tank baseplate provided cooling and platinum resistance thermometers provided a means to monitor temperature.

Near the base, are two shutters, an inner one at helium temperature and outer one at nitrogen temperature. These are normally closed and can be easily and quickly opened prior to measurement. The base of the vacuum shell has an optical window, below which is an instrument rack upon which rests a turning mirror, a null lens and a Zygo GPI phase shifting visible (633nm) interferometer. The entire assembly, tank and instrument rack is mounted on a large aluminum triangular frame which rests on three Newport Research pneumatic vibration isolation legs.

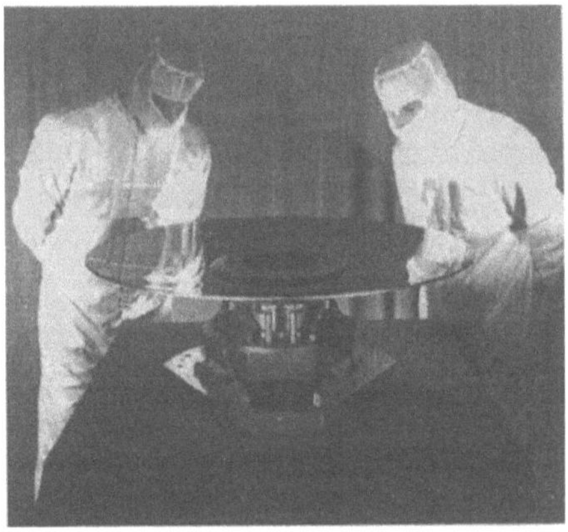

*Figure 4.* **a) The SIRTF Telescope Test Facility; b) The Infrared Telescope Technology Testbed Primary Mirror Assembly**

## 4.4. TESTING OF THE INFRARED TELESCOPE TECHNOLOGY TESTBED

HDOS fabricated the PMA, which includes the primary mirror, the bulkhead, the metering tower adapter tube, the primary mirror bipod flexures and the cooling straps and the test facility adapter, and delivered it to JPL in July, 1995. A photograph of the PMA is shown in Fig. 4.b. The initial room temperature measurements on the PMA showed an rms surface error of $0.192\lambda$ ($\lambda$= 633nm) with a peak-to-valley error of $1.56\lambda$. The dominant error feature was a series of concentric zones which resulted from form grinding in the early stages of optical fabrication. The zones were 1-2 waves in height and have subsequently been removed with further small tool computer controlled polishing at HDOS. As mentioned previously, large beryllium optics have traditionally shown "thermal hysteresis", that is, they changed shape following cycling between room temperature and cryogenic temperature. The PMA was cycled five times to 77K and three times to 5K with no evidence of hysteresis. Room temperature data recorded following these multiple cycles showed an rms surface error of 0.194$\lambda$ and the peak- to-valley error was 1.35$\lambda$. The slight change in the measured peak to valley error is not believed to be significant.

While the PMA showed no hysteresis, it did show a moderate cryogenic distortion. At 77K, the rms surface error was found to be 0.580$\lambda$ and the peak-to-valley error was 4.42$\lambda$. At 5K, the rms surface error was 0.588$\lambda$ and the peak-to-valley error was 4.30$\lambda$. It is our conclusion that there is essentially no difference between the liquid nitrogen and liquid helium test data. Furthermore, the data was highly repeatable from cycle to cycle.

Following the discovery of the cryogenic distortion in the PMA, an investigation to determine its source ensued. First, the possibility of systematic errors in the test set-up was investigated. The PMA was rotated 120° and cryo-tested again. The cryo-distortion rotated with the hardware. Then, the null lens was rotated 180° with no effect. Secondly, the PMA was decoupled from the aluminum adapter plate and the aluminum bipod flexures and cryo-tested suspended from a simple three point kinematic mount. Again, no change was observed. Following that, the primary mirror was removed from the PMA and itself cryo-tested using the same mounting scheme. The observed error in the primary mirror matched the error measured in the PMA thus indicating that the source of the cryo-distortion was in the mirror itself and not in the mounting hardware. Finally, the entire PMA was reassembled and measured once again at liquid helium temperature. The results were essentially identical to those measured earlier.

The proposed solution to the cryogenic distortion problem was to "null-figure" the mirror. This is a process which had been demonstrated experimentally in fused silica (Augason et al. 1992) and involves refiguring of the optic at room temperature incorporating the negative of the cryogenic distortion observed at the desired operational temperature such that when cooled, the correct figure is achieved. This is only possible if there is no thermal hysteresis in the mirror and had never been demonstrated on a beryllium optic. The PMA was shipped back to HDOS in February , 1996. The concentric zones were removed and the mirror was null figured in such a manner so as to have the correct shape at 5K. This process was accomplished with computer controlled polishing using small tools. The PMA was returned to JPL in September, 1996. Preliminary testing to 77K indicates that the refiguring process has been successful and that null figuring has been demonstrated on a beryllium mirror for the first time. The measured rms surface error at 77K was 0.165$\lambda$ and the peak-to-valley error was 1.38$\lambda$. Further testing to <5K will be performed, but no significant changes are anticipated.

### 4.5. Silicon Carbide Cryogenic Optical Test Flat for the STTF

The remaining pieces of the ITTT telescope assembly are in the final stage of manufacturing and will be delivered to JPL soon. Following the current PMA tests, the telescope will be assembled, aligned and tested in the STTF. The plan is to test the ITTT in the autocollimation mode utilizing a large cryogenic optical test flat (COTF). In late 1994, JPL received proposals from industry and academia to produce this optic. The principal requirement was that the COTF be 90cm in diameter and maintain an rms surface flatness of $\leq$0.07$\mu$m at 5K. The leading candidate materials were fused silica and silicon carbide.

Several key issues were considered in making the choice of material for the COTF. First, it is much easier to cool silicon carbide than fused silica due to the much higher thermal conductivity of the former. For that reason, silicon carbide seemed to be an attractive choice. However, clearly, the maturity of the fused silica technology was and still is far greater than that of silicon carbide. The tests on the 50cm RBO silicon carbide test mirror were encouraging as well as some similar data on cryogenic testing of a 25cm diameter chemical vapor deposited silicon carbide mirror (Goela et al. 1990). Furthermore, Lockheed-Martin had established a collaboration with the Vavilov State Optics Institute in St. Petersburg, Russia and produced a series of silicon carbide mirrors based on a process similar to the UTOS process. Two small mirrors including a 17cm diameter sphere and a 31cm x 21cm flat were optically tested at 6K and showed good performance (Robb et al. 1995b). In addition, a lightweighted 60cm diameter mirror was produced , though not for cryogenic applications (Robb et al. 1995a). The lightweighted mirror weighs ~5kg, shows a room temperature rms wavefront error of 0.024$\lambda$ ($\lambda$=633nm) and has a surface roughness of 10-20Å.

With this information in hand, Lockheed-Martin in collaboration with the Vavilov State Optics Institute in St. Petersburg, Russia was selected to produce the COTF. The optic and its six point aluminum mount have been fabricated in Russia and will be delivered to JPL shortly for integration into the STTF. Initial room temperature tests performed in St. Petersburg show an rms surface error of 0.07$\lambda$ ($\lambda$=633nm) and a peak to valley error of 0.4$\lambda$ for the COTF. Cryogenic optical testing, also performed in St. Petersburg, has indicated that the rms wavefront error at 4K is $\sim 0.085\lambda$ with a peak to valley error of $0.4\lambda$, thus meeting the stated requirement.

### 4.6. Future SIRTF Telescope Technology

When the COTF arrives at JPL, it will be mounted on the STTF gimbal and attached to the upper helium tank. Once the ITTT metering tower is completed and delivered to JPL it will be integrated with the PMA and the secondary mirror assembly and the telescope will be aligned. The fully integrated ITTT will then be mounted on the lower helium tank in the STTF facing up and autocollimation testing will be performed using the COTF to verify performance at 5K. If the performance is adequate and meets the prescribed error budget, the ITTT will be removed from the STTF and vibration tested to the levels appropriate for the Delta class launch vehicles. Following vibration tests, the hardware will be re-tested in the STTF to verify alignment and cryogenic optical performance following simulated launch loads.

Several possible uses are being considered for the ITTT following technology validation. There is a possibility that some elements of the hardware could be utilized in the SIRTF flight telescope. The primary mirror, for example, could be refigured for the current SIRTF design. A second possibility is to utilize the ITTT as a ground test article to validate performance of SIRTF instrument test modules or as a stimulus for the SIRTF end-to-end system tests. Finally, there is the possibility that the telescope could be utilized as part of a future mission.

## 5. Conclusions

In the near future, there is the possibility that a number of large cryogenic optical systems will be developed and launched into space to perform a variety of scientific investigations. There is currently a significant level of activity to develop the specialty optics required for such applications. A number of approaches are available to support these applications and the choice of materials and designs depends of the mission requirements. Considerable progress has been made recently in the development of large, lightweight beryllium and silicon carbide optics. In particular, the "thermal hysteresis" problem in large cryogenic beryllium mirrors has been solved. Also, the process of "null figuring" has been demonstrated in beryllium. The state of the art in silicon carbide optics is advancing rapidly and good performance at cryogenic temperature has been demonstrated in several mirrors. The future cryogenic optics needs of the space telescope and interferometry community appear to be reasonable extensions of existing technology if there is sufficient and continuing support for technology development.

**Acknowledgements:** Some of the work reported in this paper was performed for NASA by the Jet Propulsion Laboratory, California Institute of Technology. Funding for this work was provided by the NASA Spacecraft Systems Division of the Office of Space Access and Technology (Code XS) as part of the Telescope Technology Program.

## References

Arnold, J.F., Laughlin, M.J.: 1991, in OSA Space Optics Topical Meeting.
Augason, G., Young, J., Melugin, R., Clarke, D., Howard, S., Scanlan, M., Wong, S., Lawton, K.: 1992, *Proc. SPIE* **1765**, 5
Beichman, C.A.: 1996, A Road Map for the Exploration of Neighboring Planetary Systems (ExNPS) (JPL Pub. 96-22).
Chave, R.G., Nash, A.E., Hardy, J.: 1995, *Proc. SPIE* **2542**, 23
Cox, C., Harshman, J., Gumbel, H., Huff, L., Pazol, B., Triebes, K., Wolford, P.: 1994, in Silicon Carbide 50cm Light Weight Mirror Cryogenic Test Results, (JPL Report), **July 25**.
Goddard Led Study Team: 1996, in NGST Study Integration Review, B. D. Seery (Ed.), NASA Goddard Space Flight Center
Goela, J.S., Pickering, M.A., Taylor, R.L., Murray, B.W., Lompado, A.: 1990, *Proc. SPIE* **1330**, 25
Gordon, C., Augason, G., Clarke, D.S., Norris, D.D., Paquin, R.A., Kincade, J.: 1995, *Proc. SPIE* **2543**, 141
Luchik, T.S., Israelsson, U.E., Chave, R.G., Nash, A.E., Hardy, J.: 1995, *Proc. SPIE* **2553**, 547
Melugin, R.K., Miller, J.H., Young, J.A., Howard, S.D., Pryor, G.M.: 1988, *Proc. SPIE* **973**, 71
Paquin, R.A., Coulter, D.R., Norris, D.D., Augason, G.C., Stier, M.T., Cayrel, M., Parsonage, T.: 1996, *Proc. SPIE* **2775**, 480
Robb, P.N., Charpentier, R.R., Lubarsky, S.V., Tolstoy, M.N., Evteev, G.V., Khimitch, Y.P.: 1995a, *Proc. SPIE* **2543**, 185
Robb, P.N., Huff, L.W., Forney, P.B., Petrovsky, G.T., Lubarsky, S.V., Khimitch, Y.P.: 1995b, *Proc. SPIE* **2543**, 196
Young, J., Howard, S., Augason, G., Melugin, R.: 1990, *Proc. SPIE* **1340**, 111

# INTERFERO-CORONAGRAPHY USING PUPIL $\pi$-ROTATION

J. GAY
*Observatoire de la Côte d'Azur, BP 229, 06304 Nice Cedex 4, France*

Y.RABBIA
*Observatoire de la Côte d'Azur, Avenue Copernic, 06130 Grasse, France*

C.MANGHINI
*Observatoire de la Côte d'Azur, BP 229, 06304 Nice Cedex 4, France*
*All authors at Departement Fresnel, URA-CNRS 1361*

**Abstract.** A concept for interfero-coronagraphy is briefly recalled and a laboratory initial set-up is described which aims at testing feasability of the concept. First image showing extinction in the central part of the field is reported. Constraints and future use of the concept are outlined.

**Key words:** coronagraphy, interferometry, double stars, exoplanets

**Abbreviations:** OPD – optical path difference

## 1. Introduction

In the "direct imaging" approach of exoplanets, coronagraphy based on destructive interference is a promising track along which the search for faint companions of stars is a preliminary step. In this purpose, a concept has been devised by J. Gay for use with a single aperture (Gay & Rabbia, 1996). Here we present a laboratory set-up so as to check that the concept effectively works. A short reminder of the principle is given as well as the initial quick-made set-up. Constraints are briefly discussed and the scope of a next step is given in conclusion.

## 2. Principle and preliminary set-up

The concept relies on amplitude division and recombination of the coming collimated beam, just like a Michelson beamsplitting interferometer modified by applying a pupil rotation by 180 degrees on one arm and by inserting an achromatic dephasing by $\pi$ between the two arms. Both are obtained by a supplementary focusing on one arm using a cat's eye reflector (Born & Wolf, 1970; Gouy, 1890). The essential and severe constraint is that the complex transmission must remain centro-symmetric



*C. Eiroa et al. (eds.),*
*Infrared Space Interferometry: Astrophysics & the Study of Earth-Like Planets*, 187–190.

up to the recombination step. Departure from this requirement lowers the efficiency of extinction (decreasing coronagraphic gain). In the image-plane following the recombination, no light appears from any on-axis source of angular size corresponding to less than - roughly - half the Airy angle. Our initial set-up works in the visible domain, but extension toward near infrared is straightforward as soon as an IR camera is available. Also the whole set-up is compact (roughly 60cm x 60 cmx 30 cm) allowing installation at the output of most telescopes (see Fig. 1).

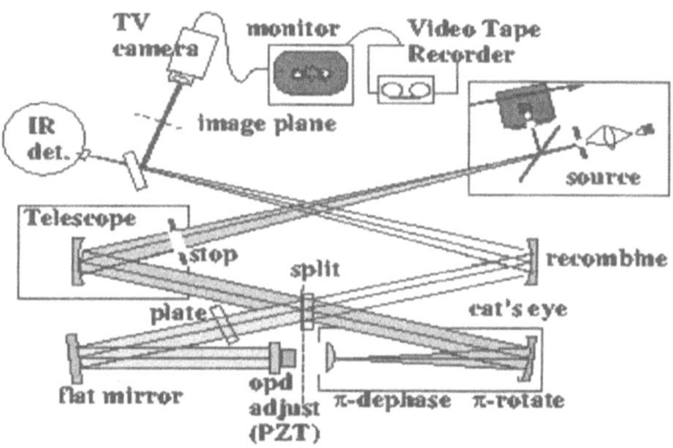


*Figure 1.* Schematic of the set-up. Light-source (up right) simulates an extended object or a double star (variable separation), an aperture stop features the telescope pupilla. A compensation plate is inserted as usually done in a Michelson interferometer. Reflexion on a cat's eye system allows both $\pi$-rotation and $\pi$-achromatic dephasing. Recombination is observed by TV Camera (up left). An IR detector helps in finding zero OPD, adjusted by motion of a flat mirror on a piezo stack.

Let us point out that when OPD is set at zero, the coronagraph prevents on-axis light to reach the final image plane, so that this light does not contribute to the observed intensity distribution. This is not like simply making a "hole" in the center of the usual light distribution. Actually, on-axis light is sent back to the sky. In this respect the reported concept might be seen as a single aperture version of a nulling interferometer (Bracewell & McPhie, 1979). Another design (double output recombination) allows to use this backward light so as to accurately maintain the central star on-axis by means of a servo system (Gay & Rabbia, 1996). Another interesting features is achromaticity, which allows a large choice for the working wavelength and/or allows to work with large spectral bandwidth.

## 3. Preliminary output

Fig. 2 shows the image of an extended source (with photocenter slightly off-axis) obtained when OPD is made zero. This image exhibits a hole featuring the solid angle within which a point-like source does not contribute to the intensity distribution in the image plane.

In spite of picture's poor quality (scanned photograph of video monitor), it is apparent that darkening is neither total nor homogeneous. This defect is attributed to aberrations, optical quality and unmatched coatings of the optical components in this quick-made preliminary set-up. Nevertheless, extinction is conspicuous thus showing that the concept works, even when centro-symmetry is not perfectly achieved. Refurbishment and improvement of the set-up in this respect are currently in progress.

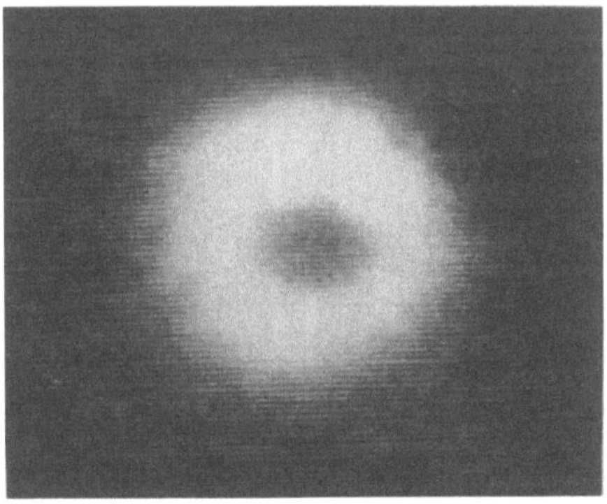

*Figure 3.* Image of an extended source,when OPD is (approximately) zero. This intensity distribution describes the angular response of the coronagraph, just like an antenna reception pattern and coordinates should be seen as angular and not linear.

## 4. Conclusion

The reported concept for interfero-coronagraphy with a single aperture is able to remove the light coming from an on-axis source. Coronagraphic gain (efficiency of extinction) is limited by phase distorsions over the aperture. Next step is to evaluate "on the sky" the effective capabilities of the instrument: plans are already made in this purpose, including the use of Adaptive Optics. An immediate scientific scope is the search for faint companions in binary or multiple sources and the study of stellar environment as well. The natural destination for this concept (and the initial motivation for devising it) is a space-based instrument such as the Hubble Space Telescope.

## References

Born M. & Wolf E.: 1970, in Principles of Optics, Pergamon, 4th edit., sect.8.8.3
Bracewell, R., MacPhie, R.H.: 1979, *Icarus*, **38**, 136
Gay, J., Rabbia, Y.: 1996, *C.R. Acad. Sci. Paris* **322, serie IIb**, 625
Gouy, L.: 1890, *C.R. Acad. Sci. Paris* **110**, 1251

# ASIX - THE ASTRO-SPAS INTERFEROMETER EXPERIMENT

A. GLINDEMANN and S. BECKWITH
*MPI für Astronomie, Königstuhl 17, 69117 Heidelberg, Germany.*

H. JOERCK
*Dornier Satelliten Systeme, 81663 München, Germany.*

C.J. SCHALINSKI
*DLR, Rudower Chaussee 5, 12489 Berlin, Germany.*

S. RÖSER
*ARI, Mönchhofstr. 12–14, 69120 Heidelberg, Germany.*

E. SCHILBACH
*Universität Potsdam, An der Sternwarte 16, 14482 Potsdam, Germany.*

**Abstract.** ASIX is a three telescope Michelson interferometer for the near infrared to fly on ASTRO-SPAS in the year 2000. ASIX is a precursor for the ESA cornerstone mission to validate and demonstrate system level functionality of an interferometer in space. After a successful mission, ASIX can be adapted within a few years to a satellite bus to be re-used as long baseline (~20m) facility instrument with a lifetime of many months. The ASIX mission will be performed in collaboration with the group of Michael Shao at the JPL, Pasadena.

**Key words:** ASIX, interferometer

## 1. Preliminaries

In September 1994, ESA's *Horizon 2000 Plus* programme was approved at the Council meeting at Ministerial Level in Toulouse. This programme intended as a roll-forward continuation of *Horizon 2000* defines the European space missions for the next 20 years. One of the cornerstone missions, to be launched in 2016, was selected to be an interferometer observatory.

Two candidates were identified for the interferometer observatory: GAIA, an astrometric interferometer as a successor to HIPPARCOS with an accuracy about 100 times better than HIPPARCOS, and DARWIN, an imaging interferometer with the main goal to detect and spectroscopically study Earth-like planets.

As a part of a collaboration with the group of Mike Shao at the JPL we propose the three-telescope Michelson interferometer ASIX for a precursor mission on ASTRO-SPAS. The ASTRO-SPAS is a reusable Spaceshuttle Pallet Satellite that is ideally suited for precursor mis-



*C. Eiroa et al. (eds.),*
*Infrared Space Interferometry: Astrophysics & the Study of Earth-Like Planets*, 191–194.

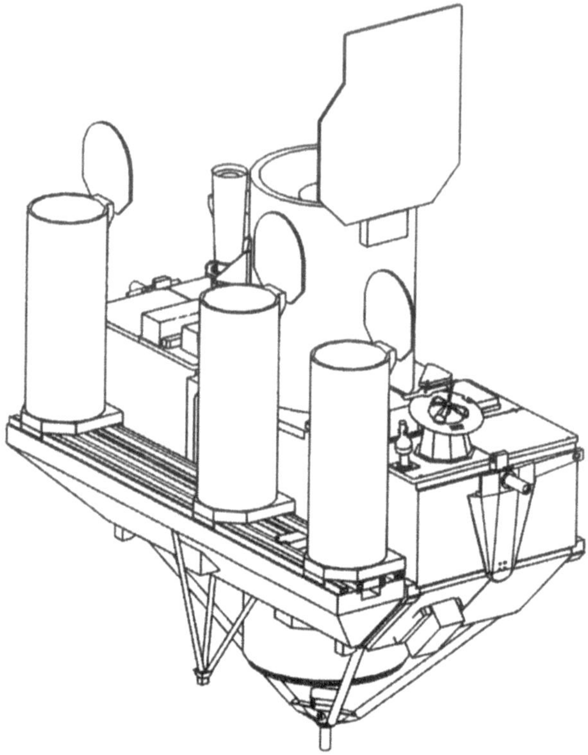

sions. Depending on the availability of the space shuttle a mission on ASTRO-SPAS could be spaceborne within a few years.

## 2. The Mission

The primary goal of ASIX is to validate and demonstrate system level functionality of an interferometer in Earth orbit. This includes both instrumental performance and mission scheduling (observing procedures) and logistics (data down link, automization). These are crucial testing items to buy down riscs and costs of future space interferometry science missions like:

- long baseline imaging systems, e.g. DARWIN (40–100m baseline),
- global astrometry programmes, e.g. GAIA and,
- extremely long baseline laser interferometers for gravity wave detection, e.g. LISA.

In order to limit the mechanical complexity the baseline has been limited to the ASTRO-SPAS width of 4m. Then, the angular resolution is between 25 and 140 milliarcsec for wavelengths between 1 and 5.5$\mu$m. The limiting magnitude is about m=10 up to 4$\mu$m and m=8 at 5$\mu$m. The limit of resolution is 4 times larger than the Hubble Space Telescope and 8 times that of ISO. On the ground, an 8m class telescope is needed to achieve the same resolution.

System level functioning of an interferometer means proving high-precision pointing, mechanical stability and accurate tracking of stellar targets. Minimum success criterion for ASIX - in addition to functioning of sub-components - is to measure interference patterns ("fringes") of stellar objects.

Technologically, the mission has to verify the key interferometer technologies in space: sub-nm metrology, active structures, beam combiner, fringe tracker, super-polished surfaces, phase closure image reconstruction. The three telescopes are mounted on a magnetic rail allowing precise pointing through the magnetic bearings, and positioning the telescopes linearly on the rail with respect to each other. The latter and the rotation of the ASTRO-SPAS with an axis perpendicular to the rail scans the $u - v$ plane. Eventually, a successful demonstration of a space interferometer brings down the cost for the ESA cornerstone missions as the know-how will then be available.

The ASIX layout as a three-element Michelson interferometer, however, allows to test further key technologies:

- implementation of closure phases (by using a third telescope) will facilitate imaging modes and data calibration
- IR detector operating between 1 and 5.5$\mu$m allows high resolution imaging of objects in a spectral range not available on the Hubble Space Telescope and only limited from the ground.

Although a 10 day test flight on ASTRO-SPAS in low Earth orbit cannot compete with dedicated science space missions, observations of selected targets in the near infrared will complement ground-based observations at shorter wavelengths. Quasi-simultaneous observations from the ground with high resolution (large telescopes using adaptive optics; speckle interferometry; near-IR interferometers as the Palomar Testbed of our JPL partners) will furthermore allow to check the ASIX measurements and test/optimise instrumental performance.

ASIX at 1$\mu$m observing wavelength has the resolution of a 40m baseline operating at 10$\mu$m, on the order of planned planet detection IR interferometry missions like DARWIN. The operation as nulling instrument (Bracewell-Interferometer) requires the insertion of a 180 degrees phase shift in one arm of the Michelson interferometer to cancel the central stellar "fringe" by destructive interference. A first test of the feasability of the nulling technique in a working space interferometer will enable the extrapolation of system performance of a future science instrument.

After a successful mission, ASIX can be adapted within a few years to a satellite bus to be re-used as long baseline (~20m) facility instrument with a lifetime of many months. Thus, ASIX has the potential to be not only a precursor but also a dedicated science instrument.

# PASSIVE COOLING OF INFRARED INTERFEROMETERS IN SPACE

T. G. HAWARDEN
*Joint Astronomy Centre*
*660 North A'ohoku Place, University Park,*
*Hilo, Hawaii 96720*
*USA*

**Abstract.** We examine three possible design approaches for radiative cooling of multiple-aperture missions of the sort required for space IR interferometry, particularly those to be sent out to 5 AU which are aimed at detecting extrasolar terrestrial planets.

- **Individually cooled free-flyers (c.f. MOFFIT):** This approach offers adaptable geometry but access to a limited fraction of the sky, perhaps less than half from elliptical trans-asteroid orbits.
- **Large semi-independent sunshade:** This may allow free-flyers access to more of the sky but has many technical challenges.
- **Structurally-linked individually-protected telescopes:** This concept can offer protection of the beam lines from all stray light and gives good sky coverage, but geometry changes may be difficult, limiting its applications to general astronomy.

A possible solution to the last problem is suggested. A unit design based on the PRISM studies at NASA Marshall Space Flight Center is outlined, together with a mission scenario

**Key words:** radiative cooling, free-flying spacecraft flotillas, independent sunshades, symmetric radiatively-cooled spacecraft.

## 1. Introduction

Since even the scientific goals of the several proposed infrared IR space interferometers are not yet fully defined, actual design details are largely indeterminate. However, all such missions must:
(i) Get cold enough to achieve their sensitivity goals;
(ii) Stay cold for long enough to carry out the mission;
(iii) Keep the optical system dimensions stable enough for interferometry, even at the low working temperature;
(iv) Protect the telescopes and beamlines from stray light in all working attitudes;
(v) Provide sufficient sky coverage to carry out the mission, especially if the target list is limited (e.g. the 80-odd nearest solar-type stars!).



*C. Eiroa et al. (eds.),*
*Infrared Space Interferometry: Astrophysics & the Study of Earth-Like Planets,* 195–203.

We discuss three possible configurations for "planet hunting" missions like the *DARWIN* and OASES proposals (c.f., e.g., Leger et al. 1995; Woolf et al. these proceedings) to detect and characterise terrestrial planets circling near-by stars. These must operate beyond the asteroid belt to reduce background emission from the zodiacal dust, if the use of very large apertures is to be avoided.

## 2. Basic temperature requirements for a *DARWIN*-like mission

The basic temperature requirements have been set out elsewhere in these proceedings as a function of the science targets of these missions. Guidelines for deriving these are presented by Thronson et al. (1995). In brief, **Telescope** temperatures must be $\lesssim$50K (if possible, $\lesssim$ 30 K) to reduce thermal background from the optics. **Instrument** temperatures must be $\lesssim$ 70K for InSb detectors working at $\lambda$ <5 $\mu$m, $\lesssim$23K for Si:As detectors sensitive to $\lambda$ < 27 $\mu$m and $\lesssim$ 18K for Si:Sb detectors working to 36$\mu$m, to reduce internal backgrounds (see Campbell et al., 1995). Finally, **Detector** temperatures must be ~30K for InSb, $\lesssim$7K for Si:As and $\lesssim$2.5K for Si:Sb devices, to constrain dark currents.

## 3. Design factors

Design guidelines for radiative and hybrid cooling to these temperatures have been set out elsewhere by Hawarden et al. (1992) and Hawarden et al. (1995); the latter also outlines the heritage of related proposals and missions.

The main design issues are the surface properties and geometries of the protective shields. Designs for reflective shields need to minimise radiation trapping between shiny surfaces. Table 1 illustrates the effects of such trapping, determined by geometry, on the temperature of a second surface. The first shield (with temperature $T_1$) has a white-painted sunward surface (probably the only option which is both long-lasting and deployable), while all internal and rear-facing surfaces are assumed to have low-emissivity reflective coats with $\epsilon_{IR} = \alpha_{IR} = 0.03$. Design geometry determines $\bar{n}$, the mean number of times a photon emitted from the rear of the first shield is reflected from the colder surface, and hence the attenuation factor at that surface. This ranges from $\alpha_{IR}$ when $\bar{n} = 1$ in open structures, up to 0.5 for large, close, parallel surfaces, for which $\bar{n}$ can be very large in the limit of high surface reflectivity (c.f. Hawarden et al. 1992). From Table 1 it is evident

Table I. Approximate temperature of a second radiation shield in different geometries (see text) for a first shield temperature of $T_1$.

| $d_\odot$ | 1 AU | 5 AU |
|---|---|---|
| $T_1$ (K) | 290 | 130 |
| Parallel Shields | 100 | 50 |
| $\bar{n} = 6$ | 70 | 32 |
| $\bar{n} = 1$ | 47 | 22 |

that two shields are likely to be needed, even at 5 AU, unless rather favourable ($\bar{\mathbf{n}} \sim 1$) geometries can be achieved.

## 4. Major constraints on interferometer designs

Several major constraints arise from the nature of interferometers and from the mode in which they are used for planet-finding:
**(i) Multiple elements:** Unit systems cannot be allowed to contribute stray heat and light to one another. This complicates their thermal design and relative positioning.
**(ii) Combination of beams:** It is a very challenging requirement to protect the beam combiner from stray radiation in full sunlight. This is a major driver in designs using free-flyer flotillas.
**(iii) Rotation:** The detection mode proposed for extrasolar planets requires the array to rotate around the pointing direction. The protection of the cold systems from external heat sources must therefore be *symmetrical*, and the beam-combiners must be protected from stray light in all working orientations

## 5. Free-flying flotilla of spacecraft

ESA's "MOFFIT" study (Bély, Laurance & Volonte, 1996; and Bély et al., these proceedings) has explored this approach. The resulting strawman design comprises 5 independent telescope units and a beam combiner. Each has a sunshade resembling a hat-brim around its service module, and is boresighted along the launch axis, producing a symmetric vehicle. In operation the flotilla would manouevre with the telescope apertures accurately coplanar to produce a rotating array pattern (Fig. 1) while avoiding any significant view of other spacecraft from points inside the telescope tubes. While not considered in the ESA

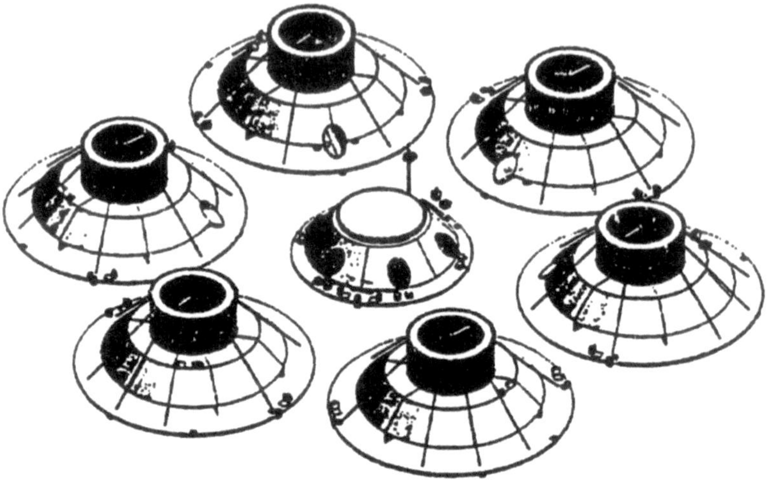

*Figure 1.* Schematic of a flotilla of free-flying satellites (from Bély et al. 1996)

study, the rotational motion required of the array by current proposals for detection of extra-solar planets by nulling interferometry can be readily accommodated by the proposed design: a 24 h rotation period at 50 m separation would require only about 0.22 g of propellant per day, much less than the 3.6 g required for the baseline 24 h observation outlined in the study. The proposed (10 mN) thruster system could maintain a 50m separation down to about an array rotation period of about 3h, at the cost of a propellant consumption rate of about 15 g $d^{-1}$.

Protecting the telescopes and their beam launchers, and also the receivers on the beam combiner unit, from direct sunlight or from views of sunlit surfaces (*c.f.* Fig. 8.6 of Bély et al. 1996) requires a large sun-avoidance angle, probably *at least* 120°. In an elliptical trans-asteroid orbit this could severely restrict sky coverage: only 25% of the sky would be accessible while at maximum solar distance (14% for the design outlined in the study; the larger value would probably require deployable sunshades). Such a mission should evidently be designed to spend as large a fraction as possible of each orbit beyond some *minimum tolerable* solar distance; circularisation by a Jupiter swing-by, or perihelion-raising by electric propulsors should be considered.

## 6. Large sunshield protecting free-flyer flotilla

A flotilla could be placed in the shadow of a very large sunshade. An asteroid has been suggested, but has the disadvantages of making power provision difficult while the spacecraft would need to accelerate continuously to resist its gravitational attraction. A large, inflated (or possibly rotating) lightweight sunshade seems a more likely option. The spacecraft would enjoy excellent viewfactors to space: T $\lesssim$45K should be readily attainable (c.f. Table 1).

However the sunshade's own rearward IR emissions (of order 0.2 W $m^{-2}$) and reflections of IR from the spacecraft, would constitute strong stray signals to be carefully excluded from the telescopes and beam receivers. To do this the flotilla craft would require to operate in planes which never intersect the edge of the sunshade.

The individual spacecraft would require only thruster operations for positioning and probably lasers for alignment information, and therefore only power need be supplied to them from the Service Module, perhaps by microwave link to avoid the torques implied by umbilical connections. Clearly, however, the approach has major technical challenges, including (*inter alia*): deployment and control of the large shield; extremely reliable station keeping and contamination by thruster exhaust, both important problems for any close flotilla; implementation of the microwave power systems; cooling the instrument and detectors aboard the beam combiner module. Perhaps most important of all is the difficulty associated with rotation of the array: it is hard to see how *any* umbilical arrangement can be tolerated during such a manoeuvre, *e.g.* when pointing near the ecliptic poles, but conversely hard to see how such a system can be operated without, at least, an umbilical connection to the beam-combiner.

## 7. Large structure linking individually-cooled telescopes (c.f. Woolf et al., these proceedings.)

Such a system is illustrated in Fig. 2. A major attraction of this approach is the possibility of permanently protecting the beamlines from stray radiation by enclosing them. This permits the protective shields of the telescopes to be arranged as cones opening along the line-of-sight, which in turn allows sun-avoidance angles to be reduced almost down to 90°, making almost half the sky accessible from a single orbital position. About 75% of the sky could therefore be reached from a trans-asteroid elliptical orbit with aphelion at 5 AU if, for example, observing remained possible down to a solar distance of 3 AU.

*Figure 2.* A "rigid" linear array of connected telescope elements.

## 8. Articulated structures

A limitation of a fixed linking structure is its lack of adaptibility to array configurations suited to observations other than the nulling mode used for planet searches. If the array configurations could be altered it would offer a much more versatile facility, capable of general purpose high-resolution imaging as well. One possibility may be to replace OASES' single linear lattice-girder with three: a central structure still carrying beam combiner and perhaps a central telescope but with second and third girders pivoted eccentrically at its ends, carrying telescope elements at their outer ends, to provide a structurally-linked five-element array with variable spacing and orientation.

The particular technical challenge of this approach would be managing the transmission and phase compensation of the beams from each unit to the combiner while retaining the "protected beamline" characteristic which is the main advantage of structurally lining the telescope elements.

## 9. PRISM: a possible unit design

A study at NASA Marshall Space Flight Centre in 1994 and 1995 developed a 1993 design concept by the writer into the PRISM proposal for a radiatively cooled spectroscopic survey telescope in a sun-synchronous Low Earth Orbit (Campbell et al.1995).

This design was derived from the conventional concentric-shell designs proposed for, *e.g.*, *EDISON*, but in it the shells decrease in length from inner- to outermost. The upper portion of each is therefore exposed. These are coupled to space, and protected from outside heating, by forward-opening conical shields or "parasols", probably of deployable fabric, attached to the upper edges of the next shells outwards and probably to the top of the spacecraft. The conical shields are similar (at least in concept) to the sunshield of the COBE spacecraft.

In the design outlined by Campbell et al., the tops of the conical "parasol" shields, and the telescope tube, are cut off at a fairly steep angle to exclude sunlight and, as far as possible, Earthlight, from their interiors in the relatively low sun-synchronous orbit proposed. In our variant the tops of the shields are coplanar with the telescope aperture, to provide symmetrical protection. The design is illustrated in cross section in Fig. 3, and a schematic perspective view without the outermost shield (attached to the spacecraft upper edge) is shown in Fig. 4.

In a free-flyer at 1 AU this design allows radiative cooling of the inner (telescope) shell and its contents (optical system, instrument bay) to $20K \leq T \leq 30K$, and will certainly offer temperatures in this range at 5 AU, even embedded in a warmer structure. Because it permits operations to within a few degrees of the ecliptic pole (down to a solar avoidance angle of $\sim$ 95 degrees) it offers excellent sky coverage.

This design concept appears suitable for radiatively-cooled unit telescopes in structurally-linked arrays, and indeed in shield-shaded flotillas, if the other problems of the latter can be overcome. It does not appear to be adaptable to individually-shielded free-flyers as it does not solve the problem of stray light in the beam lines.

## 10. Conclusion: A way forward?

More than one basic design configuration seems capable of providing a powerful and effective IR space interferometer adaptible both to planet searches and to high-resolution and imaging astrophysics. A promising route could be the assembly and checkout in LEO of a structurally-connected, reconfigurable array of perhaps 5, radiatively cooled, 1- to 2m-class telescope elements. Acceleration by low-thrust electric propulsors and an Earth-Venus-Earth gravity assist series would allow a transfer into a Jupiter-crossing (5AU aphelion) elliptical trajectory; a final gravity-assist manouevre at Jupiter could be used to establish the

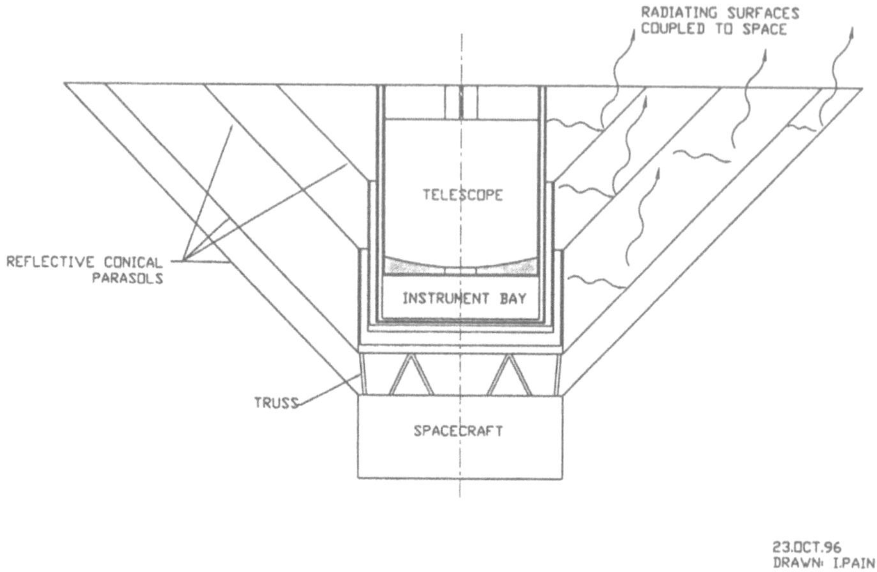


*Figure 3.* Schematic cross-section of a unit configuration based on the "PRISM" design concept (Campbell *et al.* 1995), originally intended for sun-synchronous orbit.

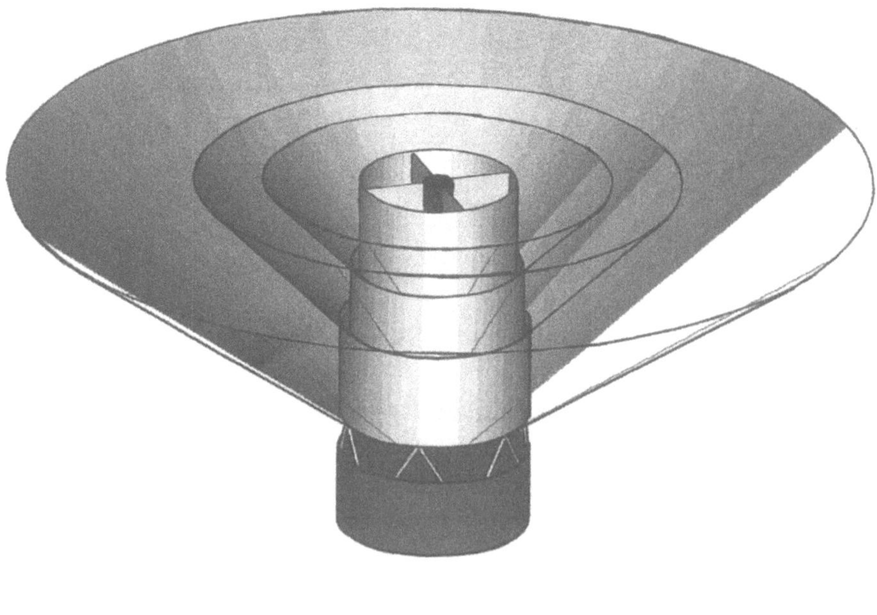

*Figure 4.* Schematic perspective view of the propsed unit configuration. Note that the outermost shield illustrated in Fig. 3 has been omitted.

observatory in an orbit lying permanently beyond the main zodiacal cloud, to offer an unparalleled opportunity to the astronomers and planetary scientists of the 21st century.

**Acknowledgements:** Some of the issues discussed here arose from points and queries raised by Thijs de Graauw, Ulrich Johann, Alain Leger and Dave Tytler at the Toledo workshop. All my co-authors from MSFC receive my especial gratitude, especially Charlie Telesco. I am particularly indebted to Chas Cavedoni and Ian Pain of the Joint Astronomy Centre for assistance with designs and diagrams, here and elsewhere.

## References

Bély, P.-Y., Laurance, R.J. & Volonte, S.: 1996, Kilometric Baseline Space Interferometry: comparison of free-flyer and moon-based versions, *Report of the Space Interferometry Study Team*, European Space Agency, SCI(96)7.

Campbell, J.W. et al.: 1995, *Proc. SPIE* **2553**, 14

Hawarden, T.G., Cummings, R.O., Telesco,C.M., Thronson, H.A., Jr.: 1992, *Space Science Reviews* **61**, 113

Hawarden,T.G., Crane, R., Thronson, H.A., Jr., Penny,A.J., Orlowska, A.H., Bradshaw, T.W.: 1995, *Space Science Reviews* **74**, 45

Leger,A., Puget, J-L., Mariotti, J-M., Rouan, D., Schneider, J.: 1995, *Space Science Reviews* **74**, 163

Thronson, H.A., Jr., Rapp, D., Bailey, B., Hawarden, T.G.: 1995, *PASP* **107**, 1099

# FLITE: FREE FLYER LASER INTERFEROMETER TECHNOLOGY EXPERIMENT

U. JOIIANN
*Dornier Satellitensysteme GmbH, Friedrichshafen, Germany.*

K. DANZMANN
*University of Hannover, Germany.*

C.J. SCHALINSKI
*DLR Institut für Weltraumsensorik, Berlin, Germany.*

R. SESSELMANN
*Dornier Satellitensysteme GmbH, Friedrichshafen, Germany.*

**Abstract.** Ambitious future space missions require extreme pointing accuracy and internal length stability, because they are based on optical/infrared interferometry. A technology precursor experiment is proposed, involving the German built ASTRO-SPAS, a space shuttle retrievable subsatellite. The objective of this experiment is the verification of key technologies and functions for an optical free-flyer interferometer, a laser gravitational wave detector (LISA) -both potential Horizon 2000plus cornerstone missions- and coherent inter-satellite laser communications (SOLACOS).

The experiment shall focus on a laser link between ASTRO-SPAS and a small, slowly departing subsatellite. Procedures common or similar in function and performance requirements for the application missions are laser-based pointing, initial acquisition and tracking (PAT), interferometric distance monitoring and control, laser optical phase locking and coherent laser communications (PSK homodyne). The free-flyer laser metrology mode for accurate internal sub-nm OPD-control and positioning of fiducial points for an optical interferometer will be verified in close distance range below 1 km. For LISA -essentially a gigant Michelson laser interferometer in space- one arm can be simulated in the range up to few km, depending on laser coherence. After increasing the distance to beyond 100 km, the laser communication mode can be tested in far-field geometry. Crucial functions and technologies (laser assembly) are verified in a realistic environment before taking the leap from laboratory to costly large science missions.

Similar configurations -to our knowledge- have not been flown in space up to now. The experiment can be performed also in a pure small satellites mission and can be expanded to include stare fringe tracking e.g. in a subsequent flight.

**Key words:** free-flyer, interferometer, Astro-Spas

**Abbreviations:** PAT (Pointing, Acquisition, Tracking), LISA (Laser Interferometer Space Antenna), OPD (Optical Path Difference), SOLACOS (Solid-State Laser Communications in Space), TPF (Terrestrial Planet Finder), FEEP (Field Emission Electrical Propulsion), FFI (Free-Flyer Interferometer), LC (Laser Communication), OPLL (Optical Phase Locked Loop), BER (Bit Error Rate), FPA (Fine Pointing Assembly), TX (Transmitter), RX (Receiver), PBS (Polarizing Beam Splitter), 4Q (Quadrant Photodiode), PN (Pseude Noise), AOCS (Attitude and Orbit Control Systems)



*C. Eiroa et al. (eds.),*
*Infrared Space Interferometry: Astrophysics & the Study of Earth-Like Planets,* 205–211.

## 1. Introduction

Some of the most rewarding space missions of the next two decades under preparation now within ESA's Horizon 2000plus program and NASA's Origins initiative are the exploration of planets orbiting other stars and the search for low frequency gravitational waves emitted by a number of exotic cosmic objects. The instruments required for such an undertaking are essentially large optical interferometers in space.

LISA, the Horizon 2000plus cornerstone mission candidate for gravitational wave detection consists of an array of freely floating spacecrafts separated by up to 5 million km and forming a giant laser interferometer. The relative distant changes caused by the gravitational wave induced strain in the space structure is interferometrically monitored to picometer accuracy in the appropriate spectral range (mHz).

The extreme angular resolution required to observe exo-planets (DARWIN, TPF) and - more general - to image small scale complex astronomical objects in the visible and near infrared dictates interferometric coupling of an array of telescopes in space, uncorrupted by the atmosphere. The telescopes (subapertures) are distributed over large distance baselines and are reconfigured during the observation. Again, their relative distances and, in addition, their orientation in space need to be monitored and controlled with nm and sub-as accuracy, respectively; this task is supported by laser metrology.

Interestingly, these technologies for large space science missions have much in common with those required for high data rate inter-satellite laser communications. This emerging technology for the communication needs of the near future requires the laser beams, carrying the information and serving as beacons for attitude control, to be pointed with sub-as accuracy and at high bandwidth (several 100 Hz) over large distances (70 000 km). For coherent optical reception technologies suitable at high data rates (1 Gb/s), the laser frequency and phase have to be controlled with sufficient precision.

Recognizing these similarities in function and requirements and being aware, that -albeit the technology has advanced tremendously on ground- no similar configuration has flown in space up to now, we propose a space-borne engineering experiment as a precursor to the large space missions.

## 2. Description of the Experiment

The basic version of the proposed experiment FLITE (Free-flyer Laser Interferometry Technology Experiment) involves two free-flying platforms in Earth orbit which are linked relative to each other by low power laser beams. Over the coarse of the mission, several distances between the platforms are established, ranging from almost contact to more than 100 km. One platform is carried as an Add-on to a larger core mission by the autonomous Astro-Spas, an already successfully flown Dornier-built carrier, released and later retrieved by the space shuttle. The other platform is a small subsatellite released and slowly floating away from Astro-Spas. The experiments will be carried out in two phases separated by a few days to establish the required separation, but each phase lasting a few hours only.

A laser link is established between both stations and several tests are carried out concerning pointing, acquisition, tracking as vital functions for all applications. Interferometric distance monitoring and laser communications is performed in a distance ranging from close to contact to beyond 100 km. The subsatellite acts subsequently as subaperture station in an astronomical free-flyer interferometer, as active amplifier mirror for a LISA laser interferometer arm, and as counter terminal for laser high data rate communications.

FLITE focuses on most critical common functions (laser link), but could be upgraded to test in addition other vital functions of the large observatories as micro-Newton thrusters (FEEP) and guide star fringe tracking for external alignment of a free-flyer astronomical interferometer. The optional experiments could be conducted in a re-flight as an Add-on to another Astro-Spas mission. In case the Astro-Spas carrier is not available, the FLITE scenario could also be converted to fit a pure small satellite experiment. Fig. 1 illustrates the mission scenario.

## 3. FLITE Test Program

The tests to be performed with the laser link are:

- Establish a bi-directional laser link by executing **pointing, acquisition and tracking modes** in near and far field telescope range. (Required by all application missions).

- **Relative attitude monitoring** (control) via the laser beacons. (The laser link defines the baseline orientation of a white light free-flyer interferometer and is collinear to the science and external reference light Coudè beams between the central beam combiner and the siderostats). Two designs will be compared: a coherent differential phase-readout on a quadrant photo diode (LISA) and a high bandwidth, high resolution fiber nutator beam sensor (LC).

- **Distance monitoring and interferometric coupling** of fiducial points on each station, respectively. (This corresponds to two func-

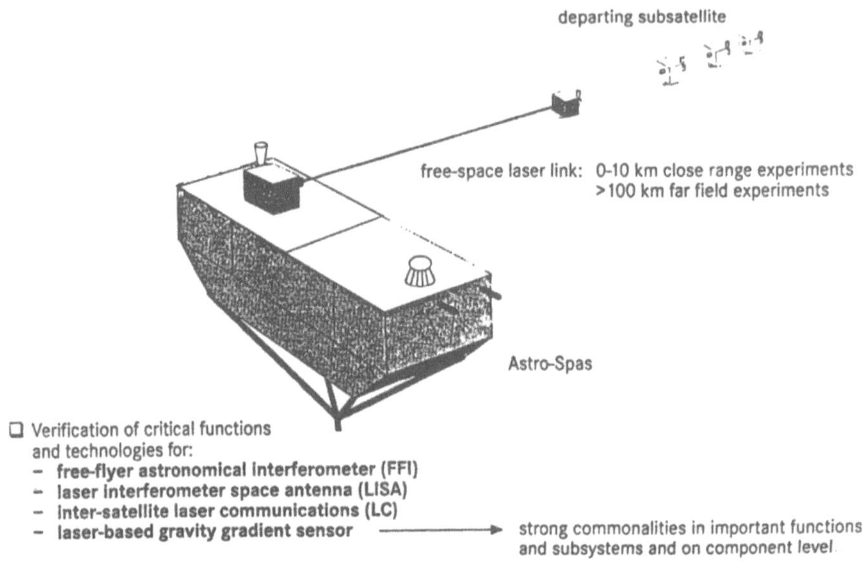


*Figure 1.* Free-flyer Laser Interferometer Technology Experiment mission scenario involving Astro-Spas and a small subsatellite

tions: first, internal optical path length difference (OPD) monitoring and control via laser fringes as required for a white light free-flyer interferometer and, second, one Michelson interferometer arm in a laser gravitational wave detector or a laser gravity gradient detector).

- **Optical phase locking** of a local oscillator laser beam to a low power received beam and subsequently **coherent phase modulated communication** by homodyne phase shift keying. (This function is required for high data rate communication itself, of course, but also for all applications for housekeeping data exchange, coded ranging and synchronization. The laser phase locking is a central function to provide an active "amplifier mirror" for the far end of a long distance Michelson interferometer arm, an essential LISA element).

- Verification of a **space-qualified solid-state laser assembly**. A prerequisite for above mentioned tests and an important step in technology for many applications in space.

In addition, **laser-based gravity gradient detectors** in low Earth orbit will benefit from the demonstrated technology. Optional experiments, enhancing the representativity of the FLITE precursor mission towards operational missions, but also increasing complexity and costs, are (as alternatives or together):

- **Test of FEEP thrusters** and distance/attitude control (a key function for LISA and astronomical interferometers).
- **Test of high resolution inertial sensors**
- **External metrology guide star**. In addition to baseline length monitoring by laser metrology, white light interferometers require extreme fine pointing and tracking of a (point-like) external reference source (star) in order to allow wavefront co-phasing (co-phase monitoring) of astronomical objects over integration time. Therefore, attitude and white light fringe tracking (monitoring) on an unresolved bright reference star lateral to experiment baseline and orbit plane shall be accomplished in this optional experiment involving two subsatellites released in opposite directions.

## 4. Preliminary FLITE Layout

Two payload boxes in form of approximately 50 cm cubes and about 70 kg mass are mounted on the Astro-Spas as an Add-on to a core instrument (e.g. the ISIS interferometer considered by JPL, USA). One box, firmly fixed to a mounting panel on Astro-Spas houses the central station, containing a fixed small optical transmit/receive telescope of less than 10 cm diameter, the master diode-pumped Nd:YAG laser, a phase modulator, several fine pointing sensors and actuators and a short stroke delay line. The other box, released by a spring at the begin of the mission is a small subsatellite, (e.g. a TUBSAT derivative) battery powered and providing its own attitude control and some limited $\Delta$v-capacity via a cold gas thruster. (TUBSAT is a University of Berlin built satellite, successfully flown in space; a new version is planned to fly end of 1996 aiming at precise attitude control to a few as : DLR-Tubsat).

The subsatellite is released at low relative speed (few cm/s). After completion of the first experiment phase the relative speed to Astro-Spas is increased (few m/s) in order to establish a distance beyond 100 km well in the telescope far field range for the second experiment phase a couple of days later.

Optical free-flyer interferometer (FFI) metrology is tested in the range close to contact up to about 1 km. LISA laser interferometer performance can be tested up to several km, depending on experiment time slot available from Astro-Spas operations. In the second experiment phase, after increasing the distance beyond 100 km, LISA interferometer arm performance and coherent laser communication are tested in the telescope far field.

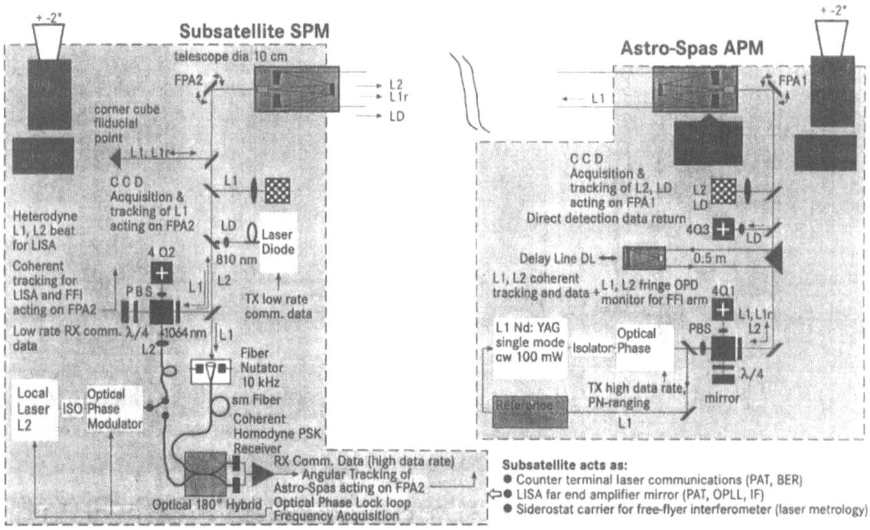


*Figure 2.* FLITE experiment functional architecture for baseline concept focusing on laser link tests. The optical front end and the laser assembly on APM are used by all part experiments in the integrated version employing functionally representative, but not fully developed components.

Fig. 2 shows functional block diagrams of the optical front-ends on the Astro-Spas payload module (APM) and the subsatellite (SPM).

### 4.1. APM (Astro-Spas Payload Module)

The laser assembly , comprised of transmitter laser, isolator, modulator and reference cavity is essential for all applications. The specifications will be driven by the most demanding, i.e. LISA for frequency and phase stability and laser communications for modulation bandwidth.

The master laser on the Astro-Spas module is a frequency stabilized single-mode laser-diode pumped Nd:YAG laser operating at a cw output power of about 100 mW. Its specifications are driven by the LISA requirements and shall be representative, except for power. The laser can be locked to an external ring cavity as foreseen in the LISA concept. The optical phase modulator is used in all applications for:

- low rate transmission between the two modules (few Mb/s)
- transmission of synchronization signals if required
- high data rate transmissions as specified for the coherent communication link (650 Mb/s)
- Pseudo Noise coded data strings for absolute distance ranging
- Phase dither for locking on reference cavity
- Phase dither for laser interferometer fringe tracking

### 4.2. SPM (Subsatellite Payload Module)

The subsatellite module is subsequently acting as:

1. A siderostat or telescope carrying subaperture in an astronomical free-flyer interferometer
3. A far end amplifier mirror for LISA
2. A counter-terminal for laser communications, receiving high and transmitting low data rates

A candidate for the subsatellite platform is the DLR/IWS and University of Berlin made TUBSAT. Its basic concept has flown successfully several times in space. A new launch is planned for 1996 to test a fine pointing attitude and orbit control system, which already would fit requirements. For FLITE it will be most probably a battery powered cube of about 50 cm side length and a mass between 50 kg and 70 kg. It will be released by spring action from the Astro-Spas FLITE payload module and shall have some limited $\Delta$v-capability from a commercial small cold gas thruster. Alternatively, all relative maneuvers possibly could also be performed by Astro-Spas itself. For a pure small satellites mission without Astro-Spas, two or optionally three such subsatellites would comprise the mission scenario.

## References

Armstrong, J.T. et al.: 1995, *Physics Today* **42**, 5
ESA SP-1135: 1990, *A Proposed Medium-Term Strategy for Optical Interferometry in Space*, 8
ESA SP-226: 1985, *Colloquium on Kilometric Optical Arrays in Space*, 4
MPQ 208: 1995, *LISA: Pre-Phase A Report*, 12
Renner, U.: 1991, *Weltraumforschung* **15**, 349
Johann, U. et al.: 1991, *Proc. SPIE* **1522**, 1
Johann, U., Sontag, H.: 1990, *Proc. SPIE* **1218**, 1

# ADVANCES IN SATELLITE DATA COMPRESSION & NOISE FILTERING BY VIRTUE OF PARALLEL COMPUTING

C. MACCONE
*Alenia Spazio - Turin Plant*
*Corso Marche 41, 10146 Torino, Italy.*

**Abstract.** Parallel Computing for space applications is a research area currently shaping up. Advances are being made that could find applications in the fairly near future. One research field is the application of Parallel Computing to Principal Component Analysis. This can be defined as a mathematical technique to reduce the dimensionality of a set of data. If the spectrum from a celestial body is represented by the flux at N wavelengths, then the spectrum can be considered as a point in an N dimensional space, with the axes consisting of the fluxes at the different wavelengths. Many such spectra then form clouds of points in the space. Principal Components Analysis finds the vectors in the space along which the data varies most significantly. These vectors are called Principal Components, with the first Principal Components being that which encompasses the most variance in the data. In this way, it is often found that a large amount of the information in the data can be retained by projecting the spectra onto a small number of Principal Components. In other words, spectra can be well reconstructed from the projections onto only a small number of principal components (for example the first ten), and when noisy spectra are being considered it is advantageous to limit the number of components used, since further projections merely reconstruct the noise. On the other hand, Alenia Spazio is now producing a string of massively parallel supercomputers, called Quadrics, that also are suited to the computation of the eigenvectors and eigenvalues of large symmetric matrices by virtue of the Jacobi algorithm. We propose to analyze large numbers of spectra by virtue of Parallel Principal Component Techniques implemented on the Quadrics massively parallel supercomputers.

**Key words:** data analysis, space applications

## 1. Introduction to the Method of Principal Components

This paper is devoted to a mathematical tool that may improve our understanding of physical phenomena: the Method of Principal Components, hereafter abbreviated as MPC. A version of MPC currently used in signal processing is called the Karhunen-Lohve Transform, hereafter abbreviated KLT. Essentially, it is something superior than the classical Fourier Transform (FT).

By adding random noise to a deterministic signal one obtains what is called a 'noisy signal' or, in case the power of the signal is much smaller than the power of the noise 'a signal buried into the noise'. Since the noise+signal is a random function of the time, one can describe it



*C. Eiroa et al. (eds.),*
*Infrared Space Interferometry: Astrophysics & the Study of Earth-Like Planets,* 213–217.

well by a statistical quantity called autocorrelation (or simply 'correlation'), defined as the mean value of the product of the values of the random process at two different instants. The correlation is obviously a symmetric function of the two instants. If one seeks for the eigenvectors of the correlation matrix, and then changes the reference frame to precisely the set of vectors, the easiest possible description of the signal+noise is achieved. This is the key-idea behind the KLT.

## 2. The KLT

The KLT is named after two mathematicians, the Fin, K. Karhunen (1946), and the French-American, M. Loève (1946), who proved, about the same time and independently, that the series expansion described hereafter is convergent. Put it this way, the KLT looks like a purely mathematical topic, but this is not the case, of course. We are going to use the language familiar to engineers and radioastronomers, and so we shall say that it is possible to represent the signal+noise as an infinite series (called K-L, or KLT, expansion) each term of which is the product of a random variable (constant in time, and thus representing the "statistical part" of the random process) multiplied by a time function (not involved at all with the "statistical part" of the random process, and thus representing the "pure time behaviour" of the random process). Assuming that the noise correlation is a known function of the two instants, it can be proved that the time functions are the eigenfunctions of the correlation, and that they can be determined by solving the integral equation whose kernel is the correlation. These time functions form an orthonormal basis in the Hilbert space, and they actually are the best possible basis to describe the noisy signal, better than any classical Fourier basis. One can thus say that the KLT adapts itself to the shape of the signal+noise, whatever it is.

Further advantage of KLT is that the random variables in the K-L expansion are orthogonal random variables. If our random function is a Gaussian process, this orthogonality of the random variables amounts to statistical independence, meaning that the terms in the KLT expansion are uncorrelated. This is certainly the most important property of the K-L expansion.

Further, since the (all positive) eigenvalues of the correlation are the variances of the random variables, any K-L expansion, when truncated to keep only the first few terms, is the best approximation to the full

K-L expansion in the mean square sense.

Finally, the mathematical theory of KLT shows that our random process need not be stationary. This too spells the difference against the classical Fourier techniques, that hold rigorously true for stationary processes only.

The 'narrowband assumption' is the rationale behind all radio filtering made thus far all over the world. Consequently, only Fourier Transform (FT) or Fast Fourier Transform (FFT) techniques were used to find the very narrow bandwith in which an unusual amount of received radio energy would indicate the presence of a signal, either sinusoidal or pulsed.

In this paper, however, we would like to maintain that the traditional usage of the FFT might sooner or later be replaced by the adoption of the KLT. In 1983, the French radioastronomer, Frangois Biraud, was the first person to describe the advantages of KLT over FFT for detecting wideband signals (Biraud 1983). Recent and very promising work on the KLT was done by Robert S. Dixon and Charles A. Klein, both with the Ohio State University in Columbus, Ohio (Dixon & Klein 1991). After aknowledging that the KLT is more general than the FT because it makes no assumption about the signal periodicity or waveform, these authors made one important new step ahead by pointing out that only the largest of the KLT output values need to be calculated, in contrast to the FT which must always calculate all the output values. This largest value is what the mathematicians call the 'dominant eigenvalue' in the solution of the integral equation whose kernel is the autocorrelation. Dixon and Klein did not attempt to prove any mathematical theorem about this fact, but they did numerical computer experiments showing that this must be the case.

It is now natural to wonder about what prevents radioastronomers from using KLT rather than FFT now. The simple answer is: the computational burden. In fact, the KLT kernel is the correlation, and, being the mean value of the product of two random variables, this kernel is not separable. Thus, in general, one cannot hope for the existence of a fast KLT algorithm. In turn, this means that the computer time required to calculate the eigenvalues and eigenvectors of a correlation matrix of order N is proportional to $N^2$, rather than to N×LOG(N), as for FFT. Nevertheless, current developments seem to be paving new ways to overcome the above difficulties: the steady improvements in computer hardware and parallelization techniques seem to lead to very fast algorithms able to get the eigenvalues and eigenvectors of large square symmetric matrix as the correlation. A comprehensive and recent book about the KLT, both non-relativistic and relativistic, is in Maccone (1994).

## 3. Parallel Computing at Alenia Spazio S.p.A.

In the last twenty years the computing power of available computers has grown dramatically. A new era in data processing is before us, mainly in those sectors where most extensively modelling and simulation techniques are applied. New computing tools and new programming techniques are required: massively parallel supercomputers and related techniques are the most appropriate tools to approach the simulation problems. High performance computing with massively parallel supercomputers is not just an evolution of the traditional electronic data processing systems: it is a real breakthrough into a new way to compute data. Quadrics Parallel Supercomputers derive from the APE Project of INFN (Istituto Nazionale di Fisica Nucleare of Italy). Following an Agreement with INFN, Alenia Spazio has acquired exclusive rights to produce and commercialize the results obtained in the APE-100 Project of INFN, a project that led to the implementation of a 100 GFLOPs massively parallel supercomputer. According to this Agreement, Alenia Spazio has undertaken the responsibility to engineer, produce and sell a line of parallel supercomputers derived from the APE-100 Project, while INFN shall continue its research and development efforts towards new generations of these machines.

## 4. Conclusion: KLT, Quadrics and Eigenquadrics

One of the most basic operations in all fields of applied mathematics is the computation of eigenvalues and eigenvectors (briefly eigensystems) of real symmetric matrices. Though many packages doing this for scalar and vector computers exist, the implementation of such a package for any parallel machine paves the way to new and unexplored applications of parallel computing. The final goal of the present paper is to foresee the possible implementation of a parallel eigensystems package for the Quadrics massively parallel supercomputer. Additionally, this package would enable the computation Karhunen-Lohve Transform. Thus Eigenquadrics can make a revolution possible: the replacement of the old FFT by the more subtle and recent KLT.

## References

Biraud, F.: 1983, *Acta Astronautica* **10**, 759
Dixon, R.S., Klein, C.A.: 1991,in USA-USSR Joint Conference on the Search for Extraterrestrial Intelligence p. 5
Karhunen, K.: 1946, *Ann. Acad. Sci. Fennicae, ser. A 1, Math. Phys.* **37**, 3
Loève, M.: 1946, *Rev. sci.* **84**, 195
Maccone, F.: 1994, in Telecommunications, KLT and Relativity, IPI Press, Colorado Springs.

# DESIGN OF INFRARED SPACE INTERFEROMETERS

J.-M. MARIOTTI
*DESPA Bât. 6, Observatoire de Paris, F-92195 Meudon Cedex, FRANCE*

**Abstract.** Infrared space observatories offer us the prospect of high sensitivity observations of the "cool Universe". On the other hand, ground-based interferometers are exploring the infrared window at very high angular resolution. An infrared space interferometer, combining these approaches would allow to tackle new and fascinating observing programs, such as the detection and the characterization of extra-solar Earth-like planets. Although the technical constraints for such a mission are certainly challenging, it appears that they could be mastered during the next decade(s). Also, an interferometer optimized for the detection of extra-solar planets (i.e. observing in the "nulling mode") could be designed in order to perform well in the imaging mode, hence expanding its observational program.

**Key words:** space projects, interferometry, infrared, detection of Earth-like planets

## 1. Introduction

The reasons for designing infrared space projects are well-known and provide the rationale for missions such as IRAS and ISO. The first one is of course that the Earth's atmosphere is opaque for ground-based sites in most of the domain spanning from the visible to the sub-millimeter wavelengths, e.g. from 35 to 300 $\mu$m. Even in the near-IR, absorption by water vapor, carbon dioxide and other gases creates large voids in the transmission curve of the atmosphere.

Between these gaps, several observing windows exist for ground-based observers but, at a wavelength of about 3 $\mu$m, starts the problem of the thermal background: the black body emissions of the telescope, the instrument and the atmosphere rapidly dominate the number of detected photons and hence provide a major source of noise, even for instruments optimized in terms of emissivity and observing from the best sites. For instance, taking a classical atmospheric model and assuming a system temperature of 275 K with an effective emissivity of 0.2 and a total throughput of 0.2, the number of photo-electrons generated by the background raises from $10^3$ $e^-/arcsec^2/\mu m$ at 1 $\mu$m to about $10^9$ at 5 $\mu$m. This problem of course worsens at higher wavelengths, for which high sensitivity observations require space-based observatories cooled at a few Kelvin.

But obviously, the main rationale for these projects is the wide range of astronomical problems that can be tackled by observations in the IR.



*C. Eiroa et al. (eds.),*
*Infrared Space Interferometry: Astrophysics & the Study of Earth-Like Planets,* 219–226.

Infrared is the natural domain for studying the emissions of the "cool Universe", i.e. from the coldest stars to the warm gas and dust. Our scope here is not to discuss these issues in details: we will mainly concentrate on one program which requires an infrared space interferometer, namely the detection and characterization of extra-solar Earth-like planets.

## 2. Space backgrounds

### 2.1. Thermal background

Space offers the opportunity to cool the experiment to temperatures unreachable to ground-based sites. The traditional approach, retained for both IRAS and ISO, is active cooling: it allows to reach extremely low temperatures but results in severe limitations for the total mass and lifetime of the mission.

Until recently, the alternative option, namely passive cooling, was thought to be limited to equilibrium temperatures reaching 70 to 100 K. But the thermal study devoted to the EDISON project (Thronson et al. 1993), has demonstrated that a careful design of the telescope structure and shields could allow to passively reach temperatures as low as 20 K.

A mission dedicated to detection and spectroscopy of Earth-like exoplanets would mainly exploit the 6–18 $\mu$m domain (Léger et al. 1996). For this range of relatively short IR wavelengths, it is not necessary to cool the whole system to such low temperatures. At about 40 K, the thermal background noise starts to be negligible with respect to other sources. A combination of passive cooling for the main components (structure, optics,...) of the interferometer and active closed cycle cooling for the very few critical components (essentially the detectors) appears hence to be a perfectly viable solution for such a mission, extending nearly indefinitely its possible duration.

### 2.2. Zodiacal light

Eliminating the instrumental thermal radiation does not unfortunately eliminate all sources of background. Indeed, the telescope is surrounded by the emission of the Solar System dust originating from the comets and asteroids, and associated with the zodiacal light. Although the column density is very small, at Earth's orbit the emission peaks in the 10–20 $\mu$m domain. At 10 $\mu$m for instance and for a bandwidth of 0.5 $\mu$m, the number of zodiacal photons collected in a coherent beam étendue is about $10^3\,\mathrm{s}^{-1}$. This represents the dominant source of background by more than two orders of magnitude.

There are two possible ways to fight the photon noise associated with the zodiacal background:

1) Since it does not depend on the diameter of the collector, increasing the size of the telescope increases only the signal and hence the Signal to Noise Ratio. In the case of the detection of extra-solar planets, Angel (1989) estimated that telescopes of about 8 meters of diameter were required for a 4-element interferometer.

2) As proposed independently by Bracewell & McPhie (1979) and Léger et al. (1995), moving the spacecraft away from the Sun reduces the temperature of the surrounding zodiacal dust, thus rapidly decreasing the background photon flux in the 10–20 $\mu$m domain. From IRAS-based models for the temperature and density distribution of the zodiacal dust, Léger et al. (1996) estimate that at 4 AU from the Sun, the background is reduced by two orders of magnitude and hence comparable to other sources. In these conditions, telescopes of the 1.5 m class are sufficient.

Using existing rockets and the gravitational assistance of the inner Solar System planets, it does not appear unrealistic to launch a spacecraft of several tons outside the asteroid belt. The main price to pay appears to be the long delay (several years) to reach the operating conditions, although this cruising period could be used to perform less demanding observations (see e.g., section 4). Also, at such distances, the solar flux is reduced by more than one order of magnitude, considerably easing the thermal design required for passively cooling the system. On the other hand, energy production and communications with Earth are more difficult, but the requirements of the interferometer are small in these domains.

### 2.3. Extra-solar sources

In the 6–18 $\mu$m domain the main sources of extra-solar background are faint stars (for the shorter wavelengths) and Infrared Cirrus (Boulanger & Pérault 1988); in average, none of them should be problematic, but since they are not isotropic, actual estimations of their levels in the direction of each source of the target list will be required.

## 3. A scientific driver: detection of Earth-like planets

The search for extra-solar planets has recently become a very hot topic with the first indirect evidences obtained by Mayor & Queloz (1995) and Marcy & Butler (1996). Although, up to now, these discoveries concern large mass planets, they demonstrate the "existence theorem" for extra-solar planets, and strongly reinforce the possibility of existence of Earth-like planets, possibly habitable or even inhabited. This topic has been the subject of several papers (see e.g., Léger 1996) and workshops (Boulder 1995) and we refer the reader to these references.

Even before these discoveries, the possibility for an infrared space interferometer to detect Earth-like planets around nearby (closer than 20 parsec) F, G and K stars, and to perform low resolution spectroscopy in search of the signatures of the main atmospheric constituants has been seriously considered, following the pioneering works of Bracewell & McPhie (1979) and Angel et al. (1986); these efforts have led to two proposals, OASES (Angel & Woolf 1996) and DARWIN (Léger et al. 1996), now under consideration at NASA and ESA.

Both projects share the same goals and the same basic features:
- they favor the infrared range, because of the reduced contrast ratio ($\sim 10^6$) between the star and the planet with respect to the visible (a few $10^9$), and because of the existence of the spectral signatures of several significant molecules, $CO_2$, $H_2O$, $O_3$, and possibly $CH_4$.
- they propose, as discussed in section 2, passively cooled spacecrafts in 5 AU aphelia orbits.
- they are based on an interferometric concept, since a 1 AU separation between the star and the planet corresponds at 20 pc to an angular separation of 0.05", i.e. the resolution of a 40 meter optical system at 10 $\mu$m. Besides, an interferometer offers the possibility to be operated in the "nulling mode", efficiently rejecting the light of the on-axis star, while efficiently transmitting the photons from slightly off-axis planets (Bracewell & McPhie 1979).

The main difference between the two concepts lies in their strategies for rejecting the exo-zodiacal light, i.e. the IR emission of the zodiacal cloud presumably associated with the exo-planetary system: while OASES relies on over-resolving the system, thus pointing to a 4-telescope linear interferometer with a $\sim 100$ m baseline (Angel & Woolf 1996), DARWIN uses an odd number of telescopes thus modulating at different frequencies the signals from the centro-symmetric exo-zodiacal light and the asymmetric planetary system (Mennesson & Mariotti 1996).

Each of these designs has his pros and cons: the linear structure of OASES is possibly easier to deploy in space and its beam-combination strategy allows to obtain a very large and deep null on-axis (Woolf 1996). On the other hand, the DARWIN concept is more efficient to disentangle in the collected signal the signature of the planets from that of the exo-zodiacal light. Indeed, a key issue for choosing a concept will be the typical level of the exo-zodiacal fluxes at 10 $\mu$m for the nearby solar-type stars: is the Solar System, with a ratio of $\sim 300$ between

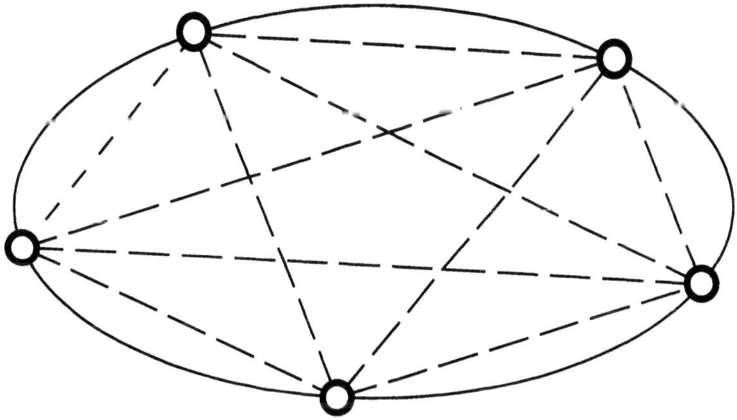

*Figure 1.* A non-redundant elliptic configuration.

the integrated zodiacal ligth flux to the Earth's flux, typical or not? This crucial question could be addressed through the interferometric connection of large ground-based telescopes (Traub et al. 1996) or a dedicated space mission.

## 4. Imaging vs. nulling

The strategy for the pupil configuration of DARWIN is to regularly distribute an odd number of telescopes (e.g., 5) on a circle with a typical radius of 25 m. A strong disadvantage of this arrangment is that it can miss a planet if its angular separation from the star is such that it never crosses the lines of maxima in the transmission map. This is why we have proposed (Mennesson & Mariotti 1996, Mennesson 1996) an alternative solution where the telescopes are located on an ellipse with typical semi-major and semi-minor axis of 25 and 12.5 m. This configuration retains most of the advantages of the previous one in terms of extraction of the planetary signal and allows to fully cover the planetary system, without the necessity of actually changing the distances between the telescopes.

An extra advantage, is that the telescopes can be distributed on the ellipse so as to form a fully non-redondant pupil, as illustrated in Fig. 1. The disposition of the telescopes allows to obtain 10 different baselines providing by rotation of the array 10 different circular tracks in the (u,v) plane. This means that the interferometer can be optimized not only in the nulling mode, but also in the more classical imaging mode, where the delays are set so as to produce a fully constructive interference condition on axis. The system could hence be used to perform other astrophysical programs, e.g. during the phase of its orbital period when it is too close to the Sun (cf. section 2).

## 5. Technical challenges

As expected for such an ambitious project, technical challenges are huge. Let us conclude by mentionning only a few of them:
- The launch and deployment in space of a structure spanning several tens of meters is a critical issue. Alternatives are an astronaut-helped assembly in Earth orbit prior to final injection, or adoption of a free-flyer concept such as the one studied by the MOFFIT team (Bély et al. 1996).
- The optical quality of the interferometer, including its pointing accuracy and stability, has to be very high in order to achieve the required rejection factor of $\sim 10^6$: most of the optical defects can be compensated through optical filtering of the beams prior to recombination (Ollivier & Mariotti 1996), while the pointing of the array and the control of the optical path differences can benefit from the high apparent luminosity of the star in the visible domain.
- The recombination set-up and the instrument has to be polarization-controlled and achromatized to a high level and for a rather large total bandwidth (Woolf 1996). Also, the requirements for detectors are stringent: the newly developed STJ technology (Peacock et al. 1996) offers a possible starting point.

During the next decade, our ability to develop the techniques necessary to control these issues (and several other) and to space-qualify them will be crucial for the development schedule of these projects. The importance of the goal, detection of Earth-like planets and possibly of extra-terrestrial life, justifies the effort.

## References

Angel, J.R.: 1989, in The Next Generation Space Telescope, P. Bély, C. Burrows and G. Illingworth (Eds.), STScI; Baltimore, p. 81
Angel, J.R., Cheng, A.Y., Woolf, N.J.: 1986, *Nature* **322**, 341
Angel, J.R., Woolf, N.J.: 1996, *ApJ*, in press
Bély, P., et al. (the Space Interferometry Study Team): 1996, "Kilometric Baseline Space Interferometry", ESA Report SCI(96)7
Boulanger, F., Pérault, M.: 1988, *ApJ* **330**, 964
Detection and Study of Earth-like Planets (Proc. Colloquium), 1995, Boulder
Bracewell, R.N., McPhie, R.H.: 1979, *Icarus* **38**, 136
Léger, A.: 1996, these proceedings
Léger, A., Mariotti, J.-M., Mennesson, B., Ollivier, M.: 1995, in Proc. Coll. Detection and Study of Earth-like Planets, Boulder

Léger, A., Mariotti, J.-M., Mennesson, B., Ollivier, M., Puget, J.-L., Rouan, D., Schneider, J.: 1996, *Icarus*, in press
Marcy, G.W., Butler, R.P.: 1996, Communication at the 187th AAS meeting, San Antonio.
Mayor, M., Queloz, D.: 1995, *Nature* **378**, 355
Mennesson, B.: 1996, these proceedings
Mennesson, B., Mariotti, J.-M.: 1996, *Icarus*, submitted
Ollivier, M., Mariotti, J.-M.: 1996, in preparation
Peacock, et al.: 1996, *Nature*, in press
Thronson, M.A., Harwarden, T.G., Penny, A.I., Davies, J.K.: 1993, EDISON proposal to ESA M3 mission, RAL, Chilton, UK.
Traub, W.A. et al.: 1996, in Proc. ESO workshop Science with the VLTI, F. Paresce and O. von der Lühe (Eds.), Garching
Woolf, N.J.: 1996, these proceedings

## 6. Questions

*M. Yanagisawa* : Is the zodiacal light background so serious, even in the celestial area far from the ecliptic plane for an Earth-orbiting interferometer?
*J.-M. Mariotti* : Yes. The contrast between the ecliptic and the pole for the background is about 3 vs. 1. We need to reduce the background more than an order of magnitude for extra-solar planet detection.

*R. Mundt* : Do you get rid of the dust emission problem also by optical filtering?
*J.-M. Mariotti* : Yes. The dust on the optical surfaces of the interferometer will be cold, so its emission should not be a problem. However it will diffuse the incoming light (e.g., from the star), creating a real problem. Fortunately, optical filtering allows to get rid of this stray light. It works very well because dust is a small scale defect which diffracts light with very wide angles (the smaller the defect on the pupil, the more efficient the filtering). In our simulations (Ollivier and Mariotti, in preparation), we assume that 2% of the primary mirror is covered by dust (a figure similar to the HST case): after filtering, the stray light is not detectable.

*T. de Graauw* : Suppose you have a slight asymmetry in the zodiacal dust around the star under study. Can you discriminate against that?
*J.-M. Mariotti* : An asymmetry in the zodiacal dust could create a local maximum of emission, thus mimicking a planet. Indeed asymmetries have been found by IRAS in the Solar System zodiacal emission (Nature, Vol. 374, April 1995). By only "imaging" the planetary system, it might be difficult to discriminate between a "true" planet and

a dust blob. However, the latter is likely to have a spectral energy distribution quite similar to that of the integrated zodiacal light. Hence, a low resolution spectrum is likely to resolve the issue.

*P. de Korte* : What would happen if you place your system near the Earth instead of at 4-5 AU away ?
*J.-M. Mariotti* : The telescope size would have to change from 1.5 m to 4-8 m diameter in order to be able to detect the planets.

# THE NEXT GENERATION SPACE TELESCOPE (NGST)

J. MATHER
*Infrared Astrophysics Branch, Code 685*
*NASA Goddard Space Flight Center*
*Greenbelt, MD 20771 USA*
*e-mail: john.mather@gsfc.nasa.gov*

**Abstract.** The Next Generation Space Telescope (NGST) is under study by NASA as a successor to the Hubble Space Telescope (HST), and the infrared missions SIRTF (Space Infrared Telescope Facility), SOFIA (Stratospheric Observatory for Infrared Astronomy), and ISO (Infrared Space Observatory). It would have an aperture > 4 m, optimized for 1-5 $\mu$m, with a goal of 8 m and 0.5 - 20 $\mu$m. It would be radiatively cooled and would be launched on an Atlas IIAS to the Lagrange Point L2 around 2006. At wavelengths longer than a few $\mu$m, it offers a speed advantage of the order of $10^6$ over a large ground based telescope.

**Key words:** infrared, space telescope

## 1. Introduction

The idea of using passive cooling to permit a large-aperture IR space telescope goes at least as far back as Tim Hawarden's 1989 proposal to ESA for an M2 mission called POIROT (Passively cooled Orbital IR Observatory Telescope, see also Space Science Reviews 61, 1992). Radiative cooling appears sufficient for large optics and instrument bays to operate at temperatures of 30-60 K, without the need for stored cryogens or closed-cycle coolers. For even lower temperatures, such as those required by long-wavelength detectors, cooling requirements can be minimized, while maintaining long mission lifetimes (>10 yr.) The most recent design of SIRTF, for example, uses passive pre-cooling of the telescope structure prior to routine science operations. By substantially reducing the stored cryogens, this design permits SIRTF to be launched on a Delta II and has substantially decreased the cost of the overall spacecraft (JPL D-12375).

IRAS and ISO have shown the enormous sensitivity advantages of cold space telescopes for observations at or beyond 3 $\mu$m, which is dominated by thermal emission from the atmosphere and optical systems for telescopes within the Earth's atmosphere. Compared to ground facilities, space telescopes also have significant advantages at shorter wavelengths for wide-field, diffraction-limited imaging.



*C. Eiroa et al. (eds.),*
*Infrared Space Interferometry: Astrophysics & the Study of Earth-Like Planets,* 227–232.

AURA's HST & Beyond Committee (1996) recognized the enormous scientific capabilities of a large-aperture, infrared-optimized space telescope and recommended strongly that such a mission be developed to follow the HST and SIRTF programs.

The Committee, chaired by Alan Dressler, recommended a passively cooled telescope with the core mission of studying the early formation of stars and galaxies at high redshift. For these studies, a diameter > 4m provides resolution comparable to HST (corresponding to scales lengths of about 1 kpc at high redshift) and the sensitivity to detect the compact, essentially unresolved star forming regions in early galaxies ($3 < z < 10$). The HST&B report also lists a wide range of scientific research which would be enabled by such a facility in both the optical and FIR, but cautions that the mission should be driven by its performance in the NIR.

## 2. Advantages and scientific investigations

The NGST would be configured as a general purpose observatory, optimized for the near IR and the study of the high redshift universe. It has major advantages in three areas:
1. Wide field diffraction limited imaging, especially at wavelengths < 1 $\mu$m, where ground-based adaptive optics are limited by a small isoplanatic patch. This is important for searches for rare objects, supernovae, large scale structures, and studies of population statistics.
2. High sensitivity at wavelengths > 2.5 - 3 $\mu$m, where thermal backgrounds at ground-based telescopes are a million times greater than the zodiacal light background. This is such an improvement that great surprises are likely.
3. Large cold aperture. The NGST would be 5 to 10 times as large as the next largest cold telescope, the SIRTF. The SIRTF is expected to be confusion-limited for near IR extragalactic astronomy.

While the high redshift universe was emphasized by the HST & B report, possible NGST scientific investigations include a wider range:

I. High redshift universe
- Cosmology, including supernova distance scale
- Galaxy formation, collisions, and evolution
- Evolution of galaxy clustering
- Star formation at high redshift
- Evolution of gas content of galaxies and intergalactic medium
- Dark matter

II. "Local"
- Galactic rotations (stellar CO bands)
- Obscured galactic regions
- Star formation
- Interstellar medium - IR penetrates dust clouds; near IR lines
- Other planetary systems, dust disks
- Planetary atmospheres, asteroid mineralogy

## 3. Study Plan

With support from the Origins office at NASA HQ, the Goddard Space Flight Center (GSFC) and the STScI began a feasibility study of a large passively cooled telescope, dubbed NGST after the 1989 STScI NGST Workshop. MSFC (Marshall Space Flight Center), JPL (Jet Propulsion Laboratory), and the Ames and Langley Research Centers are also providing resources. The goal of the study is to demonstrate the feasibility of developing a large, passively cooled telescope that meets or exceeds the HST & Beyond technical goals. The final report will be submitted to NASA HQ in August 1997, in time for the NAS (National Academy of Sciences) decade review. An interim report will be provided by Nov. 1996, in time for the development of the OSS (Office of Space Science) long-range plan. Since most of the expertise in lightweight optics and IR instrumentation lies outside NASA, U.S. industry and academic communities are playing crucial roles in the NGST study.

The study is being managed by John H. Campbell, the manager of the Hubble Space Telescope project. John Mather is Study Scientist, and he and Peter Stockman of the Space Telescope Science Institute (STScI) are leading the scientific team. Bernie Seery is the Systems Engineer and Spacecraft Manager, and Keith Kalinowski is leading the operations study team. Pierre Bely of the STScI is leading the Scientific Instrument team, and John Humphreys of Marshall Space Flight Center leads the telescope team. Participation in these teams is voluntary and open, and interested persons should contact the team leaders.

To gather the best ideas that academia and industry have to offer, NASA is funding two independent studies by industrial and academic consortia led by Chuck Lillie of TRW and Domenick Tenerelli of Lockheed-Martin. The best features of the independent studies and the NASA-led study will be merged in late August and will become the baseline mission for further definition and costing.

Details about the studies and the various engineering efforts are accessible on the World Wide Web at the address http://saturn1.gsfc.nasa.gov/ngst/

These is a working site for participants, not a general public site, and those seeking the most current information are encouraged to consult directly with the authors.

Given the outlook for NASA funding over the next decade, any mission the scale of the NGST must break the HST/Great Observatory cost curve by an order of magnitude to remain feasible. We will plan a mission that could be built for \$500 M (FY 96 funds) with a life cycle cost of \$900 M, not including data analysis. This is small compared to the HST life cycle cost, of order \$6B. The plan is designed for a launch around 2005, based on a technology demonstration to be complete by 2002. Such a schedule is clearly dependent on progress leading to approval.

It is clear that the NGST must have a significantly larger aperture than HST in order to reach the required sensitivities and NIR resolution to study the early universe. The reduced costs must come from new technologies and management approaches. To fit the mid-cost launch capabilities, the NGST will use deployable ultra light weight optics similar to those developed for ballistic missile defense requirements. Perhaps the greatest challenge will be the development of a management approach which provides the appropriate degree of NASA oversight, scientific participation, and industry incentives to provide an excellent science mission for an affordable price.

## 4. A Preliminary Concept

The top level requirements for the NGST include:

- Aperture > 4 m
- Diffraction limited at 2 $\mu$m
- Wavelength range 1 - 5 $\mu$m; goal 0.5 - 20+ $\mu$m
- Broadband imaging, celestial background limited sensitivity
- Moderate resolution spectrometry
- Lifetime 10 years
- Guest Observer mode
- Entire sky visible each year
- Significant sky visible for long periods

Other possible but less likely capabilities include:

- High resolution spectrometry
- Coronagraph or equivalent
- UV or far IR coverage

The appended drawing shows a preliminary version of the NGST mission. Although significant study must be undertaken, it is likely that the final design will include the following features:

- Launch on an Atlas IIAS class expendable launcher,
- Mass less than 2818 kg,
- Direct insertion to orbit around the Lagrange L2 point, roughly in line with the Sun and Earth, and 1.5 million km away,
- Segmented, deployable primary mirror (nominally 8m dia.) made of either SiC or Be, or with a nickel or glass meniscus on a reaction plate,
- Deployable secondary (1m dia.) made of a similar material,
- Adjustable mirror, either primary or a conjugate, for figure corrections,
- Large, multilayer sunshield deployed by inflation, to shade all parts of the spacecraft over a reasonable range of pitch and roll, and reach telescope temperatures $< 60$ K (30 K goal),
- Passively-cooled instrument bay containing large format imagers, spectrographs, wavefront sensors, and fine guidance sensors,
- InSb detectors from 0.5 or 1 to 5 $\mu$m, passively cooled,
- Si BIB detectors for 5-30 $\mu$m, cooled by stored helium or active coolers,
- Compact and shielded spacecraft bus, assembled from standard components, isolated thermally and structurally through an active structure from the instrument bay and telescope,
- Active vibration isolation between reaction wheels and structure,
- High bandwidth communications (about 1 MHz) to 1-2 ground stations from L2,
- Solar array power, perhaps as part of the sun shade.

Some parts of this concept entail technologies that are not fully developed, though much has been accomplished by military funding, ground-based adaptive optics research at observatories, and the general advance in computers and industrial capabilities. It is clear that a mission based on the above concept could be made to work. To reduce cost and risk, we are planning a technology demonstration program that may include a small space mission, depending on the risk assessment.

`Acknowledgements`: This study is supported by the Origins Program in the Astrophysics Division at NASA Headquarters.

## References

Dressler, A. *et al.*: 1996, HST and Beyond, Committee Report. Avail. Space Telescope Science Institute, Baltimore, MD

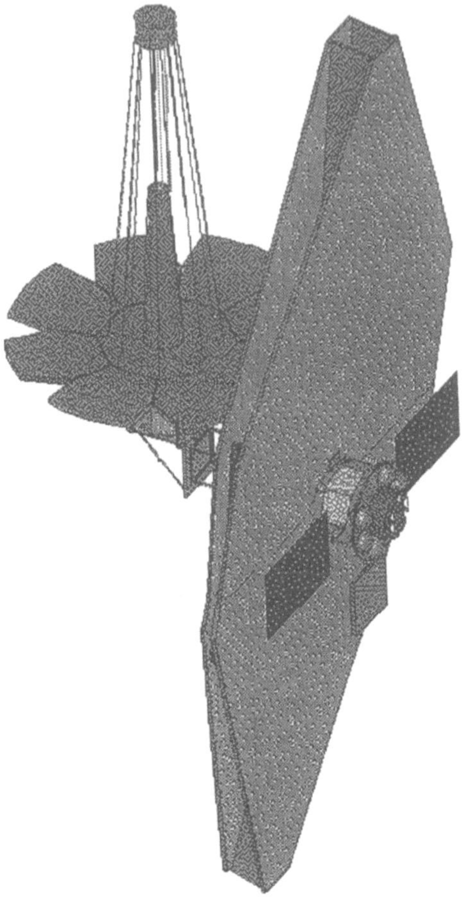

*Figure 1.* One concept for Next Generation Space Telescope.

# HIGH ACCURACY OPTICAL VISIBILITIES ON LONG BASELINES: FIRST RESULTS AND PROSPECTS

G. PERRIN
*Observatoire de Paris, DESPA, F-92195 Meudon, France*

V. COUDÉ DU FORESTO
*Max-Planck-Institut für Astronomie, Königstuhl, D-69117 Heidelberg, F.R.G.*

J.-M. MARIOTTI
*Observatoire de Paris, DESPA, F-92195 Meudon, France*

S.T. RIDGWAY
*National Optical Astronomy Observatories, Tucson, Arizona 85726-6732, U.S.A.*

N.P. CARLETON and W.A. TRAUB
*Harvard-Smithsonian Center for Astrophysics, Cambridge, Mass. 02138, U.S.A.*

**Abstract.** In this paper we report the first results of a fiber optics beam-combiner at the focus of the IOTA interferometer. The spatial filtering of the atmospheric turbulent optical modes by single-mode fiber optics enabled us to measure fringe contrasts with precisions far superior to that of classical all-optics beam-combiners. This demonstrates the potential of single-mode fibers for light recombination in imaging interferometers with high-dynamic range. A potential application is expected to be the direct detection of planet-like objects around stars.

**Key words:** atmospheric effects, interferometers, stars, extra-solar planets

## 1. Introduction

In actual optical interferometers, the accuracy of visibility calibration is limited by the non-stationarity of the atmospheric turbulence. The best reported values are of the order of one percent, i.e. consistent with the accuracies obtained in speckle-interferometry, a technique indeed limited by the same phenomenon.
For several years (Coudé du Foresto & Ridgway 1991), we have developed the concept of single-mode interferometry which provides visibility moduli measurements that are independent of atmospheric turbulence, thus opening the way to interferometric image reconstruction with very high dynamical range in the optical domain. We report here the first results obtained at 2.2 $\mu$m and with a baseline of 21 meters.



*C. Eiroa et al. (eds.),*
*Infrared Space Interferometry: Astrophysics & the Study of Earth-Like Planets*, 233–239.

## 2. Experimental set-up

Recombination is achieved at 2.2 $\mu$m with single-mode fiber optics (Perrin et al. 1995), the Fiber Linked Unit for Optical Recombination (FLUOR), set at the focus of the Infared-Optical Telescope Array (IOTA). IOTA is located at the Smithsonian Institution's Fred Lawrence Whipple Observatory on Mount Hopkins in Arizona, and is a collaborative project of the Smithsonian with Harvard University, the University of Massachusetts, the MIT (Lincoln Lab) and the University of Wyoming. IOTA is currently a two-element interferometer with 45 cm entrance pupils (Carleton et al. 1994). The available baselines range from 5 to 38 m. It is operated both in the visible and in the infrared. First fringes were obtained at 2.2 $\mu$m with an all-optics beam-combiner (Dyck et al. 1995).

## 3. Results

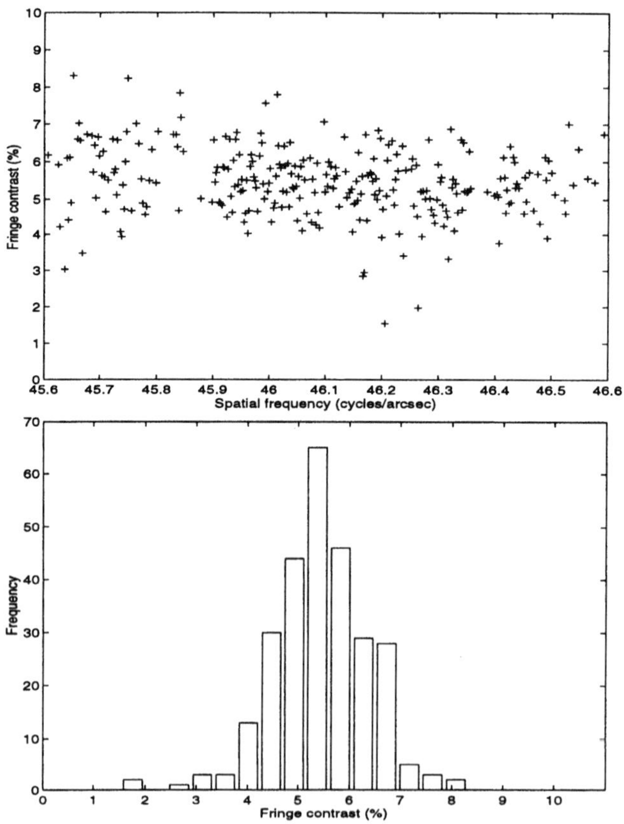


*Figure 1.* Fringe contrast measurement on Betelgeuse.

Use of single-mode fibers provides a major advantage over classical techniques. Single-mode fibers are perfect spatial filters and perfect waveguides. A beam injected into a single-mode fiber keeps a constant spatial distribution (the fundamental mode) and phase changes are only determined by the propagation along the fiber. After transmission through the atmosphere, the wavefronts of the two beams are corrugated by the atmospheric turbulence. In all-optics interferometers, turbulence induces fluctuations of contrast that depend upon seeing conditions. Conversely, the beam injected in the fiber is transversally fully coherent. The beams are then recombined without the usual loss of coherence due to the distortion of the phase of the two wavefronts. The expression of a monochromatic interferogram recorded with FLUOR is:

$$I = P_1 + P_2 + 2C\eta\sqrt{P_1 P_2}\cos(\phi + \alpha) \tag{1}$$

where $C$ and $\alpha$ are the modulus and phase of the coherence factor, and $\phi$ is the modulated phase difference between the two beams (phase modulation being produced by scanning the interferogram through the zero optical path difference with a delay line). $P_1$ and $P_2$ are the fluxes injected in the two input fibers. The interferometric efficiency $\eta$ of the set-up is fully determined by experimental parameters, such as mismatch of polarizations, losses in the X-couplers, frequency response of the detectors,... This term is extremely stable and can be calibrated out by observing with the same set-up a source of known visibility, e.g., an unresolved star. The process of spatial filtering by the fiber trades phase fluctuations on the wavefront for intensity fluctuations of the guided beam. The intensity fluctuations are monitored with the two photometric outputs $P_1$ and $P_2$ while, in a classical set-up, the interferometric efficiency also depends on the instantaneous corrugations of the wavefronts at the time when fringes are recorded. Full calibration requires calibration of the instrumental visibility (visibility measured on an unresolved source through a non turbulent atmosphere) and of the statistical loss of coherence caused by corrugated turbulent wavefronts. The latter calibration is seeing dependent and a large amount of data is needed if high accuracy is aimed for. Besides, the turbulence is continuously evolving and its statistics are not stationary, leading to biased contrasts measurements.

For a single-mode interferometer with optical waveguides, the contribution of turbulence to contrast loss (the atmospheric transfer function) is directly deduced from the two photometric signals. Measurement of the instrumental visibility thus allows full calibration of the interferogram and yields an unbiased estimate of the visibility (Coudé du Foresto 1996). Fig. 1 is an example of contrast measurement with FLUOR. The values displayed are estimates of $\eta C$ after photometric calibration.

The maximum projected baseline is 21.2 m. The source (Betelgeuse) is totally resolved and is sampled at spatial frequencies that stand beyond the peak of the second lobe of its visibility function, hence the low values of the contrast. Contrast fluctuations have two causes. Firstly, a global tendency of contrast to decrease with spatial frequency is clearly visible and is compatible with the expected decrease of the visibility function due to the super-synthesis effect. Secondly, contrast measurements are sensitive to detector, photon and optical-path-fluctuations noise. In the infrared, detector noise is predominant over photon noise which can be neglected for all astronomical sources. Atmospheric optical path fluctuations (or piston) are the first order of turbulent modes. This is the only mode left unfiltered by the single-mode fibers. It is a fluctuating delay between the two beams. It could be reduced down to a negligeable level by scanning fringes more rapidly than the 300 Hz fringe frequency used for the acquisitions reported here. From standard models of turbulence it is shown that for average conditions piston is attenuated by more than a factor of ten for scanning frequencies of the order of 3 kHz (Perrin 1996). The lower view displays the histogram of the measurements for the 258 recorded interferograms. Almost all the contrasts were found to lie in the range $4-7\%$. The final measurement and the attached error are the mean and the statistical error of the samples. Hence $\eta C = 5.47\% \pm 0.06\%$.
We have plotted the statistical error contrasts (normalized to one acquisition) versus magnitude in the K photometric band for all the observed sources (Fig. 2). The points are well fitted by the power law (full line):

$$\epsilon(\%) = 21.31 \times 10^{0.38\,\mathrm{m_K}} \propto \Phi^{-0.95} \qquad (2)$$

where $\Phi$ is the brightness of the star. The dominant term is thus detector noise for which the error is exactly proportional to the reverse of the source flux. Piston noise accounts for most of the error for bright sources hence the -0.95 exponent in the power law instead of the -1 expected exponent.

Imaging is not yet possible with two telescopes, yet sources with simple, *a priori* known or assumed, geometries can be easily studied by directly fitting the visibility data by the Fourier transform of the model. High accuracy optical visibilities give access to some interesting fields of astronomy. We discuss such a case in the following section.

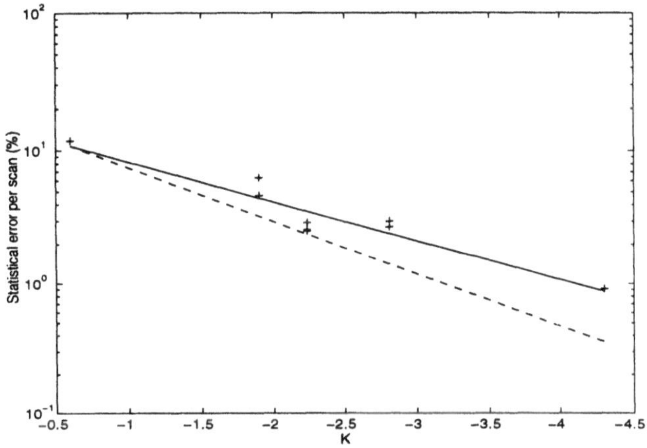


*Figure 2.* Statistical error per interferogram vs. K magnitude. The full line is a fit to the error by a power law. The error is mostly due to detector noise.

## 4. Prospects for high accuracy visibility measurements

High precision on visibility translates in terms of high dynamic range in the reconstructed image. Dynamic ranges as high as $10^4$ would be attainable with infrared interferometric imagers with configurations similar to that of the radio Very Large Array. Besides, detection of very faint details around stars does not seem unrealistic. Let consider the recent detection of a planet-like object around the solar-type star 51 Peg (Mayor & Queloz 1995). 51 Peg is a G2V star located 14 pc away from our Sun. The companion is found to be a Jupiter-like planet orbiting 0.05 AU from the star. From theoretical models of giant planets (Guillot et al. 1995) it is possible to estimate the physical parameters of a planet this close to its parent star. For a predicted mass in the range 0.5-2 $M_{Jupiter}$ and after an 8 Gyr evolution, the planet has a size quite similar to that of Jupiter with a radius between 1.2 $R_{Jupiter}$ and 1.4 $R_{Jupiter}$. Because of its proximity to the primary, the temperature of the planet is relatively high with $T_{eff} \simeq 1250$ K. The radius of 51 Peg is a little bigger than the Sun ($R_\star = 1.29\, R_\odot$), its temperature being $T_{eff} = 5773$ K. In the K band the brightness ratio of the two objects derived from the Planck law is thus $\beta \simeq 7 \times 10^{-4}$. The visibility function of the system is the weighted sum of the visibility functions of each element. We consider that the planet is unresolved (the required baseline would be 10 km). Let us expand the modulus of the visibility function to the first order with respect to $\beta$ :

$$\left|V(\vec{S})\right| = \left|V_\star(\vec{S})\right| (1-\beta) + \beta \cos(2\pi \vec{S}.\vec{r}) \quad (3)$$

where $\vec{S}$ is the vectorial spatial frequency, $V_\star$ is the visibility function of 51 Peg alone and $\vec{r}$ is the position vector of the planet. The second term in the above equation is oscillatory with a frequency equal to the spacing between the planet and the primary. In first instance, assuming a uniform stellar disk, the first zero in the visibility function of 51 Peg occurs for a baseline of 1 km at 2.2 $\mu$m and the expansion is thus valid for baselines shorter than a few hundred meters. One may wonder if other classical features of the star can mimic the planet signature. Star spots and active regions of solar-type stars may produce a signature of about 0.001 magnitude in the visibility function (von der Lühe et al. 1996). But, these features are very sharp details with a maximum separation of one stellar diameter. Their influence on the visibility is thus a very smooth function at spatial frequencies lower than the first zero of the stellar disk. On the other hand, this first lobe contains 9 periods of the oscillatory term due to the planet. Hence the signature of the planet in the Fourier space can be easily distinguished as long as it is over the noise. We assume that the raw visibilities for 51 Peg can be calibrated against those of another solar type star, e.g., 85 Peg (G2 V). Taking into account the errors on the visibilities of 51 Peg and its reference, the number of interferograms required for a 3 $\sigma$ detection of the planet signal is thus of the order of 2000.

We hence suggest that it is possible to detect the planet with a fiber optics single-mode experiment. For the detection to be credible it is necessary to sample two periods of the oscillation. This requires a 300 m baseline length. The current configuration of the experiment does not meet these requirements. Nevertheless, other interferometers exist or are to be built in the coming years with sub-kilometric and kilometric sizes which could house a fiber optic combiner.

## 5. Conclusions

The fiber optic recombination technique in optical interferometry has proved to be efficient to measure spatial frequency components of bright stars with high accuracy. Future improvements will improve its sentivity, allowing observations of astronomical sources with high contrast ratios, e.g., the 51 Peg system. This technique will provide its full advantage when fiber optics recombiners will couple large telescopes equiped with adaptive optics systems (Mariotti et al. 1996).

## References

Carleton, N. P., Traub, W.A., Lacasse, M.G.: 1994, in *Proc. SPIE* **2200**, 152

Coudé du Foresto, V., Ridgway, S. T.: 1991, in High Resolution Imaging by Interferometry II, J.M. Beckers, F. Merkle (Eds.), Garching, p. 731
Coudé du Foresto, V.: 1996, *A&AS*, in press
Dyck, H. M., Benson, J.A., Carleton, N.P., et al.: 1995, *AJ* **109**, 378
Guillot, T., Burrows, A., Hubbard, W.B., et al.: 1996, *ApJ* **459**, L35
Mariotti, J.-M., Coudé du Foresto, V., Perrin, G., Zhao, P., Léna P.: 1996, *A&AS* **116**, 1
Mayor, M., Queloz, D.: 1995, *Nature* **378**, 355
Perrin, G., Coudé du Foresto, V., Ridgway, S.T., et al.: 1995, *Proc. SPIE* **2476**, 120
Perrin, G.: 1996, *A&A*, submitted
von der Lühe, O. et al.: 1996, in Stellar surface structure, IAU Symposium 176 K.G. Strassmeier (Ed.), in press

# THE COAST PROJECT

J. ROGERS
*MRAO/Cambridge University, Cavendish Labs, Madingley Road*
*Cambridge. CB3 OHE U.K.*

**Abstract.** The Cambridge Optical Aperture Synthesis Telescope (COAST) is the first independent element Interferometer to produce an astronomical image. Resolutions of the order of 20 milliarcseconds (mas). are routinely obtained and as the the instrument is further developed this is expected to improve to 1 mas. In terms of resolution COAST is already superior to the present generation of space instruments although the sparse aperture filling mens that the instrument at present is limited to objects brighter than 7th. Magnitude. Foreseeable improvements should, however, increase this by 2 to 3 orders.

**Key words:** optical ground interferometry

## 1. Introduction

Ground based, optical, telescopes having an aperture greater than about 10cm do not produce diffraction limited images because of atmospheric distortion. Adaptive optics systems which continually monitor and compensate for atmospheric perturbations have been developed to eleviate this problem. Such systems must individually correct 10 to 20cm ($r_o$) sub-areas of the main mirror in times of the order of 10msec ($t_o$). On a large telescope this means that hundreds of sub apertures must be separately controlled in real time. The error signals are provided either by the object itself or a laser beacon, both optically faint objects. Unfortunately, the intensity of the faint object must be divided by number of regions to be controlled. This problem may ultimately govern the size of monolithic telescopes. For observing faint objects there is no substitute for the collecting area of the primary mirror. The resolving power of a large telescope can, however, be mimicked by an array of smaller telescopes linked together. This principle of aperture synthesis has been used, for decades by radio astronomers, but is considerably more difficult to implement in the optical regime largely due to the mechanical precision and stability required. A second radio astronomy technique that can be applied at optical wavelengths is that of Closure Phase. The latter is a powerful procedure for reducing atmospheric perturbations on the image and requires three, diffraction limited, telescopes observing the object simultaneously. Closure Phase applied to an optical array effectively means that the fringe pattern produced when the in phase light from the telescopes is combined is not distorted by atmospher-



*C. Eiroa et al. (eds.),*
*Infrared Space Interferometry: Astrophysics & the Study of Earth-Like Planets,* 241–246.

ic effects but simply moved around the image plane. COAST utilises both Aperture synthesis and Closure Phase techniques, as does Aperture Masking (Haniff et al. 1987, Buscher et al. 1990, Rogers et al. 1994) which can be used on large, conventional, telescopes to enhance their resolving power. Images of giant stars with resolutions approaching 30 mas have been obtained by Aperture Masking on a 4m telescope. Such resolution has revealed surface features and non circularity in some of these stars. Aperture Masking was initially developed to test and develop the basic principles and data analysis algorithms for COAST and involves placing a small mask in a collimated beam formed from the telescope output beam. An array of holes penetrate the mask, thus converting the large telescope into an assembly of co-mounted, diffraction limited telescopes. It is possible to take this process a stage further with large aperture telescope arrays such as the Keck pair and VLT array. Baselines shorter than the diameter of a single mirror can be produced as already described, whilst a number of other baselines can be generated by combining light from different portions from two of the large telescopes. This technique would reduce the number, or even remove the need for "outrigger" telescopes in such systems.

## 2. The COAST Array

The resolution of Aperture Masking is ultimately limited by the diameter of the host telescope. An array of separated small telescopes was an obvious next step and thus the concept of COAST (Baldwin et al. 1994) emerged. From the outset the philosophy was to build the simplest possible system which would work and give valuable operational experience. Ideally COAST would have been situated at a high altitude site with good seeing. As no such site exists within the U.K. it was considered that a remote site would greatly increase development time and costs. Thus, COAST was always intended to be a precursor to a more powerful instrument which will be outlined later in this paper. In order to understand the capabilities of COAST, a brief technical outline follows. It consists of a "Y" shaped array of four independent telescopes which can be moved between prepared bases up to a maximum separation of 100m. Each telescope comprises of a 50cm siderostat flat, feeding a fixed, horizontal, 40cm f5.5 Cassegrain telescope. The light is reduced to a 2.5cm beam by the secondary mirror, which is then intercepted by a piezo controlled mirror that performs tilt/tip corrections and directs the starlight into the optics building for path compensation and beam combining. A corrugated steel tunnel 32m long and 6m wide forms the optics building. It has been covered with approximately 300 tonnes of Earth, sown with grass and thick insulating end walls. This arrangement provides an extremely stable thermal environment without the use of air conditioning. Externally the building has the same emissivity and thermal inertia as the surrounding field and thus, does not give rise to significant convection. In addition, its shape minimises wind turbulence. Equalisation of the stellar light paths is performed by a two stage path compensator. The coarse correction is performed by a trolley running on 25m long rails. Mounted on the trolley is a voice-coil system which exerts the fine control on the retro-reflector. Positional informational for the dual servo system is provided by a laser interferometer. Furthermore, the retro-reflector can be modulated by a triangular waveform which provides a linear sweep through the fringe pattern. A different rate is chosen for each of the four telescopes so that there is a unique fringe frequency for each of the baselines. The four light beams emerging from the path compensators are each split at a dichroic, wavelengths shorter than 650nm are used for acquisition and auto-guiding. Wavelengths longer than 650nm are fed to either of the two independent, pupil plane beam combining and detection systems. In the visible light system, up to 900nm, "Avalanche Photon Diodes" (APDs) are used as detectors, whilst a NICMOS III chip is used for the near infrared system which will operate up to 2500nm. In both cases a similar arrangement of beam-splitters and mirrors is used to combine the four input beams into four output beams each with a quarter of the light from each telescope contained within it. Since beam combination takes place in the pupil plane the fringes must be detected temporally by sweeping the path compensation trollies through distances of a few wavelengths. The detectors, therefore, need only to be single pixel devices. Even the NICMOS chip with its numerous pixels is used as a simple four element device.

Fig. 1 shows fringes obtained from Capella between two telescopes only. Notice that the mean level has been subtracted from this plot. Subtracting the difference between two detector outputs performs this task, since due to the beam-splitter arrangement they are 180° out of phase. Fringes are at maximum near zero trolley deviation if the sweep was centred about the packet. The large peak in the power spectrum near the origin is due to atmospheric scintillation. Clearly separated from this, at 700Hz, the trolley modulation frequency is the peak due to the fringes. An estimate of the square of the visibility amplitude can be calculated from the area under this peak. It can be seen that the signal to noise ratio is very high, in this example of the order 10000:1. To determine the closure phase between three telescopes it is necessary to modulate the input beams at three different frequencies in order to isolate the corresponding visibilities. Using Earth rotation synthesis,

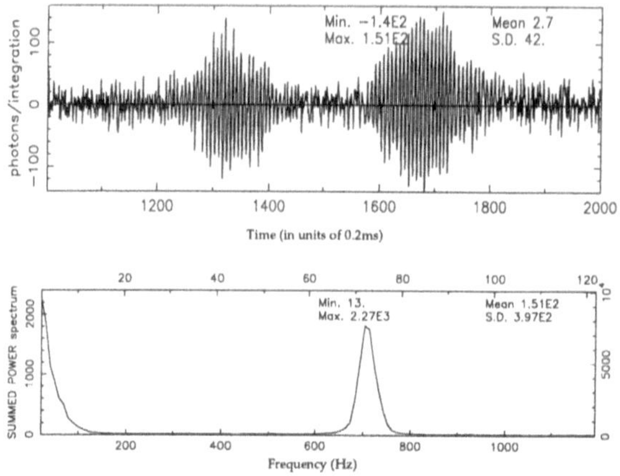


*Figure 1.* Interference fringes measured from Capella at 830nm (top) together with the power spectrum obtained from 100 seconds of such data (below). Two sweeps of the fringes past the APDs are shown. The scale on the vertical axis of the power spectrum is arbitrary and the zero frequency is not plotted.

similar data obtained over a period of time can be used to build up an image of the source. Using a maximum base line of 6.1m an image of Capella (Fig. 2) with a resolution of the order 20 mas has been obtained (Baldwin et al. 1996). The two component stars of the Capella system are clearly resolved although they are separated by only 6 light minutes yet they are 40 light years distant. These are the first true images from an array of separated optical telescopes demonstrating the validity of the technique and also its ability to work even at poor sites.

## 3. Future Developments

The basic concepts of COAST have proved to be sound and any future interferometric array we would consider building would largely be a development of the present system. It would be primarily an imaging instrument, probably using 6 telescopes at any time. A limiting factor is that beam combination system presently used produces the same number of output beams as telescopes, each containing an equal portion of light from corresponding telescopes. Thus, 6 telescopes appears to be the optimum giving sufficient light whilst offering improved phase closure constraints. In order to provide rapid imaging capability the total array will consist of 12 to 18 telescopes, the 6 active telescopes being a switchable subset of this. This feature will almost certainly

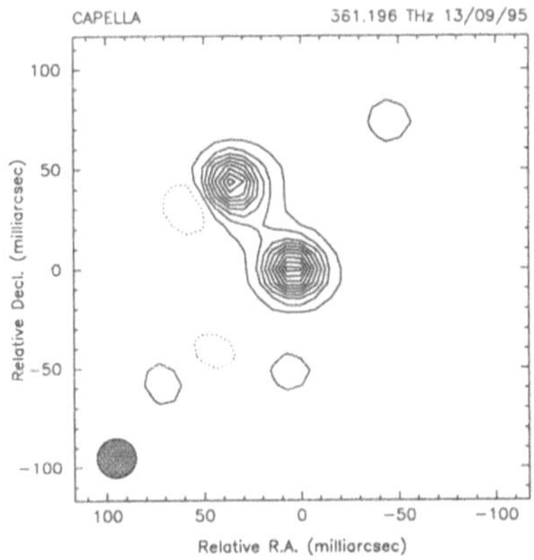


*Figure 2.* A map of Capella obtained using COAST. The beam used in the reconstruction is illustrated in the bottom left corner of the map. The contours are drawn from -5% to 95% of the peak flux with an interval of 10%.

prove necessary as experience has already shown that high resolution observations often reveal short time scale phenomena. It is envisaged that the instrumentation would initially allow operation from the visible regime up to 3500nm. The telescopes themselves would be similar in concept to the present instruments but canted upwards at about 20° to improve sky coverage and modified to accept mirrors of 1 to 1.5m diameter. A system for replicating these optical components would probably be necessary as their expense dominates the overall cost of the project. It is envisaged that the array will eventually have a base line of at least 500m which will require evacuated tubes for relaying the light to the beam combining laboratory. Vacuum delay lines would negate the need for dispersion compensation and allow simultaneous fringe measurements spanning the visible to near infrared spectrum. Technically the evacuation of the light path system is the most radical departure form the present COAST design. Furthermore, much work will be required to automate as fully as possible the entire observing procedure in order to make the instrument comparatively easy to use and hence, time efficient. A technique to the automatic alignment of the optical paths would be a priority. This task would be eased by combining a number of components in miniaturised "Bulk Optics". Although COAST has demonstrated that optical interferometry is a powerful tool for reducing atmospherical effects, if the new telescope is to function at its full potential it must be located on a new site having optimum seeing conditions for an interferometer and a high occurance of clear nights. No such location exists within the U.K. and potential sites have yet to be investigated. COAST was built with a budget of 1.2M U.S. dollars but due to the international aspect and scope of this future project we estimate the cost to be an order of magnitude higher. This project would almost certainly have to be a collaborative effort to raise the necessary funds and to acquire a suitable site. The cost of COAST2 is of the the same magnitude as a conventional, 4m class telescope but will have sub-mas resolution. As such we feel that it would prove remarkable value and almost certainly provide a major step forward in observational astronomy.

## References

Haniff, C.A. et al.: 1987, *Nature* **328**, 694
Buscher, D.F. et al.: 1990, *MNRAS* **245**, 263
Rogers, J. et al.: 1994, *Proc. SPIE* **2200**, 541
Baldwin, J.E. et al.: 1994, *Proc. SPIE* **2200**, 118
Baldwin, J.E. et al.: 1996, *A&A* **306**, L13

# PROSPECTS FOR DIRECT IMAGING FROM THE GROUND

D. G. SANDLER *
*ThermoTrex Corporation,*
*10455 Pacific Center Court, San Diego, CA 92121*

**Abstract.** NASA's Extra-solar Neighboring Planetary Systems (ExNPS) roadmap recommends a decade of vigorous exploration from the ground during the design and development of the space infrared nulling interferometer. An element of the ground observational program will be discovering and analyzing nearby extra-solar planetary systems through direct imaging of giant planets. To accomplish this goal will require extending the current capabilities for direct detection of faint sources to a new domain of very high contrast and sensitivity, made possible only through the technology of adaptive optics (AO). In this paper, we will discuss the potential of AO for discoveries from the ground, and show that a new advanced form of AO on large telescopes will yield direct images of Jupiter-like planets around nearby stars.

**Key words:** adaptive optics, Jupiter-like planets

## 1. Introduction

The recent discoveries of Jupiter-mass planets around nearby stars by measurement of stellar reaction motion (Mayor et al. 1995) may be viewed as the beginning of a new era for ground-based astronomy. The next step will be to confirm discoveries and make new ones by imaging the reflected light of giant planets. However, the challenge is severe: from a distance of 10 pc, Jupiter would be $10^9$ times dimmer than the sun, and at 0.5 arcsec separation would be swamped by the strong glare of the central star. To recover the diffraction-limited planet image, peaked above the surrounding halo and concentrated in few pixels on the detector, the stellar halo must be suppressed by several orders of magnitude, and speckle noise caused by residual wavefront errors must be severely reduced and randomized to allow efficient smoothing of the halo through long exposures. These requirements can in principle be met by the technique of adaptive optics (AO) (Beckers 1994).

To illustrate the above points, we show in Fig. 1 an image at $\lambda = 2.2\mu$m of Gliese 229, taken recently at the Multiple Mirror Telescope by the Steward Observatory AO group. This image clearly confirms the existence of the brown dwarf Gliese 229B (Nakajima et al. 1996) at 7.7 arcsec separation from the primary star, at a star-to-companion bright-

* Also with Center for Astronomical Adaptive Optics, Steward Observatory, University of Arizona, Tucson, AZ 85721



C. Eiroa et al. (eds.),
*Infrared Space Interferometry: Astrophysics & the Study of Earth-Like Planets*, 247–254.

ness contrast ratio $R \sim 3000$. The AO system used for the observation controls only the image motion over each of the six, 1.8 m primary mirrors, using a fast beam-combiner controlled by simultaneous centroid measurements of the six images of the star (Gray et al. 1995). For this simple low-order system, the function of AO is to increase the signal-to-noise ratio (SNR) by peaking the companion signal, through concentrating its energy into a smaller area in the image plane, by about a factor of 5 for 0.65 arcsec seeing.

The strength of the background in the vicinity of the brown dwarf is still the uncorrected atmospheric halo. A Jupiter at 10 pc would be 14 times closer to the central star, and some 300,000 times dimmer than Gliese 229B. In this case, AO must perform a double function, both peaking the planet and drastically reducing the atmospheric halo at inner radii well within the atmospheric seeing disc.

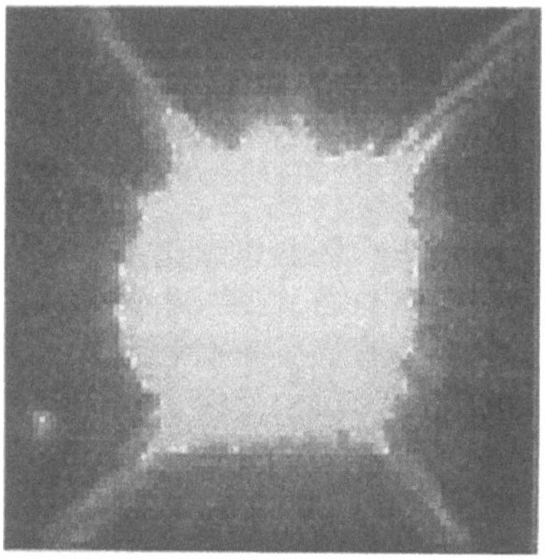

*Figure 1.* Image of brown dwarf Gliese 229B at 2 microns wavelength, taken at the MMT. Low-order AO was used corresponding to tip/tilt correction over the six MMT mirrors.

## 2. Conventional adaptive optics for astronomical imaging

AO sharpens astronomical images by measuring the wavefront across the pupil of the telescope using a wavefront sensor (WFS) and applying an equal but opposite correction using a deformable mirror (DM) containing actuators distributed across its surface. The WFS typically measures local slopes or curvature of the wavefront from a reference star, over portions of the pupil (subapertures) of width $d \sim r_0$, where $r_0$ is the atmospheric coherence length at the imaging wavelength. For good astronomical sites, $r_0 = 15 - 20$ cm at visible wavelengths, and increases to $r_0 \sim 1$ m at $\lambda = 2\mu$m. The first astronomical AO systems to yield impressive sharpening of science objects (Graves et. al. 1995; Rousset et. al. 1994) have used partial correction ($d > r_0$), optimized for use of natural field stars which may be outside the isoplanatic patch to obtain a sufficiently bright reference. Full diffraction-limited correction of the wavefront to the diffraction limit (Strehl ratio $S \geq 0.5$) requires $d \leq r_0$,and hence bright reference stars, typically $m_v \leq 10$ for correction in the near infrared. Since the probability of finding such a bright star within ~10 arcsec of an arbitrary science target is low, the use of laser guide stars (LGS) to project artificial beacons (Fugate et al. 1994) has been adopted for many new astronomical observatories. Astronomical AO using a beacon created by laser excitation of naturally occurring sodium atoms in the Earth's mesosphere at 90 km promises to provide diffraction-limited imaging in H-K bands on the new generation of 6.5-10 m telescopes (Sandler et al. 1994).

Development programs for sodium LGS AO are underway for the 6.5 m MMT (Sandler et al. 1995) and Keck II (Gleckler and Wizinowich 1995) telescopes, and similar systems are planned for several other large telescopes. The MMT design is unique in its use of an adaptive secondary mirror as the corrector element, which transmits the corrected beam through a dichroic beamsplitter directly to the infrared science detector, with shorter wavelengths reflected into the AO sensors. The use of an adaptive secondary greatly simplifies the optical design and maximizes throughput of the system.

The 64 cm diameter secondary mirror will have 300 actuators to control atmospheric turbulence, with nominal update times of 1 millisec for closed-loop control. The design is based on a thin 2 mm glass shell whose shape is controlled using voice coil actuators. The shape of the thin shell is referenced to a thicker, stable glass surface, with 100 $\mu$m gap between the two surfaces. The static figures of both surfaces match those of an ideal telescope secondary mirror in the absence of atmospheric aberration. Capacitive position sensors are used to monitor the precise separation every 0.1 millisec, thus providing extremely accurate correction by avoiding hysteresis.

The design of the new MMT system employs many other unique features, intended to make the system ideal for astronomical AO. These include a 4 W continuous wave laser; the latest in fast low-noise CCD's for WFS detector; and infrared quad cell to measure overall tilt from laser-sharpened field star images, thus increasing sky coverage for tilt field stars. Soon after first light for the new MMT in late 1997, the AO system will yield diffraction-limited images at $\lambda \geq 1.6\mu$m for any object in the sky. Since the design is optimized for the lowest emissivity, the AO instrument for the new MMT will be ideal to search for brown dwarfs cooler and closer in than Gliese 229B. In addition, the MMT AO system will provide very strong correction at $\lambda = 10\mu$m wavelength, providing an established architecture for use with a ground-based interferometer composed of two corrected large telescopes. Candidates include configuring the twin Magellan telescopes or Keck I and II as an interferometer, and the Large Binocular Telescope. A nulling interferometer using AO was first proposed by Woolf and Angel (1995), to image from the ground the zodiacal emission around nearby stars.

## 3. Properties of the corrected image halo

We focus here on correction with a relatively bright central star, with prospective planets within the isoplanatic patch at the imaging wavelength. In this case, the dominant residual errors after AO correction are fitting errors (of high spatial frequencies, arising from the finite number of actuators), temporal lag errors (with a broad spectrum of spatial frequencies, arising from atmospheric winds and the finite control bandwidth of the AO loop), and photon noise errors (random Poisson noise arising from the finite number of photons available for wavefront sensing). Fig. 2 shows cross-sections of image profiles for several representative cases which are relevant to the search for brown dwarfs and extra-solar planets. Proceeding from top to bottom along the intensity axis, first the case of normal adaptive optics at $\lambda = 2\mu$m is shown. The halo is suppressed by a factor of 10 at 0.5 arcsec compared to the uncorrected seeing halo. The corrected halo joins with the seeing halo at 1 arcsec radius, corresponding to the cut-off radius $\lambda/d$ for adaptive correction, with $d = 0.35$ m. The AO system gain, $G$, is the ratio of peak intensity to mean halo level. From Fig. 2 we see that $G \sim 10^4$ for radii $< 1$ arcsec; the value $G = 10^5$ at 1.5 arcsec is in agreement with the value reported by Lena (1996) for AO correction at the CFHT.

The integration time $T$ required for detection at planet signal-to-noise ratio $SNR_p$ is given by

$$T = \tau \left(\frac{R}{G}\right)^2 \left(\frac{SNR_p}{SNR_h}\right)^2 \tag{1}$$

where $\tau$ is the effective speckle lifetime, which defines the integration time for short exposures such that the residual speckles in the image halo are uncorrelated from frame to frame. The quantity $SNR_h$ is the the inverse of the rms level of fluctuation about the halo at the separation of interest, normalized to the mean halo level. It contains contributions from residual speckle noise and photon, detector, and back-

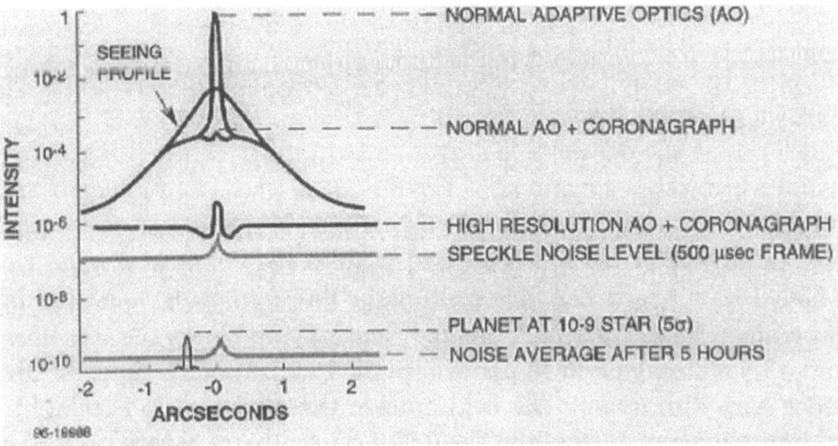


*Figure 2.* Cross section of theoretical stellar image profiles for varying degrees of AO correction.

ground noise in the image. As an example of normal adaptive optics at $\lambda = 2\mu$m, for $\tau = 5$ millisec and $SNR_h = 1$,we find that a faint object with $R = 10^6$ at separation $< 1$ arcsec can be detected at the $5\sigma$ level for a $T = 20$ min integration.

## 4. High resolution AO for extra-solar planet detection

To reach $R = 10^9$, a much larger gain is required. Angel (1994) investigated the fundamental limits imposed by photon noise and AO errors on detection at this level, and found that direct detection should be possible by an AO system on a $D = 6.5$ m telescope using a DM with $\sim 10,000$ actuators operating at 2 kHz update rate. His method thus relies on very strong suppression of the stellar halo by extending correction to smaller scale sizes and faster time scales than normal AO. Using analytic formulas, he obtained $G \sim 10^6$ for an $m_R = 4 - 5$ central star, and $SNR_h$ of order unity. Assuming random AO errors, with completely decorrelated speckle from frame to frame ($\tau = 0.5$ millisec), from equation 1 a detection can be made in $T = 3.5$ hours. Angel also pointed out that amplitude errors caused by scintillation, neglected in normal AO because they degrade the Strehl ratio by only a few percent, must also be at least partially corrected because they decrease the value of $SNR_h$.

A schematic of this "super AO" system for planet detection is shown in Fig. 3. Image motion correction and correction of the strongest, low-frequency atmospheric aberrations is performed by a 300 actuator DM, which for the new MMT we take to be the adaptive secondary mirror. Correction of residual fast-changing, fine-scale structure required to suppress the stellar halo is made by a high-resolution deformable mirror of low stroke. Scintillation correction is performed by an approach which places one deformable mirror away from the pupil, using an algorithm which takes advantage of the full set of phase and intensity measurements to control both phase and amplitude errors. Wavefront phase and amplitude errors are measured using stellar light in a band from 0.65-0.9 by a Mach-Zehnder white light interferometer, operated in a closed-loop servo system. A tapered, central stop blocks the star image, and a Lyot stop is used to perform pupil apodization to suppress high-order Airy diffraction. The field outside the central stop is relayed to an infrared array, imaging in the 0.95-1.35 $\mu$m band, where $r_0 = 40$ cm for 0.7 arcsec seeing.

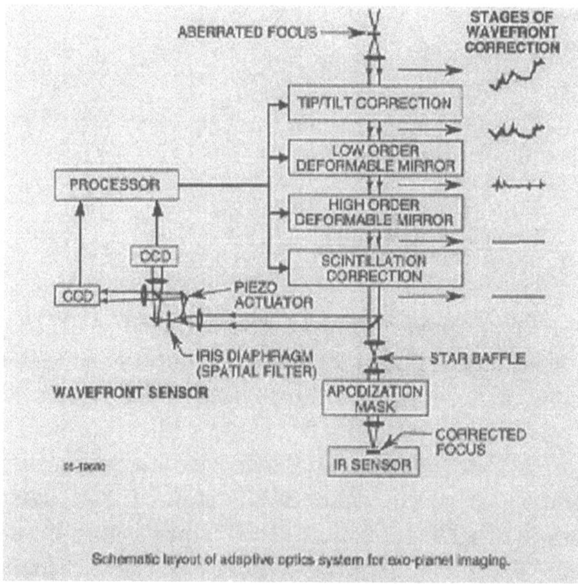


*Figure 3.* Schematic of high resolution AO system for extra-solar planet detection.

Stahl and Sandler (1995) have performed detailed simulations of the performance of "super AO", applied to the new 6.5 m MMT and Magellan telescopes. They took $d = 5$ cm, corresponding to 13,000 correction segments for fine scale correction across the aperture. Scintillation correction was modeled an intensity mask which clips all pixel intensities above a given threshold value., with forty percent of the light available to the imaging camera is lost in the scintillation clipping.

For the above parameters, the system gain is predicted to be $G = 0.4 \times 10^6$, with 0.03 of the energy in a broad halo of width 4 arcsec. The behavior of the corrected image halo is pictured in Fig. 3. The simulations show that the predicted gain should be achieved by the system, and the residual speckle noise is below the level of photon noise. However, in order to obtain the necessary speckle lifetime of $\tau = 0.5$ millisec, Stahl and Sandler had to introduce a predictive algorithm which uses adjacent phase measurements about a given subaperture and measurement of the effective wind speed to overcome time lag errors. Because this filter operates over very small scales compared to $r_0$, the algorithm was found to be robust. The upshot of is that Stahl and Sandler found that a Jupiter-like planet can be detected in a $T = 5$ hour integration. Fig. 4 shows a simulated 5 hour exposure of our solar system as viewed from 8 pc, with Jupiter clearly appearing at 5 sigma above the noise.

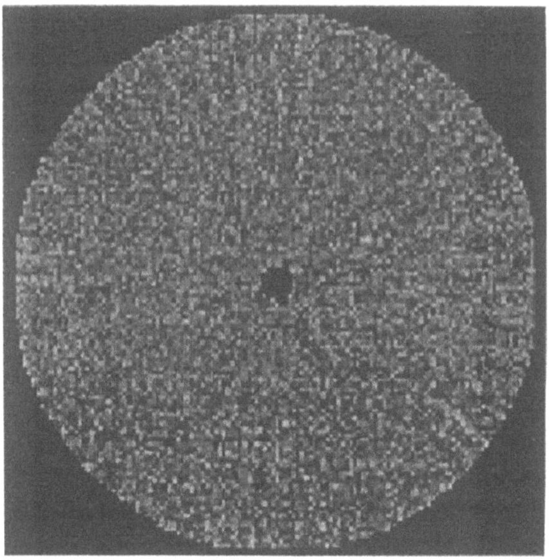

*Figure 4.* Simulation of 5 hour exposure of a twin solar system seen from 8 pc, using high resolution AO on the new 6.5 m MMT.

## 5. Future directions

The hardware requirements for "super AO", while challenging, are reachable in the near term by drawing on and extending the latest AO technology. It is clear that to measure a planet signal at 1/1000 of the background halo, very accurate calibration and control of all systematic errors will be required. Intensity spikes from stellar diffraction are smoothed by broadband imaging, and reduced dramatically by rotation of the image during the 5 hour exposure. The next step is to test high-resolution adaptive correction and control of systematic errors in the field, which we plan with R.Q. Fugate at the Starfire Optical Range 3.5 m telescope. After successful experimental tests, the full scale instrument will be built for a survey at the new MMT and Magellan telescopes. Within a decade, it should be possible to build a two-telescope interferometer with "super AO", which should reach to Uranus and Neptune-like planets, thus providing an extraordinary tool to probe extra-solar planetary systems.

**acknowledgements:** This work was supported by grant F49620-94-1-0437 from the Air Force Office of Scientific Research to the Center for Astronomical Adaptive Optics at Steward Observatory.

## References

Angel, J.R.P.: 1994, *Nature* **368**, 203
Beckers, J.M.: 1993, *ARAA* **31**, 13
Bruns, D., Sandler, D., Martin, H., Brusa, G.: 1995, *Proc. SPIE* **2534**, 130
Fugate R.Q. et al.: 1994, *JOSA* **A11**, 310
Graves, J.E. et al.: 1994, in Active and Adaptive Optics, ESO Conference and Workshop Proc. No. 48, Garching, p. 47
Graves, J.E., Roddier, F., Roddier, C., Northcott, M.: 1995, in Proc. OSA/ESO Conference on Adaptive Optics, ESO, Garching, in press
Gray, P.M. et al.: 1995, *Proc. SPIE* **2534**, 2
Lena, P.: 1996, private communication
Mayor, M., Queloz, D.: 1996, these proceedings
Nakajima, T. et al.: 1996, *Nature* **378**, 463
Rousset et al.: 1994, in Active and Adaptive Optics, ESO Conference and Workshop Proc. No. 48, Garching, p. 65
Sandler, D., Stahl, S., Angel, J.R.P., Lloyd-Hart, M., McCarthy, D.: 1994, *JOSA* **A11**, 925
Sandler, D., Lloyd-Hart, M., Martinez, T., Gray, P., Angel, J.R.P., Barrett, T., Bruns, D., Stahl, S.: 1995, *Proc. SPIE* **2534**, 372
Stahl, S., Sandler, D.: 1995, *ApJ Letters* **454**, L153
Woolf, N.: 1996, these proceedings
Woolf, N., Angel, J.R.P.: 1995, in Adaptive Optics, 23, OSA Technical Digest Series (OSA, Washington, DC), p. 46.

# ACTIVE COOLING SYSTEMS

F. SCARAMUZZI
*CNR/IFSI and ENEA/ERG/FUS 00044 Frascati, Italy*

**Abstract.** Space cryogenics tries to solve the problems connected with cooling down components in space missions. Very low temperatures are required in particular to cool radiation detectors used in astrophysics. Up to now the cryogenic problems have been solved by using a reservoir containing a cryogen, say liquid helium, on board satellites. A new trend is now making progress, consisting in trying to "make the cold" on board: this could result in less weight and volume to put in orbit, and in longer time duration of the missions.
The first step to cool on board can be performed by means of passive radiation, a subject that has been addressed in another paper in this Workshop. Next step is "active cooling", which is the subject of this paper. Various techniques and associated problems are discussed, and the state of the art evaluated.

**Key words:** cryogenic techniques, active cooling

## 1. Introduction

The use of cryogenic techniques on space missions, and in particular on those dedicated to astrophysics, is now well established. Optical components must be cooled, in order to reduce thermal noise, with temperatures down to a few kelvins; radiation detectors may require temperatures as low as a few millikelvins, in order to reach high sensitivities. Thus, all the range of temperatures considered under the name of "cryogenics", i.e., from about 100 K to the millikelvin range, is of interest for the devices used in space science . This field is known presently under the name of "space cryogenics" (Scaramuzzi 1991): its main differences from general cryogenics are the typical space requirements of reliability and automatic operation, the ability to operate with small amounts of power, and in most of the instances the ability to function in the absence of gravity.
The technique preferred up to now to reach low temperatures consists in having on board of the satellite a cold bath, which goes normally under the name of a "cryogen". The most advanced cryogen, liquid helium, was used for the first time on the IRAS (InfraRed Astronomic Satellite) mission in 1983: it assured the temperature of 1.8 K to the IR detectors, plus some other ancillary cooling to the optics, for about 9 months. Since then two other missions have been launched, in which liquid helium is on board, and are still in operation: COBE (COsmic Background Experiment) and ISO (Infrared Space Observatory). In order to envisage the limitations of that technique, it is useful



C. Eiroa et al. (eds.),
*Infrared Space Interferometry: Astrophysics & the Study of Earth-Like Planets*, 255–265.

to consider some of the design parameters of such missions. Consider as an example ISO: roughly 2000 liters of liquid helium are necessary to ensure the operation of the mission for a period of about 2 years. This means that a volume of more than 2 cubic meters, a weight of about 2 tons (including the dewar) have to be launched in order to have a mission that is nevertheless limited in time.

The obvious alternative solution to the cryogens consists in producing the refrigeration on board. In this way, it is possible to plan longer missions, with smaller volumes and weights to be put in orbit. Of course, other problems arise, such as the power requirements, the reliability of the devices and the mechanical noise that is sometimes produced by the refrigerating circuits. This will be the subject of this presentation, with the aim of evaluating the state of the art of the field and the possible perspectives.

A first step in cooling a system on board a satellite can be made by sing "passive cooling", i.e. the cooling obtained by irradiating energy in the empty space. This topic has been the subject of a communication in this workshop, and will not be treated here. We will assume that it is possible to realize a passive refrigerator with a reasonable cooling power at roughly 70 K.

The development of "active coolers" has been a very active field in recent years. There is a topical conference, called the "International Cryocooler Conference", which is held every two years, with steadily increasing attendance and number of communications: it is a very good forum for this subject. The last meeting was the 8th, held in June 1994 in Vail, Colorado, USA. Its proceedings (Ross 1995) contain about 100 papers in about 1000 pages, and give an almost complete picture of the field. (Of course, not all of the papers refer to space cryogenics, but it can be seen that a substantial percentage of them are oriented in that direction.) Thus, it is hard in a short communication to give a thorough account of such an active field: necessarily, this will be a brief review, with the aim quoted before.

In chapter 2 an analysis of the relevant parameters for the coolers will be performed, in the light of the requirements of space applications: this will help in evaluating particular solutions that will be examined in the following. This exam will be structured into temperature intervals: in chapter 3 the coolers in the range 70 - 10 K will be considered; in chapter 4 the problem of reaching 4 K will be faced; in chapter 5 we will touch the issue of reaching temperatures lower than 4 K; and in chapter 6 the range of the millikelvin temperature will be addressed. In chapter 7 conclusions will be drawn.

## 2. Relevant parameters for coolers

**a.** One of the most important features to be considered for an active cooler to be operated in space is the performance of the thermodynamic cycle. This is relevant mostly for the evaluation of the amount of power at high temperature that is necessary in order to have the required cooling power at the operating temperature (high temperature can mean room temperature, or the temperature of the preceding stage in case of cascade refrigerators). The ratio of these two quantities is usually referred to (rather improperly) as the efficiency, and is expressed in watts (at high temperature) over watts (at low temperature): we will use this term in the present paper.
**b.** Reliability is a paramount feature for long missions. This parameter is very difficult to evaluate when durations of 5 to 10 years are considered, since testing for such long periods is practically impossible. Thus, evaluation of the probability of a failure becomes the only way for guessing the reliability. However, some generalizations can be made: for example, a solution with moving components is *a priori* less reliable than one without.
**c.** The elimination of mechanical noise is a stringent request in some missions and is clearly tied to the presence of moving components. It has to be noted, however, that there has been a substantial improvement in the noise performance of mechanical coolers, by adopting mechanically equilibrated solutions.
The three requirements described above often compete with each other: normally, techniques with good efficiency have poor noise and reliability performance, and viceversa. Thus, the problem has to be considered case by case, looking for the best compromise.
In addition to the above items, it is also convenient to consider the possibility of obtaining high performance (which means: very low temperatures, very low noise, and so on) not continuously, but for finite periods of time. This means that the apparatus will be performing at its best only for a certain percentage of the useful time: then, a "recycle" can be made, and the best conditions restored. This type of solution could allow advanced parameters, but with a limited duty cycle, in cases where the continuous solution would pose impractical requirements on the system.

## 3. Coolers in the range 70 - 10 K

We will consider three different families of refrigerators operating in this temperature interval:

**a.** Mechanical coolers. The most developed type is based on the Stirling cycle. The Stirling cycle is one of the most efficient in thermodynamics. On the other hand, these refrigerators have moving cryogenic components, plus the compressor, which can affect both the reliability and the noise level: however, important improvements have been made toward having long-life, low-vibration units. A large variety of prototypes have been made, with different number of stages, correspondingly different minimum temperatures, and with different cooling capacities. Here we will quote, just as examples, a single-stage unit, producing cooling powers of the order of 1 W at temperatures around 50 K, which has been space qualified (Horsley et al. 1995), and two-stage units, producing cooling powers of the same order at temperatures around 30 K, all with quite interesting performance.

**b.** Pulse tube refrigerators. These refrigerators, with a thermodynamic cycle not much dissimilar from Stirling, have the advantage of working without a piston or a displacer. Thus no moving cryogenic components are used, the only source of mechanical noise being the compressor: the mechanical noise level can thus be substantially reduced. The development of these coolers has been quite impressive in the last five years, reaching efficiencies not far from the Stirling refrigerators. There is a large variety of projects, with very interesting results, for single and multi-stage devices. Small single-stage coolers, giving about 1 W at 80 K, have been produced and space-qualified (Tward et al. 1995). Temperatures as low as 20 K have been obtained in double-stage coolers. A multi-stage cooler reaching temperatures lower than 4 K is reported as well (Matsubara and Gao 1995).

**c.** Sorption coolers. Absorption (chemisorption) and adsorption (physisorption) are used to make pumps and compressors in which no moving parts are required. In particular, sorption compressors are used as the drivers for a JT cooler. The compressor is a container filled with the active material, the ad(ab)sorbant: when heated, it desorbs the ad(ab)sorbed gas, thus generating high pressure; when cooled, it ad(ab)sorbs the gas, thus reducing the pressure. In Fig. 1 the scheme of a particularly simple cooler of this type is shown. The compressor C1 is heated and develops high pressure, which drives the gas through the two heat exchangers HE1 and HE2, where the gas is cooled before expanding in the JT valve, reaching the minimum temperature TL. The cold low pressure gas, after passing through HE2, is ad(ab)sorbed by the compressor C2, which is cooled by closing the heat switch HS2, in contact with the high temperature source, at the temperature TH. Heat exchanger HE1 is also in good thermal contact with the source at TH. More complex designs introduce valves in the flow circuit, in such a way that compressors C1 and C2 can reverse their functions,

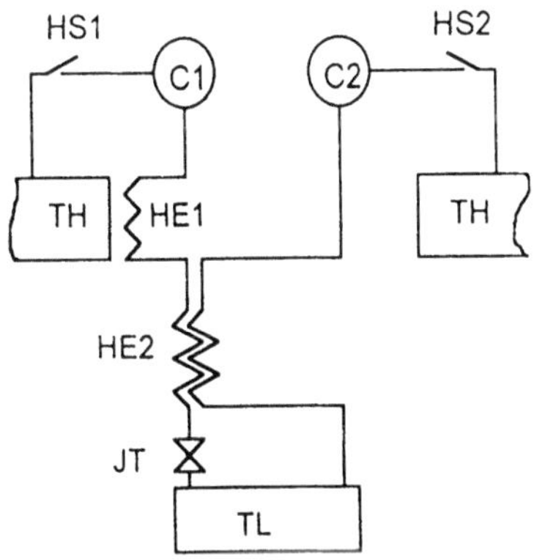


*Figure 1.* Schematic drawing of a single-shot sorption cooler.

the latter becoming the pressure driver. In this way they can go on indefinitely producing refrigeration. Doubling this circuit, and operating the two out of phase, permits smoothing out the operation of the refrigerator, by avoiding the effect of the dead times due to heating or cooling of the compressors. It is particularly interesting to quote, in this class of coolers, the refrigerators realized at JPL, using hydrogen as the circulating gas, and its absorption on metals to produce hydrides as the basic process. A prototype, using 70 K as the high temperature and reaching 10 K has been accomplished in a single shot version. The name of the project is BETSCE (Brilliant Eyes Ten-kelvin Sorption Cryocooler Experiment): it has been ground tested, and it is ready for space qualification (Bard et al. 1995). In another project, for a cryocooler to operate on board a stratospheric balloon for more than one month, a prototype functioning continuously between 70 and 25 K is presently being tested, providing 480 mW at 25 K, while consuming 216 W at room temperature and 1 W at 65 K (Wade 1996). These coolers are characterized by a very high degree of reliability and by the total absence of mechanical noise: the latter feature makes them particularly suitable for those experiments in which very weak signals have to be detected. On the other hand, the thermodynamic cycle is not very efficient.

Most of these coolers operate between room temperature and the design temperature. If a certain cooling capacity at intermediate temperatures is available, for example from a passive radiator, an improvement on their efficiency can be obtained, for example, by using a "cryocooler coldfinger interceptor" (Johnson and Ross 1995).

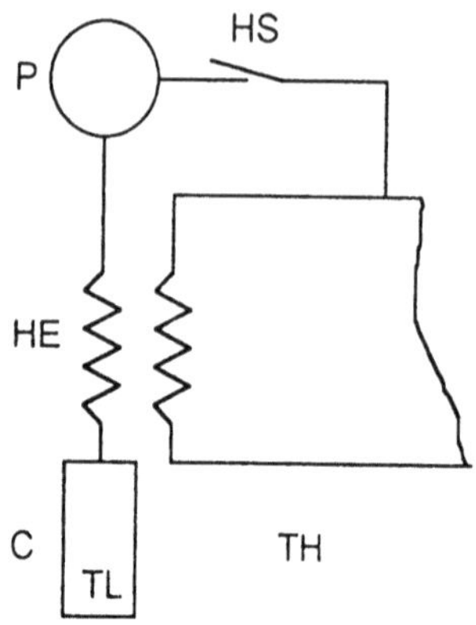


*Figure 2.* Schematic diagram of a single-shot refrigerator using an adsorption pump

## 4. How to reach 4 K

Reaching 4 K on board satellites can be considered an important step, since it means being able to liquefy $^4$He at a reasonable pressure. Lower temperatures can then be obtained by reducing the vapor pressure of liquid $^4$He: this will be addressed in next chapter.

The extension of mechanical or pulse tube refrigerators down to 4 K has been up to now considered impractical, since they use regenerators, and the heat capacity of known materials suitable as regenerators goes rapidly to zero with temperature. Nevertheless, new materials are being developed, with promising results. Presently, however, the most straightforward way, once a cooler producing temperatures of 30-10 K is available, consists in adding a JT stage, using $^4$He (or $^3$He) as the fluid, and thus liquefying it: this implies the addition of a compressor, and can provide continuous cooling at about 4 K. This has been done with success by coupling the JT stage to Stirling coolers: temperatures as low as 2.5 K have been reached (Orlowska et al. 1995).

Turning once more to the problem of eliminating moving components from active coolers, in order to increase their reliability and decrease the mechanical noise, it is possible to do so here too using a sorption cooler, and mounting it in cascade with a hydride absorption cooler (Wade 1996, Johnson and Ross 1995). At these low temperatures helium must be used as the circulating gas, and thus for the compressor it is necessary to use adsorption. Many materials have been investigated as potential adsorbants, among which one of the most commonly used in cryogenics is activated charcoal. This raises many problems. Onc of the most dramatic is the fact that the heat of adsorption is very large, much larger than the latent heat of evaporation, at least for the first couple of layers adsorbed. This feature is particularly serious, and,

together with the poor performance of JT cycles, leads to very low efficiencies. Designs and evaluations of such a technique exist (Chan et al. 1981, Tward 1985), but no real development has been pursued up to now. Nevertheless, if small cooling powers are required, this is a very interesting solution to be considered.
Worth quoting also are efforts based on the use of Adiabatic Demagnetization Refrigerators (ADR) in this temperature range. A recent result is reported, in which 27 mW are produced at 2 K in a single shot mode (167 s), starting from 10 K (Kashani et al. 1985).

## 5. Cooling below 4 K

In satellites with liquid $^4$He on board the operating temperature is necessarily lower than 2.17 K, the lambda point, since only using the peculiar properties of superfluid helium has it been possible to accomplish a reliable phase separator between liquid and vapor, able to function in the absence of gravity (the "porous plug"): normally the temperatures used are in the range 1.5-1.8 K. Here, however, we are concerned with making our "cold" on board and avoiding cryogens. For this range of temperatures we will assume that we are able to liquefy $^4$He, in a first approximation at pressures close to 1 bar, so that a temperature of 4 K is our starting point: in chapter 4 we have shown different ways to obtain this result. However, by reducing the vapor pressure in equilibrium with a liquid $^4$He bath, temperatures as low as 1 K can be reached; doing the same on a liquid $^3$He bath, it is possible to reach temperatures around 0.3 K. Thus a straightforward method to cool below 4 K, and down to 0.3 K, consists in having liquid helium baths on which to pump. In the absence of gravity, in order to solve phase separation problems, such baths can be constituted by the liquid sticking on a sponge-like structure, where gravity is replaced by Van der Waals forces (Kittel 1981).
Pumping on liquid helium baths can be accomplished once more with the help of adsorption. In Fig. 2 the scheme of a simplified single-shot sorption pump, say for $^4$He, is shown. The pump P contains the adsorbant, say activated charcoal, and is connected through a low conductivity pipe to the cold point, the cell C, passing through the heat exchanger HE, which is in good contact with the high temperature (TH) bath, say 4 K. Heating the pump, and keeping the heat switch HS open, allows the temperature of the adsorbant to rise, say at 40 K, so that $^4$He gas is desorbed: it takes place in HE liquefaction and the liquid is thus stored in cell C. At this point, the heater is shut down and the switch HS is closed, so that the pump P cools down and starts

pumping on the liquid, thus reducing its vapor pressure: the cell then cools down, reaching the minimum temperature TL, of about 1 K. If $^3$He is used instead, the minimum temperature reached is 0.3 K, but it is necessary to start from a lower temperature for TH, since the critical point of $^3$He is lower than 4 K: a $^4$He circuit such as the one described before can be used to perform this task, in cascade with the $^3$He cycle. The device described here is single-shot: once all the liquid has been evaporated, the cell heats up. However, the cycle can be repeated and duty cycles higher than 90% can be obtained. Coolers working on this principle have been made and operated with success on stratospheric balloons, where there is no need to worry about absence of gravity (Palumbo et al. 1994). However, experiments in the laboratory have proved that it can work in zero $g$ (Lounasmaa 1974). It is possible also to imagine continuous designs, working, as for the compressors, with two or more similar circuits operating out of phase with each other .

## 6. Lower and lower temperatures

Temperatures lower than 0.3 K would be useful for some radiation detectors in space. It is possible to reach these temperatures in more than one way. We will quote here three of them:

**a.** The method used on the ground for continuous cooling down to a few millikelvins is the dilution refrigerator (Lounasmaa 1974): when a mixture of liquid $^3$He and $^4$He is cooled below 0.8 K, a phase separation occurs and the liquid separates in a $^3$He-rich component and a $^4$He-rich component, which, in the presence of gravity, sit one upon the other, the $^3$He-rich above, being the lighter. Forcing $^3$He to pass from the $^3$He-rich to the $^4$He-rich solution generates cooling. This already shows the difficulty of operating such a circuit at zero $g$, since gravity plays an important role. It has been suggested that gravity could be replaced by an electric field gradient, and laboratory tests have proved that this is feasible (Israelsson et al. 1988). Recently, however, a clever solution has been suggested, in which mixing and dilution take place dynamically in a cryostat into which cold $^3$He and $^4$He gases are sent, and then are recovered, after performing their task, as a mixture. This has been proved to work well independently of gravity, and is proposed as a possible solution for future missions (Benoit and Pujol 1994, Benoit et al. 1994). Naturally, in space the mixture cannot be used again. Thus, the duration of the experiment will be limited by the amount of gas stored: a reasonable figure for the rate of consumption is 2 liters NTP of $^3$He per day for a refrigeration of about 100 nanowatts at about 100 millikelvins.

**b.** Cooling by an ADR is also possible and can be made without worrying about absence of gravity. A brilliant single-shot prototype has been developed, in which part of the demagnetization process is used to compensate the heat inlets during the experiment, thus resulting in a constant temperature of about 100 mK all along the run, lasting more than 8 hours (Timble et al. 1990, Bernstein 1991). The system has been space qualified with a rocket launch and has been used on stratospheric balloons.

**c.** A third method has been proposed recently, which is based on the use of a normal-insulator-superconductor tunnel junction (Nahum wt al. 1994): forcing the hottest electrons to pass from the normal metal to the superconductor produces cooling as low as 100 mK starting from 0.3 K. The method is fascinating, also because it only needs the application of a small current to the junction. Conversely, the main disadvantage is the extremely low cooling power developed by a single junction, which is in the order of femtowatts: attempts to produce many junctions operating in parallel, and thus multiplying the cooling power by a significant figure, are in course.

## 7. Conclusions

It is worth trying to get a feeling on the performance of the different active coolers described in the preceding chapters, by means of the "efficiency", as we have defined it at the beginning of this paper, i.e., the ratio between the power to be spent at high temperature and the cooling power obtained at the operation temperature. This is not an easy task, since not all the papers published on this subject contain quantitative information about the coolers' peformance.

Nevertheless, in Fig. 3 a plot is presented, in which some data taken from the literature are reported, as a function of temperature - the high temperature is the room temperature. On the same graph the performance of a perfect Carnot refrigerator working with a high temperature of 300 K is also reported, as a continuous line. It is obvious that the difficulty increases steeply when approaching very low temperatures.
It is clear that it begins to be possible to get rid of cryogens on board satellites. For the range of temperatures down to about 30 K the use of mechanical refrigerators is already feasible, still keeping in mind the problems posed by reliability and mechanical noise. The solution for the cryostats able to reach lower temperatures begins to be at hand, but more effort is needed.
An interesting field is the development of the chain of refrigerators

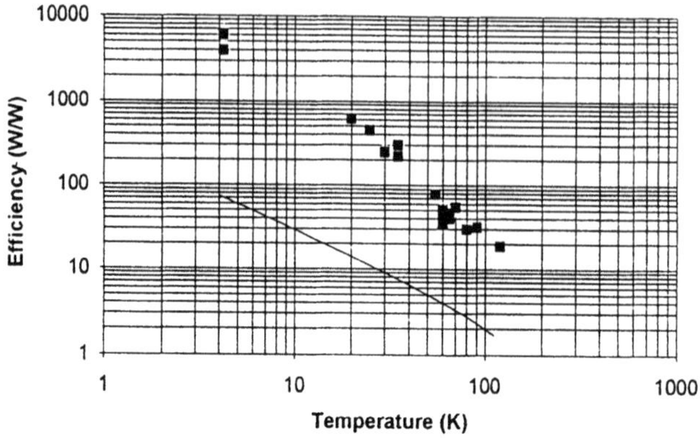


*Figure 3.* The efficiency of cryocoolers as a function of temperature. The continuous line is the performance expected for a Carnot cycle.

based on the use of sorption cooling, which is the only way to obtain very high reliability and substantial absence of mechanical noise. Here the state of art is quite good for temperatures down to 10 K in a single-shot mode. The accomplishment of continuous systems seems very promising, together with the development of a stage able to go down to 4 K. This, as mentioned above, poses very tough problems as far as power consumption is concerned, but it is worth proceeding on this line, considering that, mostly for astrophysical measurements, the requirement of very low temperatures for periods of many years, coupled with the absence of mechanical noise, can become important.

## References

Bard, S., Wu, J., Karlmann, P., Cowgill, P., Mirate, C., Rodriguez, J.: 1995, *Cryocoolers* **8 (ref.2)**, 609

Benoit, A., Bradshaw, T.,, Jewell, C., Maciaszek, T., Orlowska, A., Pujol, S.: 1994, in 5th European Symposium on Environmental Control Systems, Friedrichschafen, Germany

Benoit, A., Pujol, S.: 1994, *Cryogenics* **34**, 421

Bernstein, G., Labov, S., Landis, D., Madden, N., Miller, I., Silver, E., Richards, P.: 1991, *Cryogenics* **31**, 99

Chan, C.K., Tward, E., Elleman, D.D.: 1981, *Advances in Cryogenic Engineering* **27**, 735

*Cryocoolers* **8 (ref.2)**, 647 (1995, Chapter Magnetic Refrigerators and Low-Temperature Regenerators)

Horsley, W.J., Hicks, E.F., Kiehl, W.C., Simmons, D.W., Taylor, D.J., Wells, E.E., Wells, J.A.: 1995, *Cryocoolers* **8 (ref.2)**, 23

Israelsson, U.E., Jackson, H.W., Petrac, D.: 1988, *Cryogenics* **28**, 120

Johnson, D.L., Ross, R.G. Jr.: 1995, *Cryocoolers* **8 (ref.2)**, 709

Kashani, A., Helvensteijn, B.P.M., McCormack, F.J., Spivak, A.L., Kittel, P.: 1995, *Cryocoolers* **8 (ref.2)**, 637

Kittel, P.: 1981, *Physica* **108B**, 1115

Lounasmaa, O.V.: 1994, Experimental Principles and Methods Below 1 K, Academic Press London and New York

Matsubara, Y., Gao, J.L.: 1995, *Cryocoolers* **8 (ref.2)**, 345

Nahum, M., Eiles, T.M., Martinis, J.M.: 1994, *Appl. Phys. Lett.* **65 (24)**, 3123

Orlowska, A.H., Bradshaw, T.W., Hieatt, J.: 1995, *Cryocoolers* **8 (ref.2)**, 517

Palumbo, P., Aquilini, E., Cardoni, P., de Bernardis, P,. Masi, S., Scaramuzzi, F.: 1994, *Cryogenics* **34**, 1001

Ross, R.G. Jr. (Editor): 1995, Cryocoolers 8, Proc. 8th International Cryocooler Conference, Plenum Press

Scaramuzzi, F.: 1991, in Infrared and Submillimetric Astronomy from Space, (Proc. International School of Space Science), F. Melchiorri and P. Encrenaz (Eds.), p. 495

Timble, P.T., Bernstein, G.M., Richards, P.L.: 1990, *Cryogenics* **30**, 271

Tward, E.: 1985, Proc. 3rd Cryocooler Conference, NBS Special Publication 698, R. Radebaugh, B. Louie and S, McCarthy (Eds.), p. 220

Tward, E., Chan, C.K., Raab, J., Orsini, R., Jaco, C., Petach, M.: 1995, *Cryocoolers* **8 (ref.2)**, 329

Wade, L.A.: 1996, private communication

## 8. Questions

*D. Tytler* : Are there problems with temperature stabilization ?
*F. Scaramuzzi* : As far as I know, there are no serious problems. I can quote the $^3$He cryostat that we sent on a stratospheric balloon which performed very well for many hours. Dr. Hawarden quotes the figure of $10^{-5}$ stability in the experiment BETSCE at JPL, where a temperature of 10 K was kept for 20 minutes.

*B.W. Jones* : Your presentation was based on passive cooling to 70 K. What are the implication of passive cooling to, say, 20K?
*F. Scaramuzzi* : My choice of 70 K, as a working hypothesis for passive cooling as a first stage in a cascade of refrigerators, took into account the fact that the "cooling power" of such a stage is a strong function of temperature ($\sim$ T$^4$). Thus, at 70 K, the cooling power is 1 W m$^{-2}$ of radiation: this is still a challenging condition. At 20 K it would be of the order mW, hardly useful in any cryogenic system.
*T. Hawarden* : At 20 K, a blackbody field carries only 9 mW m$^{-2}$ - a big radiator for very little cooling. Below 20 K, there is little to offer.

# SPACE INTERFEROMETRY MISSION

M. SHAO, S. UNWIN, A. BODEN and D. VAN BUREN
*Jet Propulsion Laboratory, California Institute of Technology*

S. KULKARNI
*California Institute of Technology*

**Abstract.** The Space Interferometry Mission (SIM) will be a 10-m optical interferometer in Earth orbit. SIM will enable a wide variety of science programs in Galactic and extra-galactic astronomy, including the calibration of cosmic distance indicators, and will address questions related to the origins and prevalence of planetary systems. SIM is designed primarily as an astrometric mission, providing high-throughput wide-angle astrometry, with an expected noise floor of about 4 $\mu$as for bright targets, and about 1 $\mu$as over small angles. Accurate parallaxes and proper motions will be available for a large number of stars of various types, providing data for fundamental programs in the stellar dynamics of star clusters and our Galaxy as a whole. SIM will have seven light-collecting siderostats (effective diameters 0.3m) spaced along the main boom, to allow synthesis imaging with a resolution of 10 mas, by rotating the spacecraft about the line of sight. It is NASA's first space science interferometer, and in addition to enabling a strong astronomical science program, it is designed to demonstrate several key technologies needed for future missions in NASA's Origins Program, such as the Next Generation Space Telescope (NGST) and Terrestrial Planet Finder (TPF). Interferometric nulling will be used by TPF to provide coronagraphic-type extinction of a bright star, allowing detection of faint companions. SIM will implement an achromatic null to a level of $10^{-4}$ and will study dust disks around main-sequence stars, and image the disks around young stellar objects.

**Key words:** space, interferometry, astrometry, imaging

## 1. Instrument and Mission

SIM is a 10 meter baseline visible-wavelength interferometer capable of global astrometric measurements with 4 $\mu$as accuracy and narrow-angle astrometric measurements on the order of 1 $\mu$as (Table 1 summarizes the principal instrument and mission parameters). In the course of its approximately 5-year mission, it will determine the distances and velocities of stars to unprecedented accuracy. There are many basic questions in galactic astrophysics which can be addressed with these new fundamental data on the distribution and kinematics of stars.

As well as performing unique science programs, SIM will serve as a technology pathfinder for a line of future astrophysics missions. The baseline concept for SIM is derived from the Orbiting Stellar Interferometer (OSI) concept developed at the Jet Propulsion Laboratory (Rayman et al. 1992). The SIM architecture (Fig. 1) represents a



*C. Eiroa et al. (eds.),*
*Infrared Space Interferometry: Astrophysics & the Study of Earth-Like Planets,* 267–278.

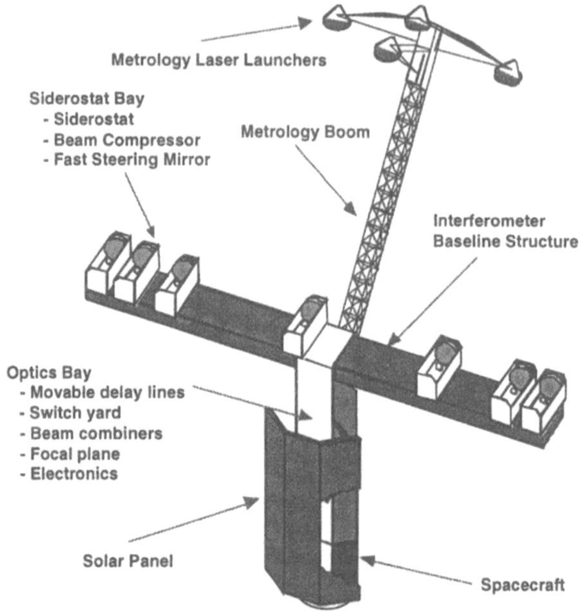


*Figure 1.* Overview of the SIM spacecraft.

paradigm shift from the era of rigid monolithic telescopes. Instead, it uses active optics distributed across a lightweight flexible structure which maximizes resolution per unit mass carried into orbit. This is a necessary step forward for future low-cost large-aperture space astrophysics missions.

SIM leverages on a number of past, current and upcoming activities to significantly reduce mission cost and risk. It benefits from experience with ground-based interferometers, such as the Mark III systems at Mt. Wilson, and the currently operating (dual star) Palomar Testbed Interferometer, which have provided some of the first interferometry technology demonstrations using starlight. The Interferometer Technology Program (ITP) at JPL focuses on technologies needed for space. Ground testbeds, like the Micro-Precision Interferometer (MPI) testbed, are demonstrating end-to-end operation of an interferometer on a large flexible, flight-like structure. Advances in the development of integrated modeling tools are enabling high-fidelity prediction of interactions between SIM's subsystems, and hence predict its performance on orbit.

The SIM beam combiners comprise three independent units, plus one spare, which are fed by delay lines to equalize the paths and allow

Table I. SIM Instrument and Mission Parameters

| | |
|---|---|
| **Instrument** | |
| Baseline | 10 meters |
| Wavelength range | 400 – 1000 nm |
| Spectral resolution ($\Delta\lambda/\lambda$) | 100 |
| No. siderostats | 7 |
| No. delay lines | 4 |
| No. beam combiners | 4 |
| Aperture diameter | 0.30 meters |
| Astrometric FOV | 10deg x 10deg |
| Imaging FOV | 2.4 x 0.4 arcsec |
| Detector | Si CCD and APD |
| **Mission/Flight System** | |
| Orbit | 900 km Sun-synch. |
| Orbit Period | 103 min |
| Launch Vehicle | Delta-II 7920 |
| Mass | 1800 kg (79% margin) |
| Power | 1030 W (63% margin) |
| Lifetime | 5 years |
| **Science Performance** | |
| Astrometry (wide-angle) | 4 $\mu$as (1-$\sigma$) on 20 mag in 10 hrs |
| Astrometry (narrow-angle) | 1 $\mu$as on 15 mag in 3 hrs |
| Imaging resolution | 10 mas @ 500 nm |
| Number of baselines | 20 |
| Imaging sensitivity | |
| Point source | 25 mag in 1 hr |
| Extended source | 20 mag/pix in 1 hr |
| Nulling | $10^{-4}$ |
| **Project Timeline (tentative)** | |
| Technical Design Review | Oct 1996 |
| Prelim. Non-Advocate Rev. | July 1997 |
| Non-Advocate Review | Mar 1999 |
| Preliminary Design Review | Sep 1999 |
| Critical Design Review | July 2000 |
| Indep. Assessment Review | Nov 2001 |
| Launch | Dec 2002 |

white-light fringe detection. Two combiners are used as 'guide interferometers', taking light from pairs of siderostats which view bright guide stars. Delay measurements from these guide interferometers are fed forward to the 'science interferometer', which performs all astrometric and imaging fringe measurements. Through switchyard mirrors located near the center of the siderostat boom, any combination of siderostats can be routed into the beam combiners, providing a total of 20 baselines for imaging, and flexibility in the event of component failure. There are two detectors in each beam combiner: a CCD detector records a low-resolution dispersed fringe, and a fiber-fed APD records the white-light fringe, using light from the second half of the beam splitter element.

## 2. SIM Science Objectives

The science program for SIM is very broad, and involves many areas of astrophysics. Some principal science objectives include:

- Calibration of stellar and 'standard candle' luminosities used in cosmic distance scale and globular cluster measurements
- Precise studies of galactic dynamics including rotation curves and halo object motions
- Obtaining distances to nearby spiral galaxies via rotational parallaxes
- Searching for planets down to an Earth mass
- Measuring the apparent astrometric motion in gravitational microlensing events
- 10 mas resolution synthesis imaging of circumstellar disks around young stellar objects (YSOs)
- Imaging the narrow-line regions of active galactic nuclei (AGN)
- Demonstrating nulling by imaging the dust around $\beta$ Pictoris to within ~0.1 AU of the star, to search for gaps or structure that may be caused by the presence of planets

In this Section, we summarize the science topics in tabular form, in the observational categories of microarcsecond astrometry, and milliarcsecond imaging. In the following Section we present a brief discussion of some selected topics in more detail.

### 2.1. Microarcsecond Astrometry

The majority of science topics addressed by SIM will use its microarcsecond astrometric capabilities. SIM will measure parallaxes and proper motions about a factor of 1000 better than can be achieved from the ground, and a factor of about 250 times better than the best available astrometric catalog, from Hipparcos (e.g. Lindegren 1995).

Wide-angle astrometry (over the $10 \times 10$ deg siderostat field of view) with SIM is expected to have a 'noise floor', (representing of uncalibrated systematic errors) of about 4 $\mu$as $(1-\sigma)$ for single measurements of bright stars. By observing a reference grid over the whole sky repeatedly during the mission, astrometry over the whole sky should achieve at

Table II. SIM Astrometric Program Highlights

| Science Topic | Objective | Target Description | Magnitude Range | Comments |
|---|---|---|---|---|
| Dark Companions | Astrometric Detections of Planets, BD, WD, MACHO Candidates | Nearby Stars, LMC, SMC, and Bulge Field Stars, Possible MACHO Companions | Planet astrometry: 2 –12 Bulge MACHOS: 15 – 18; LMC: 18 - 22 | Detect Gas Giants; BD and WD companions, MACHO Astrometric Effects |
| Luminosity Calibration | Calibrate/Constrain Astrophysical Models | MS Stars, Cepheids, RR Lyrae, PN, Nova/CV | MS: 4 – 12 Cep: 7.5 – 11 PN: 10 – 18 | Calibrate MS Stars, Standard Candles, (O (1%) range error) |
| Rotational Parallaxes | Single-step Distances to Nearby Spiral Galaxies | Spiral Galaxies to 5 Mpc | M31: 16 – 18 M81: 18 – 20 | M31 Distance to ~1 % |
| Binary Systems | Mass Distributions of Binary Constituents | MS Binaries, CV (White Dwarfs), XRB (Neutron Stars, Black Holes) | MS: 6 – 12 CV: 12 – 18 | Unequivocal Mass Distributions |
| Globular Clusters | Calibrate Cluster Distances/Ages, Cluster Dynamics | Globular Clusters (47 Tuc, w Cen, M15, etc.) | 6 – 14 | Cluster Age Determination, Dynamics (Crowded Field Operation) |
| Open Clusters | 3-D positions and 3-D velocities | F5 - M0 Dwarfs in Nearby Clusters | 14 – 18 | Measure internal motions to 1% |
| Galactic Structure | Rotation Curve Spiral Arm Halo Kinematics | Early Main Sequence at 2–20 kpc, Halo K Giants | 10 – 18 | Spiral Arm Densities, Dark Matter Distribution |
| AGN | BLR Fluctuations, Transverse Peculiar | AGN Within 100 Mpc | 13 – 20 | Correlate Astrometric and |

Table III. SIM Imaging Science Summary

| **Target Classification** | **Objective** | **Typical Surface Bright (mag $arcsec^{-2}$)** | **Total Int Time (hrs)** |
|---|---|---|---|
| Globular Clusters | Core Resolution, Dynamics | 8 – 12 | 6 – 24 |
| Main Sequence ($\beta$ Pic) nulling | Superplanets, Exo-Zodiacal Emission | 12 – 15 | 24 – 36 |
| YSO Disks (GM Auriga) nulling | Resolve Disk Densities and Features | 12 | 24 – 30 |
| AGN NLR (NGC 2110) | Resolve NLR Emission Line Features | O($10^4$) photons $m^{-2}s^{-1}arcsec^{-2}$ | 24 – 30 |
| Jets in AGN | Counterparts to radio jets | 17 – 21 | > 48 |
| Distant Galaxies | Early Galaxy Morphology | 16 – 18 | 24 – 96 |

least this accuracy. Certain science programs - such as searching near-by stars for evidence of planets down to masses approaching an Earth mass - will use bright reference stars within about 1 deg to reduce the astrometric error to around 1 $\mu$as. For faint objects, photon counting statistics contribute to the noise. Fig. 2 shows the expected wide-angle astrometric accuracy as a function of stellar magnitude and integration time, assuming the 4$\mu$as noise floor, 0.3m siderostat apertures, and silver-coated optics. Table 2 summarizes the astrometric program which SIM will undertake; specific targets or target classes are identified.

### 2.2. Synthesis Imaging

SIM is also an imaging mission. With a resolution of 10 mas in the optical - exceeding that of HST by a factor of four - it represents a logical next step in space-based high-resolution imaging. Future high-resolution instruments, such as TPF, will use interferometry with 'dilute' apertures. By combining different siderostat pairs, and rotating the spacecraft about the line of sight, the instrument samples the Fourier transform (u,v) plane, from which the image brightness distribution can be deduced. Sensitivity depends on how extended the target is. For a target occupying 10 pixels (each 10 $\times$ 10 mas), the surface brightness

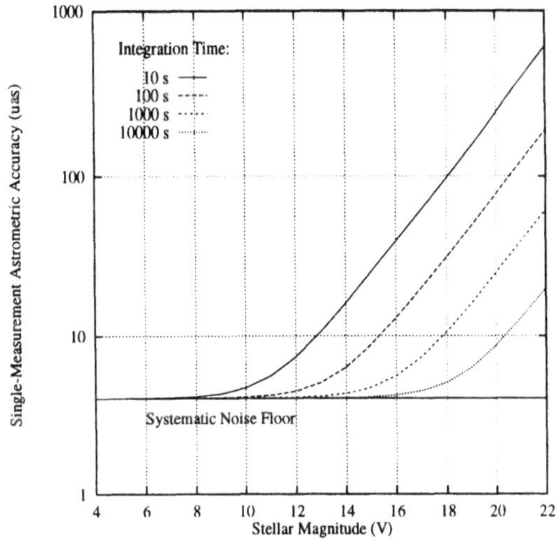


*Figure 2.* SIM Single-measurement astrometric performance.

sensitivity (SNR = 10) for a 1000 s observation is 13 mag arcsec$^{-2}$. Science projects using SIM's imaging capability are given in Table 3.

SIM will demonstrate the use of a nulling beam combiner. This will be implemented via a polarization flip in the beam combiner, achieving an achromatic null to about $10^{-4}$. This capability will allow astrometry of faint objects near to bright targets, and imaging of extended regions around bright stars.

## 3. SIM Origins Science

The science program for SIM is very broad, as Tables 2 and 3 indicate. As an element of NASA's Origins Program, SIM is an important precursor for future missions, such as NGST and TPF, both in technology and beginning to explore some of the science areas which those missions will address in detail. In this Section, we briefly outline some of the programs which relate to the Origins goals.

**Planets around Nearby Stars.** Which nearby stars have planets? What are those planets like? How pervasive are planets and planetary systems around nearby stars? What statistical information can be gained about the formation of these systems, and what can we infer about how these systems are formed? These are some of the questions which SIM will begin to address.

There have been significant recent developments in the nascent field of extra-solar planetary research. Companions with masses in the planetary range of 1 to 10 Jupiter masses have been found around nearby stars (from radial velocity measurements): 51 Peg (Mayor and Queloz 1995) and 70 Vir (Marcy and Butler 1996). Equally significant is the direct detection of a genuine brown dwarf, the companion of the near-by star Gliese 229 (Nakajima et al. 1995). Spectroscopy (Oppenheimer et al. 1995) reveals strong methane absorption bands. Such bands are not seen in other stars, but are common to large giant planets (e.g. Jupiter). Planets have also been detected orbiting pulsars (via pulsar timing).

The extrasolar planet candidates discovered to date are not easily explained with the standard paradigm for planet formation, derived from our solar system, since all these extra-solar planets are 'peculiar' in some way. To make planetary science into a comparative field we need more data, specifically a census of planetary bodies around stars in the solar neighborhood. SIM will contribute in a unique manner to this census. Its astrometric precision coupled with high throughput enable us to search for rocky planets around the nearest few stars and intermediate-mass planets (Uranus) over the nearest 1000 stars. None of these bodies are accessible to ongoing or future (ground-based) radial velocity studies.

The astrometric detection of dark companions to luminous primaries is achieved by measuring the transverse reflex motion of the primary as it orbits the common center of mass of the system. Two dimensional astrometric information uniquely defines the orbital inclination of the companion, and a distance uniquely defines the stellar mass, thereby uniquely defining the mass of the companion. Astrometric companion detection is done via narrow-angle astrometry, where SIM's performance will be between 0.6 and 1 $\mu$as.

In searching for dark companions to stars nearer than 100 pc, SIM will be sensitive to reflex motions of amplitude greater than 0.02 solar radii and periods shorter than twice the mission length (10 years). This corresponds to planets with masses down to 1% of Jupiter orbiting solar-mass stars, so SIM will easily detect gas giant and brown dwarf-type companions around a very large number of candidate stars – enough to establish highly significant mass distributions. Detection of Earth-mass planets is one of the most challenging tasks for SIM. With an estimated narrow-angle astrometric accuracy of 0.6 - 1$\mu$as astrometric performance, SIM is capable of detecting Earth-mass companions around the nearest 12 - 90 stars (as extracted from the Gliese catalog, Shao 1996). This program has important implications for the Terrestrial Planet Finder mission.

**Characterizing Brown Dwarfs.** What is the frequency of brown dwarf companions? What is the nature of the dark companions observed thus far, and surely to be discovered in the near future? What is the frequency of large gas planets?

SIM will be able to study brown dwarfs and other sub-stellar companions to bright stars. In addition to detecting reflex motion (the signature of a brown dwarf is much larger than that of a planet), SIM will be able to detect light from brown dwarfs directly.

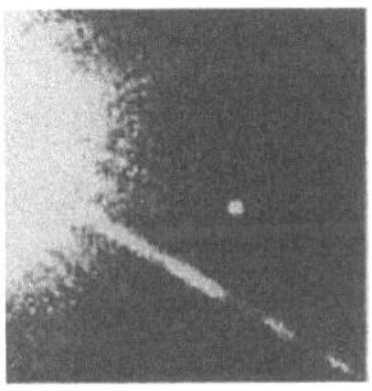

*Figure 3.* HST Image of the Brown Dwarf GL229B (Nakajima, Kulkarni, Durrance and Golimowski. Courtesy STScI and NASA).

SIM's nulling capability is the only way to directly detect the majority of brown dwarfs inferred from indirect (radial velocity or astrometric perturbations) techniques. In studies of such systems, SIM and HST will play complementary roles – SIM will enable us to explore and study brown dwarf companions with orbital separations less than 1 arcsec, while HST can image systems at larger separations (see Fig. 3). Using Gliese 229B as a prototype, the absolute magnitude in the range 0.8 - 1.0 $\mu$m is between 17.5 and 15 (Nakajima et al. 1995). SIM need only achieve a contrast of $10^4$:1 to image and measure the flux from Gliese 229B-like objects. It will be interesting to see how the spectra of brown dwarfs depend on effective temperature. Gliese 229B with Teff = 900 K peaks at 1 micron (strong methane absorption bands dominate) whereas Jupiter ($T_{\rm eff}$ = 120K) peaks at wavelengths longer than 10 $\mu$m (the true emitted spectrum) SIM will determine the optical spectrum of brown dwarfs at 1% spectral resolution. SIM can synthesize images of such compact objects efficiently; an image can be synthesized in roughly six hours.

**Stellar Debris Disks.** What is the nature and distribution of circumstellar material around main sequence and pre- main sequence stars? What are the implications for the formation of stars and planetary systems? Can the presence of planets be inferred from structure in these disks?

The IRAS satellite revealed that many main-sequence stars are surrounded by disks of circumstellar dust, with stars such as Vega and $\beta$ Pictoris having up to $10^4$ more material than in our solar system. The exact distribution of the dust in these star systems is poorly known, but the R-band surface brightness of $\beta$ Pic (Golimowski et al. 1993) is consistent with there being approximately $10^4$ more material in orbit than around the Sun, with $\beta$ Pic being 10 times brighter than the Sun

SIM can measure the total brightness of the $\beta$ Pic disk and of other systems within a range of distances from the central star that is critical for planet formation. Toward $\beta$ Pic, the nulling system will measure the visible exo-zodiacal emission at distances from about 0.2 AU to about 3.5 AU. This is just the range of distances where the Terrestrial Planet Finder will be searching for planets, so that expanding our knowledge of the inner zodiacal clouds of even a few stars is very important. The minimum data that SIM will obtain are total flux measurements of the exo-zodiacal cloud toward dozens of $\beta$ Pic analogs. By combining nulling with rotational synthesis, and also by combining different wavelengths, SIM will generate u-v coverage sufficient to search for structures such as gaps or density enhancements due to possible presence of planets.

**Imaging YSO disks.** Many young stellar objects (YSOs) – and T Tauri stars in particular – are accompanied by optically thick circumstellar disks. Perhaps 50% of YSOs exhibit an infrared excess indicative of the reprocessing of visible radiation by these disks, and HST/WFPC2 has resolved several of these disks (e.g. Orion Nebula, by O'Dell and Wen 1994; HH30 by Stapelfedt 1996). Circumstellar disks are thought to be critical to star formation and to be the progenitors of planetary systems, but their physics is poorly understood (Bodenheimer 1995).

Koerner et al. have imaged the disks around several T Tauri stars in millimeter wave including classical T Tauri star GM Auriga (Koerner et al. 1995a,b). This star has a large circumstellar disk extending out to 1000 AU with a mass of roughly 0.1 solar masses. This same disk has been imaged with HST/WFPC2 (Stapelfedt 1996). Surface photometry as close as 200 mas to the central star measures V-band surface brightness from 14.5 to 17.5 arcsec$^{-2}$, sufficient for imaging with SIM even with low contrast. A 10 mas resolution image of a small portion of the disk with a minimum SNR of 10 could be synthesized roughly 30 hours, resolving features as small as 1.4 AU. This resolution may be sufficient to detect a central 'clearing' if such a clearing - perhaps indicative of planet formation in the disk- exists.

**The Cepheid Distance Scale.** Establishing the distance scale is a fundamental problem in cosmology. Current estimates of the Hubble parameter are in the range of 45 - 90 km s$^{-1}$ Mpc$^{-1}$, large enough to strongly affect models for the Universe's evolution and force some rather uncomfortable potential discrepancies with inferred globular cluster ages (Chaboyer et al. 1996). Of crucial importance in setting the distance scale is an accurate calibration of Cepheid luminosities.

SIM will improve this situation directly, by measuring precise parallaxes of Galactic Cepheids, improving extinction estimates for the nearest Cepheids, and directly determining the distance to M31 and its Cepheid population. By directly measuring the distances to galactic Cepheids by trigonometric parallax as accurately as 0.5%, their luminosities will become known to 1%. This astrometric program is quite simple, and observing time modest, involving 10 measurements of 50 stars within 5 kpc.

Once astrometric distances are available at such precision, extinction along the line of sight to Cepheids becomes the dominant source of luminosity error in the P-L diagram. SIM can address this problem as well. If one simultaneously determines the flux, spectrum, and angular size to a high accuracy, then the extinction is readily found. By scanning the fringe null in delay, the angular diameters (expected to be a few mas) of the brightest and closest Cepheids can be determined to 0.5 - 1%.

**Rotational Parallax Distances.** SIM can measure the distances to nearby spiral galaxies directly, using rotational parallax, giving an extragalactic calibration of the Cepheid distance scale. HI or stellar observations yield the radial velocity field. SIM measures the transverse velocity field, averaging over many stars to reduce the effects of non-circular velocities. The scale factor between these measurements is just the distance to the galaxy - from the Earth's orbit diameter to a galaxy distance in a single step. For the nearest spiral galaxy, M31 at a distance of 0.77 Mpc, SIM can measure the associated proper motion with 3% errors for a single star, or about 1% using ~25 stars. About 2 days of observing time are required. This technique is useful out to at least M81 (5 Mpc).

## 4. SIM Infrared Capability

The SIM project is evaluating the possibility of extending the wavelength coverage into the near IR (but not beyond about 2 $\mu$m, since none of the optics are cooled). This would allow astrometry of stars which are obscured by dust, which is an area of particular interest because star (and planet) forming regions are often dusty environments. Kinematics of stars at the Galactic center is another promising near-IR project for SIM. With nulling, SIM could directly detect planets in the 51 Peg class. Such planets should have spectra which peak between 1 - 2 $\mu$m, with a star/planet ratio $\leq 10000$. A NICMOS-3 detector, cooled to around 80K, would have a read noise of $\sim$15 $e^-$ (multiple reads), sufficient to provide photon-noise limited performance for all but the faintest (>20 mag) stars. Detector cooling would be via a sorption cooler or turbo-Brayton refrigerator, providing low vibration and modest power requirements.

**Acknowledgements:** This work was performed at the Jet Propulsion Laboratory, California Institute of Technology, under contract to the National Aeronautics and Space Administration.

## References

Bodenheimer, P.: 1995, *ARAA* **33**, 199
Chaboyer, B., Demarque, P., Sarajedini, A.: 1996, *ApJ* **459**, 558
Golimowski, D.A., Durrance, S.T., Clampin, M.: 1993, *ApJ* **411**, L41
Koerner, D.W., Chandler, C.J., Sargent, A.I.: 1995a, *ApJ* **452**, L69
Koerner, D.W., Sargent, A.I.: 1995b, *AJ* **109**, 2138
Lindegren, L.: 1995, in IAU Symposium No. 166, E. Hog, P.K. Seidelman (Eds.), Dordrecht: Kluwer, 55
Marcy, G.W., Butler, R.P.: 1996, *ApJ* **464**, L147
Mayor, M., Queloz, D.: 1995, *Nature* **378**, 355
Nakajima, T., Oppenheimer, B.R., Kulkarni, S.R., Golimowski, D. A., Matthews, K., Durrance, S. T.: 1995, *Nature* **378**, 463
O'Dell, C.R., Wen, Z.: 1994, *ApJ* **436**, 194
Oppenheimer, B.R., Kulkarni, S.R., Matthews, K., Nakajima, T.: 1995, *Science* **270**, 1478
Rayman, M. et al.: 1992, FY92 Study Team Progress Report, JPL internal document D-10374.
Shao, M.: 1996, in NASA Conference Detection of Earth-like Planets Around Other Stars, H. Thronson (Ed.)
Stapelfedt, K.R.: 1996, personal communication.

# THE INFRARED IMAGING SURVEYOR (IRIS) PROJECT

H. SHIBAI
*The Institute of Space and Astronautical Science (ISAS)*
*Yoshino-dai 3-1-1, Sagamihara, Kanagawa 229, Japan*

**Abstract.** InfraRed Imaging Surveyor (IRIS) is the first Japanese satellite project dedicated to infrared astronomy. IRIS has a 70 cm telescope cooled down to 7 K by using superfluid helium assisted by two-stage Stirling coolers. The expected hold time of the super-fluid helium is one year. Two focal plane instruments are planned; the InfraRed Camera (IRC) and the Far-Infrared Surveyor (FIS). The total spectral coverage is 2 to 200 microns. The major scientific objectives are to investigate birth and evolution of galaxies in the early universe by survey of young normal galaxies and starburst galaxies.

**Key words:** infrared satellite, infrared astronomy, infrared survey, galaxies, cosmology, cooled telescope

IRIS is a Japanese space project solely dedicated to infrared astronomy promoted by the Institute of Space and Astronautical Science (ISAS) under collaboration with scientists of universities and institutions in Japan. The IRIS project is expected to start in 1997 and to be launched by an M-V rocket in 2002.

## 1. Scientific Objectives

It is considered as the most important target of current astronomy to know how galaxies formed and how they have evolved. Astronomers have already found exotic objects, such as AGN's, Seyfert galaxies, QSO's, which may have played important roles in the early universe. However, we have only limited information about normal galaxies and starburst galaxies in the early universe, although they are the majority in the vicinity of our Galaxy at present. IRIS will be able to observe the integrated starlight of the normal galaxies even at high redshifts and even with heavy extinction due to dust, by near- and mid-infrared deep survey. On the other hand, the process of the starburst (burst of star formation) must play an important role in the early universe, much more so than that in the present universe. IRIS is expected to detect a large number of starburst galaxies in the early universe by a far-infrared survey.

## 2. IRIS Satellite

Table I summarizes the IRIS mission. The telescope is a Ritchey-Chretien type reflector whose apperture is a 70 cm diameter. The mir-



*C. Eiroa et al. (eds.),*
*Infrared Space Interferometry: Astrophysics & the Study of Earth-Like Planets*, 279–281.

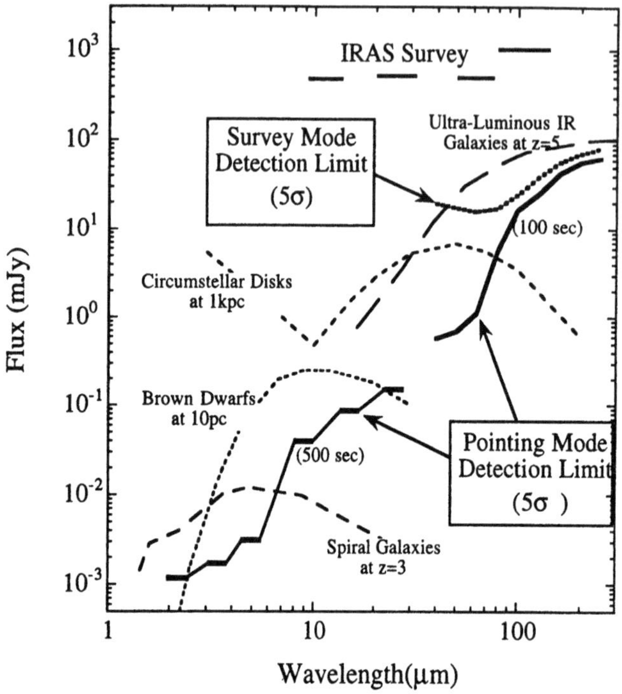


*Figure 1.* Detection Limits of IRIS

rors are made out of SiC or beryllium for weight reduction. All of the telescope mirrors, support structures, and a baffle are cooled down to 7 K. The required image quality is less than 2 arcsec. This is comparable with the minimum pixel size of 1.2 arcsec for the InSb array and the diffraction limited image size. IRIS adopted a hybrid cooling system. It uses a relatively small amount of liquid helium as well as Stirling coolers and a radiation cooling system. The major portion of the input heat to the cold part is absorbed by the mechanical cooler instead of helium evaporation. As the result of this, the amount of liquid helium can be significantly reduced.

## 3. Focal Plane Instruments

IRIS has two instruments on the focal plane of the telescope. The IRC aims for very deep imaging-surveys in the wavelength region from 2 to 25 microns. The major scientific objectives are to detect young normal galaxies in the early universe and to survey brown dwarfs in the nearby space. It uses an InSb 512x512 array and a Si:As 256x256 array with several wideband filters and grisms to measure spectral energy distributions as well as prominent line emissions. The spatial resolution is nearly limited by the pixel size of 1.2 arcsec in the shorter wavelength region and by diffraction of the telescope aperture in the longer wavelength region. The FIS is designed for an IRAS-type all sky survey in the wavelength region from 50 to 200 microns. The major scientific output will be an all sky survey in the far-infrared region with better sensitivity compared to IRAS and detection of far-infrared line emission from distant star-burst galaxies. It uses stressed and unstressed Ge:Ga array detectors for point source detection with diffraction limited resolution. A Fourier transform spectrometer will be incorporated for detection of line emissions from distant starburst galaxies. Fig. 1 shows the detection limits of IRIS.

## References

Shibai, H., Murakami, H.: 1996, *Proc SPIE* **2744**, in press.

Table I. Summary of the IRIS Mission

| | | | |
|---|---|---|---|
| Cryogenics | LHe 150 liters plus Stirling coolers | | |
| Optical system | Primary mirror aperture 70 cm at less than 7 K | | |
| | Ritchy-Chretien F/6, Focal length 4.2 m | | |
| | Image quality better than 2 arcsec | | |
| Instruments | IRC (Near/Mid IR Camera) | | |
| | InSb 512x512 | 2.2, 3.8, 5 micron | 1.2"/pixel |
| | Si:As(BIB) 256x256 | 9, 15, 25 micron | 2.3"/pixel |
| | Grism spectrometer (R = 50) | | |
| | FIS (Far-Infrared Surveyor) | | |
| | Ge:Ga 6x20 | 50-100,60,75,100 micron | 30"/pixel |
| | Stressed Ge:Ga 4x12 | 100-200,170 micron | 50"/pixel |
| | Fourier spectrometer (R = 200) | | |
| Orbit | Sun-synchronous orbit at 700-900 km launched by M-V rocket | | |
| | flies along day/night border | | |
| Observations | Continuous survey mode | | |
| | constant scanning along a great circle perpendicular to the Sun | | |
| | full sky coverage in a half year | | |
| | Pointing mode | | |
| | deep survey with a fixed attitude (10 minutes) | | |
| | for IRC imaging and FIS spectroscopy | | |
| Attitude control | Pointing accuracy better than 30 arcsec | | |
| | Pointing stability better than 1 arcsec/minute | | |
| Data rate | Generation rate 100 kbps | | |
| Mission life | More than one year with liquid helium | | |
| | Near-infrared observation can be continued after its exhaustion | | |
| Total Mass | 875 kg (wet) | | |

# PLANET FINDER OPTIONS II

*The Development Of Methods For Seeing External Planet Radiation*

N.J. WOOLF
*Center for Astronomical Adaptive Optics*
*Steward Observatory, University of Arizona*
*Tucson AZ 85721, USA*
*e-mail: nwoolf@as.arizona.edu*

**Abstract.** This paper describes the history of the development of schemes for observing the radiation from external planets. It explores the roles of imaging, division of wavefront interferometry and nulling interferometry for this task. A linear nulling interferometer is proposed for imaging planetary systems and obtaining spectra of the planets.

**Key words:** external planets, imaging, interferometry

## 1. Introduction

The spoken version of this paper dealt both with the development of concepts for observing external planets, and a particular solution, a linear nulling interferometer. However, there is both a technical version of this discussion available (Angel and Woolf 1997), and a popular one (Angel and Woolf 1996). Also, Beichman (1996) has published a report of the ExNPS study, and this gives a different perspective on the development of the concept. For those reasons, the printed version of this paper deals with the development of instrument concepts, with references to the journals where the ideas were proposed, and does not deal here with the details of the linear interferometer concept. It is still evolving.

The problems of seeing the radiation from external planets are two-fold. The planet is very faint compared with the star, and angularly very close to the star. As examples if we consider a system resembling the Solar System, and at a distance of 10pc (within which there seem to be reasonable prospects for finding such a system), then an Earth-like planet would be 0.1 arcsec away from the star and Jupiter would be 0.5 arcsec away. At visible wavelengths, the "Earth" would be $10^{10}$ times fainter than the star and "Jupiter" $10^9$ times fainter. In a broad IR band centered near 10 microns, both the "Earth" and "Jupiter" would be about $10^7$ times fainter than the star. In terms of the quantity of



C. Eiroa et al. (eds.),
*Infrared Space Interferometry: Astrophysics & the Study of Earth-Like Planets*, 283–294.

radiation to be detected, the limits are also disconcerting. At that distance, the visible radiation of Jupiter is like a 27.5 magnitude star and Earth, 29th magnitude, that is at or beyond the current limit of detection with large ground based telescopes. For the infrared the situation is worse. The star itself is near the current limit of normal photometric observations in the 10 micron window from the ground. This is due to both the thermal radiation of the telescope and atmosphere and to emission variations of the atmosphere. These are reasons for making IR observations from a cooled telescope in space. Additional reasons are that the infrared spectral features sought in an Earth-like planet would be in regions absorbed by the terrestrial atmosphere.

## 2. Basic Optical Options

I recall discussions by Bob Danielson in the 1960's of the possibility of using a controlled, moving occulting disk in space, and using its Fresnel diffraction. This seemed a way for using a moderate sized, imperfect telescope at visible wavelengths to see Jupiter sized planets, but the extension to Earth-like planets required an extremely large telescope. The technical problems seemed immense (cf., Huang 1973). The alternative is to do it all with the telescope.

We would like to image an external planetary system, and so we start by considering a perfect optical telescope. The problem is that a star image center is surrounded by a diffraction disk, and comes to its first zero at a radius $1.22\lambda/D$, where D is the primary mirror diameter. However, the diffraction pattern rises again to the peak of the first ring, and some 16% of the total energy finds its way into an extended pattern of rings. The range of radius over which the null is less than $10^{10}$ or $10^{7}$ of the central energy is both minute, and varies rapidly with wavelength, so that to use it, the imaging would need to be combined with high resolution spectroscopy, a task of immense technical challenge. Apodizing to reduce the diffraction pattern will be discussed later.

To get this minimum 0.1 arcsec from the star at visible wavelengths requires a telescope of 1.2 meters, and at 10 microns wavelength the corresponding diameter is 24 meters. The infrared telescope size is beyond reasonable budget, and the optical image of the planet is extremely faint. Any stray light scattered by optical imperfections of the telescope would appear as speckles, mimicking a planet. Some speckle averaging would be possible by rotating the telescope around its optical axis, but while there seemed possible an approach to the IR problem by using optical radiation to perfect the telescope figure, the tolerances for even better surface quality at a 20 times shorter wavelength seemed unmanageable. On the other hand, an interferometer size of about 20 meters did seem manageable, and so it was appropriate to consider IR interferometry.

Within interferometry there are two options. The first is division of wavefront - the interferometer is optically equivalent to pieces of a telescope mirror. Each object produces bright and dark fringes in the focal plane at a place corresponding to its image. If the planet bright fringes fall between the bright fringes of the star, then the fringe minima will not go to zero, and from the fringe contrast, the planet brightness can be determined. But the observation of extreme contrasts at small angular separations is extremely difficult. Both imaging and division of wavefront interferometry suffer from this problem.

If instead of division of wavefront interferometry, we put the radiation from two spaced apertures into the two inputs of a Michelson spectral interferometer, adjusted for zero path difference, the two outputs will both correspond to the radiation from the sky. At each of the two output beams, the radiation from the sky will be partially transmitted, with a transmission that varies sinusoidally across the sky in one direction, and is constant along lines perpendicular to that direction. The spacing of the maxima is $\lambda/S$ where S is the separation of the apertures. If the interferometer is pointed exactly at a star, there will be from one output beam a full transmission of starlight, but in the other beam, the star's radiation will interfere and be canceled. In that beam at an angular distance $\lambda/2S$ from the star however, the transmission will be at its first maximum. A planet there will have its radiation transmitted. This is the concept of a nulling interferometer, and it offers the prospect of seeing a planet while at the same time canceling out the radiation of the star. Note that in general the Airy pattern of one of the apertures will not resolve the star and the planet. Some special modulation scheme will be needed to tell where the source of the radiation is.

## 3. The Bracewell Nulling Interferometer

Ron Bracewell (1978) published the first suggestion for applying this concept to the detection of a planet. He wanted the planet to star ratio to be as favorable as possible, and so he suggested looking for the radiation of Jupiter-like planets well beyond their Planck maxima, where both the planet and star were on the Rayleigh Jeans tail of their radiation. For such a case, at perhaps 40 microns, the radiation of the planet is only about $10^4$ times fainter than the star. Because this wavelength is not transmitted by the atmosphere, the device would need to be in space. The planet would be detected by rotating the device about the axis to the star. The radiation from the planet would be modulated as it went from high transmission to the null and back again. Further references to this work are given in Angel and Woolf (1997).

In considering the application of this concept to detecting Earth-like planets, one problem is that the sinusoidal transmission fringes have a very sharp null, so sharp that if maxima are at 0.1 arcsec away from the star, then 0.5 milliarcsec away, the radiation transmitted is about 1 part in 4000. But 1 milliarcsec is the typical angular diameter of a star at 10pc. Closer star will have a worse situation. The interferometer will transmit too much star radiation. For the detection of Jupiters at 5 times greater distance from the star, the star transmission will be 25 times less, and the planet is brighter, 2- aperture star light nulling works, but for Earth-like planets it is inadequate.

## 4. The Angel Apodizing Ring

It became apparent to Angel, Cheng and Woolf (1986), that the spectrum of an Earth-like planet in the 10 micron region was very interesting because of the broad atmospheric absorption bands. In particular, the presence of Ozone, which produces one of these features is on Earth, believed to be a result of life processes. Therefore we wanted to find a way to be able to observe the radiation of the planet.

Angel reasoned that the difficulty of using a conventional aperture telescope is that the null between the central maximum and the first bright ring is so sharp. This is because the wavefront amplitude is changing roughly linearly with distance from the star. As it passes through zero, the intensity, the square of the amplitude goes through zero, but to get an extended minimum, not only should the amplitude go through zero, but it should simultaneously have a zero slope at that point. To produce that slope, Angel considered a ring around the aperture, at such a distance that it also produced a null in the same place as the central mirror disk, but the amplitude would be changing in the opposite direction. Then the ring width could be tailored so that the rate of change of amplitude was equal and opposite to that from the disk. Thus there would be an extended null around the central maximum in which a planet could be detected. Because the null was now extended, the system permitted a relatively broad bandpass $\lambda/10$, and so a series of filter images could study the planet spectrum for features. It was assumed that the equivalent of the disk and ring would be obtained by using a telescope with a cold aperture mask, and that different mask sizes could be used for different wavelength bands. In order to see the planet's atmospheric features, it would be necessary to have the telescope above Earth's atmosphere in space.

One loss that came with the design is that the first minimum was pushed about twice as far away from the diffraction core center. Thus instead of discussing a 24m telescope for observation of planets at 10pc, we were discussing a 60m telescope! The concept may still valid if there is need to collect a substantial amount of energy from a planet, e.g for relatively high resolution planetary spectroscopy for observations out to 10pc. The design at present does not carry the concept far enough. A second ring of telescope aperture is needed to slow the rate of growth of brightness out from the center of the dark band around the star image core. The concept's advantage for spectroscopy is the very large collecting area of the telescope. It is a truly distant future Space Telescope.

## 5. Wavefront Irregularities and Optical Solutions

Every scheme to detect planets around stars has to deal with issues of optical imperfections in the observing system. If the system is a telescope or an interferometer, the starlight produces a diffraction pattern, and the goal is to see the planet when it is in a null in the starlight pattern. But any such null will be filled in by radiation scattered by the wavefront errors and amplitude irregularities across the entrance pupil of the system. These can be corrected by adaptive optics, in which a an optical surface has a shape modified by actuators as e.g. discussed by Angel (1994) and Stahl and Sandler (1996).

The surface irregularities have a scale $d$ set by the projected spacing of actuators in the pupil plane. If D is the mirror diameter, then the scattered light is spread over an area larger than the diffraction core by a factor $(D/d)^2$. The fraction of starlight in this scattered radiation will be 1- exp( $-\sigma^2$ ), or roughly $\sigma^2$ . Here $\sigma$ is the wavefront rms error in radians of phase. Typically for a large mirror, the area factor will be at most $10^4$. Thus to observe an object $10^9$ times fainter than the star, we would like the exponential term to be little more than one millionth. This would imply for visible radiation an rms error of about 0.1 nm. This seems nowhere near attainable. In proposing a search for Jupiter-like planets from the ground, the Center for Astronomical Adaptive Optics does not propose this, but rather to make $\sigma$ about 0.2 and to gain sensitivity beyond that by averaging for about $10^6$ speckle lifetimes.

For an infrared telescope in space, to search for planets $10^7$ times fainter than a star, we could have $\sigma$ about 0.03, and this would correspond to about 0.1 wave of visible light. Thus very large telescopes in space seem reasonable for observing planets in the infrared without major problems from scattered radiation. On the other hand a division of wavefront interferometer would have a much smaller area gain, and so does not seem a practical concept.

These same considerations would cause severe problems for a nulling interferometer also - if it tried to work with an image. Instead, nulling interferometers are proposed to work single mode, either by putting the radiation through a single mode optical fiber, or by passing it through a diffraction limited hole. The effect is to reduce the problem to one of adjusting single parameters of amplitude and phase. These helpful ideas and actualization have been reported by Shao and Colavita (1992) and Mariotti et al (1995). If a large mirror is required for light collection, it is still possible to collect star and planet radiation in a single mode by using rectangular or ellipsoidal optics, so that the long direction of the diffraction pattern reaches out as far as the planet.

## 6. From Telescope to Interferometer

Angel (1990) discussed options for a Next Generation Space Telescope (NGST), and showed that in the IR, a 16m telescope could only see Earth-like planets out to a distance of 2.5pc. Therefore he started to consider options for dividing a 16m aperture into a nulling interferometer. Shao (1990) also considered this possibility, and described a Bracewell-like nulling interferometer with large mirrors. In his discussion, Angel discovered the problems of the Bracewell nulling interferometer for looking at Earth-like planets, and so he considered reconfiguring a 16m NGST as four 8m sized apertures in a square or diamond configuration.

The advantage of such a two-dimensional arrangement of the aperture is that the null has its transmission varying not as $\phi^2$ but as $\phi^4$, where $\phi$ is the off-axis angle. This provided a broader deeper null in which the star could be hidden from view. He noted however that the telescope would respond to both the radiation from Solar System zodiacal dust, and radiation from the external system dust. Because of its photon noise he found that the radiation from solar system dust was limiting the performance. Although the 4-mirror interferometer concept allowed in principle for the mirrors to be smaller, 8m mirrors would be required for their diffraction pattern to sufficiently cut down the solar system dust radiation.

## 7. The Léger Practical Scheme

Even now we struggle to devise concepts for a single 8-meter mirror telescope as NGST. Four such mirrors seemed impractical. However, Léger et al.(1996) noticed and Bracewell had before, that out at some distance 3-5 AU from the sun, zodiacal dust would be cooler, and its 10 micron flux substantially lower. With such lower fluxes it became possible to consider an interferometer using smaller mirrors, and it became apparent that 1m-class mirrors would be adequate. Originally a 4-mirror cross arrangement following Angel (1990) was proposed. Since IR telescopes with such mirror sizes have already been launched, the launching of an interferometer with four of them did not seem unreasonable. Further study seemed to show that there were no obvious major show-stoppers, though the technical problems of nulling interferometry in space, and the precision needed pose acute problems (see Paper III in this series, in this same volume). Recent suggestions have included an oval arrangement with five mirrors. Léger's proposal marked a turning point, in that for the first time it became possible to plan for a real device to look for Earth-like planets.

One further comment about this scheme, is that if the interferometer with four large mirrors were operated at a large distance from the Sun, it would be an alternative configuration for high resolution spectroscopy of external planets.

## 8. The ExNPS development

When in 1995 NASA decided through JPL to set up teams to explore options for making images of the surfaces of planets, it was seen as an opportunity to put together a group with a strong background in this area. In one team, we combined Angel, Bracewell, Léger, Mariotti and Woolf. Our analysis of the problems of making meaningful images of the planets around other stars demonstrated to us that the difficulties were so immense that no technique could be selected at this time to give meaningful progress towards that goal. Indeed it seemed more likely that it was a mission for a space probe to another star than that it was a task for "local" interferometry. However, it seemed possible to divert interest from imaging of planet surfaces, to imaging planetary systems, watching planets revolve around their star. Then the information that the public imagined to come from imaging could instead be obtained by spectroscopy. We could probably tell whether simple life had developed on a planet.

The late Bernard Oliver strongly impressed on Angel and the writer of the advantage of exploring options of multiple rings around a mirror as in the 1986 concept, and the writer kept pushing to explore long thin mirrors to get a good diffraction pattern in 1-dimension, while keeping down the overall size and weight of the telescope. So we considered long apertures with apodizing apertures off the end. Then we realized that this was very like an interferometer!

With these concepts together, it was obvious that the Bracewell interferometer could have its null broadened by using a second smaller in-line mirror pair, so phased that it went through a null in the same place, but with opposite amplitude changes. In order for the interference pattern to be regular and repetitive, the spacing of the outer pair had to be an integer number (n) of times that of the inner pair. And the same time the maximum amplitude from the outer pair had to be 1/n times that from the inner pair. The example with n=3 was able to produce totally coherent maxima. The example with n=2 was 2/3 of the length for the same angular resolution, and gave satisfactory performance. It was adopted as a baseline planet finder concept, and dubbed OASES (Outpost for Analysis and Spectroscopy of Exo Systems). As compared to the square or diamond shaped arrangement of the apertures, this had the advantage of a null varying as $\phi^6$ rather than $\phi^4$.

The signal from a single planet would consist of a series of "blips" when a transmission fringe crossed over a planet (see Fig. 1). Planets at larger radial distances would produce more and narrower blips. It was found that the analysis of the data from such a rotating interferometer could be obtained by cross-correlating the interferometer output with that expected for a planet at every possible combination of position angle and radial distance. Although the result for a single wavelength provided a number of optional planet locations, when the signal from the interferometer was spectrally dispersed, and the signal at each wavelength analyzed separately and the results added, a good picture of the planetary system appeared. All this is described in Angel and Woolf(1997). Because the different wavelengths are separated as a necessary part of the observations, the device not only images the complete planetary system, but also obtains separate low resolution spectra of each of the planets. The system multiplexes system imaging with spectroscopy of each planet.

Residual annoyances with this design were two-fold. First, the signal from the planets had a 180° position angle ambiguity. This could be resolved by using a natural slight precession of the interferometer, but the ambiguity was resolved at a lower Signal/Noise than was the planet signal itself. Another ambiguity occurred in separating the

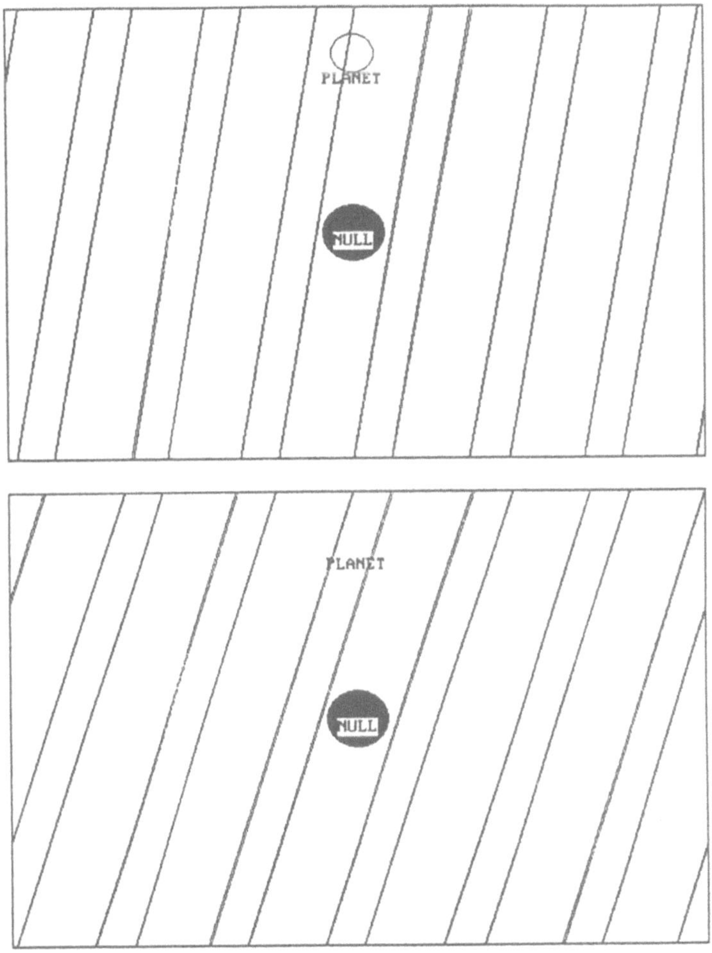


*Figure 1.* The rotating transmission fringes change the planet signal

signal of an innermost planet from the signal due to the smooth component of zodiacal dust in the external system. This could be resolved by watching the inner planet make an appreciable movement in revolving around the star. Now, since the meeting linear nulling interferometers are better understood (Woolf and Angel 1996, Paper I of this series), and optimized linear interferometer configurations are available that resolve these problems.

## 9. Exo-zodiacal radiation

Perhaps the most disconcerting aspect of the problem is that the analysis shows the limiting factor for observations is the brightness of the zodiacal radiation around the other star. In calculations to date this has been assumed to be identical with the solar system value. But the solar system value is the only one known at this level. The lowest level measurable around other stars to date is about 100 times the solar system value. About 15% of stars seem to have that strong an emission. The time required to make an observation is proportional to the amount of exo-zodiacal emission, and the times predicted are already long. A further complication includes the possibility that exo zodiacal emission may be clumped, and for dusty systems, the emission in one clump may exceed the radiation of planets themselves (Backman 1997).

A scheme has been proposed by Woolf and Angel (1995) for observing exo zodiacal dust emission from the ground, using the Large Binocular Telescope. This telescope is uniquely suited to this task because of the close spacing of its elements (14m), and because of the very few optical elements in the beam combining path. This system could detect total emission as small as a few times the solar system value for nearby stars. However, the best way of distinguishing planets from zodiacal dust clumps seems to be by the spectrum, obtained with an interferometer at 5AU, that is by the same system that would also hopefully detect the planets by their absorption spectrum.

## 10. Where next?

The problems of obtaining images of planets have some areas such as the development of radiatively cooled telescopes which seem comparatively straightforward. Other aspects are considerably harder. For an example, for focal plane designs look at paper III by Woolf, Angel and Burge in this volume. Another difficult area is the achieving of mechanical tolerances in the relative placing of the interferometer elements, the precise pointing at a star, and moving from star to star.

Although it does seem possible to obtain images of planetary systems, and the spectra of their planets, the long range science goals of such a program still leave us with major concept challenges. The scientific study of Earth-like planets, and especially the search for evidence of whether there is life on them suggests a need to not only observe oxidizing molecules in the atmosphere, such as ozone, but also reducing ones such as methane. Whereas a crude spectrum of an exo-Earth could be obtained with a interferometer using several 1m class mirrors, in orbit (and needing about two months of observation), if we want to observe methane, we are asking for at least 10 times higher spectral resolution, and perhaps up to 5 times higher signal to noise. That would require between 10 times and 250 times as much collecting aperture. If the smaller number is adequate, we could consider a second generation interferometer in which each 1m mirror of the current version is replaced by an NGST. That is the simpler, optimistic solution. Another option is to search for methane using a large Angel-ring telescope in a 1AU orbit. It is clear that the search for, and study of external Earth-like planets will push telescope/interferometer technology beyond the range of our current vision.

## References

Angel, J.R.P.: 1990, in The Next Generation Space Telescope, P.Bely, C.J. Burrows (Eds.), STScI, Baltimore, p. 81
Angel, J.R.P.: 1994, *Nature* **368**, 203
Angel, J.R.P., Cheng, A.Y.S., Woolf, N.J.: 1986, *Nature* **322**, 341
Angel, J.R.P., Woolf, N.J.: 1996, *Scientific American* **274**, 60
Angel, J.R.P, Woolf, N.J.: 1997, *ApJ*, in press
Backman, D.: 1997, *Committee Report*, in preparation
Beichman, C.: 1996, A Road Map for the Exploration of Neighboring Planetary Systems (ExNPS), NASA: JPL
Bracewell, R.N.: 1978, *Nature* **274**, 780
Huang, S.S.: 1973, *Icarus* **18**, 339
Leger, A., Mariotti, J.M., Mennesson, B., Ollivier, M., Puget, J.L. Rouan, D., Schneider, J.: 1996, *Icarus* **123**, 249.
Mariotti, J.M., Coude de Foresto, V., Perrin, G., Zhao, P., Lena, P.: 1995, preprint
Shao, M.: 1990, in The Next Generation Space Telescope, P.Bely, C.J. Burrows (Eds.), STScI, Baltimore, p. 160
Shao, M., Colavita, M.M.: 1992, *Advances in Astronomy and Astrophysics* **30**, 457
Stahl, S.M., Sandler, D.G.: 1995, *ApJ* **454**, L153
Woolf, N.J, Angel, J.R.P.: 1995, in Adaptive Optics, *OSA Technical Digest Series* **23**, 44

## 11. Questions

*P. Bely:* To satisfy the 1:2 amplitude ratio between the two telescope pairs for the OASES configuration, can you not use telescopes of different sizes.

*N. Woolf:* It is indeed in principle possible to use telescopes of different sizes. It is a practical issue whether the cost of making two different sizes makes it worth while. Remember, the amplitude varies as the square root of the intensity, and so the telescope mirror diameter for the smaller ones is only about a factor 1.2 smaller.

*T. Hawarden:* The proposed launch into an elliptical orbit to 5 AU wherein observing occurs near aphelion means that you can only look at half the sky with that mission. Would it be worth looping around Jupiter to circularize?

*N. Woolf:* Orbit options are still far in the future. For a start, how would we set up this giant device? The history of deploying civilian mechanical devices in space is not currently encouraging. Angel has suggested that maybe the first deployment should be near Earth, for simple astrophysical observations. Astronauts would then be available to repair the device. Finally, solar electric propulsion could send it out to distant places. Actually, although we show illustrations with Jupiter, to indicate the distance from the Sun, Jupiter itself provides a difficult environment, with radiation belts, and probably relatively dense dust clouds around it. I would prefer the spacecraft to go in the opposite direction! And if we do need to look at only half the sky, it would be good to look at that part of the ecliptic where the majority of the Earth's telescopes can observe. It should be sent towards Cancer.

# PLANET FINDER OPTIONS III

*Focal Plane Instrumentation*

N.J. WOOLF, J.R.P. ANGEL and J.M. BURGE
*Center for Astronomical Adaptive Optics*
*Steward Observatory, University of Arizona*
*Tucson AZ 85721, USA*
*e-mail: nwoolf@as.arizona.edu, rangel@as.arizona.edu, jburge@as arizona.edu*

**Abstract.** Schemes for interferometers to observe systems of planets orbiting their star, and to obtain the spectra of these planets are in a state of concept development. The simplest of these schemes use four telescopes either in a diamond shaped or linear configuration. These preliminary designs are used as a basis to explore solutions to the problems that will be encountered in the development of focal plane instrumentation.

**Key words:** external planets, imaging, interferometry

## 1. Introduction

The focal plane instrumentation of the planet finder has to null the star, take the spectrum of the residual radiation and record the fringe pattern at each wavelength band in the mid-IR spectrum. It also has to produce signals that control the pointing of the telescopes and the spacecraft. These tasks require an unprecedented consistency of performance. As with other complex devices, the difficulty is not in using it, but in setting it up. That is the focus of this paper.

## 2. The Sequence of Discussion

The two fundamental aspects of the interferometer configuration set the discussion sequence. The concept is to use destructive interference of light from different apertures to cancel or "null" starlight and not have fluctuations in the residual transmission of starlight that become disturbing noise signals. Then also, the signal from a planet is very weak, and inefficiency of optics would need to be compensated by increase in telescope sizes. The initial goal is to observe systems for an image of the planet configuration in about 6-12 hours, and to observe the spectra of the planets in about 50 days. These considerations drive the design.

The mission constraints of needing of order 100 stars to be observed for planetary systems, and a goal of seeing planets out to about 1 arcsec



*C. Eiroa et al. (eds.),*
*Infrared Space Interferometry: Astrophysics & the Study of Earth-Like Planets*, 295–307.

from the star, when linked with detector considerations define a range of rotation periods of the interferometer of about 1 to 6 hours.

Next, the requirements of phase and amplitude control for adequate nulling lead to a need for single mode operation. Then, single mode operation while preserving high constancy of amplitude places a requirement for star tracking to about 1 milliarcsec.

Phasing is affected by spacecraft pointing as well as by the internal spacing and path lengths of the planet finder. For this reason, while internal control may be maintained by laser interferometry, the path length control needs to use the star to provide the signal. This in turn sets a requirement for a shorter wavelength interferometer system with a substantially common path with the mid-IR system that acquires planet data.

Phase and amplitude control need to be maintained to a consistency of $\pm$ 0.03%, and to a specification of $\pm$ 0.5%. One of the biggest areas of concern for the specification is the chromatic performance of beamsplitting. If relief is sought from the specification over the full mid IR bandwidth by dividing the spectrum into zones, then the edges of those zones must also have equality of amplitude and phase for the separate signals. If the spectrum needs to be divided into zones typically $2\mu$m wide for achromatism and amplitude control, that division requires extreme consistency in selection of zone edges. Single mode operation at high efficiency might also require division into spectral zones before beam combining. There are also concerns that the two linear polarization components might need to be separated. The benefits of not having to separate by polarization are modest, and can be compensated by optimizing for each polarization. But the benefits of not also having to divide spectrally are very great, though we do not yet know the trade-off options. .

Then the IR spectrum needs to be divided into bands of about $0.5\mu$m width for detection of signals, recording and analysis. Design factors for this spectrograph and the detectors are reported. Finally, the optical sequence of operations is discussed as a precursor to optical design.

## 3. Rotation Period and Detector Noise

The planet signals will be very weak as reported by Beichman (1996) and Angel and Woolf (1997). Indeed, new detectors will be needed with lower readout noise and dark current. In order to increase the signals from the planets, it would be helpful if data were obtained slowly. To allow for the fringe motion of a Saturn-like planet, about 1000 separate data points will need to be recorded per revolution of the interferometer. In order for each of these to contain an integration of at least 4 seconds, the device cannot rotate faster than once per hour. On the other hand, for the brighter stars, it would be good if decisions whether to make further observations could be made within several hours. This sets several hours as a lower limit to the rotation rate. With this range of the rotation rate, it is possible to see how the control of pointing and nulling is achieved.

Using numbers from Beichman, for four 1.5m telescopes, the typical photon arrival rate for an Earth-twin at 13pc will be 15 per second in an octave, or 1.12 $t\eta$ in a band of $\lambda/20$ as required below, where $\eta$ is the optical efficiency and $t$ the observation time for one data point. The noise will, (if all else works as planned), be due to background exo-zodiacal radiation at 225 $t\eta$ in a single band if it is like the solar system (about 1/3 will be nulled). This will have an associated photon noise of 15 $(t\eta)^{0.5}$. We hope that the efficiency will be quite high, but for current detectors with noise about 30 events per readout, unless the optical efficiency were near perfect, or readout could be at a slower rate than once every few seconds, detector noise would dominate. Slow rotation alows a greater integration time and reduces the amount of required detector development.

If we make rotation substantially slower, then the total observing time searching for systems with Earth-like planets becomes significant. For example, if we make one rotation every 6 hours we use 25 days in a preliminary survey. The time in slewing from object to object is likely to be no less, and we cannot afford it to be much more. But we would like to spend most of a mission in observing spectra of planets. To observe spectra would typically take about 50 days of observation per planetary system, so in one year, 100 stars could be observed for the presence of planetary systems, and six systems observed spectroscopically. The mission would probably last a few years. Much slower rotation would clearly have an impact on the number of objects observed. We shall consider a range of the rotation rates of 1 hour to 6 hours for the control of pointing and nulling.

## 4. Precision Needs for Phase and Amplitude Matching

The reason for having an interferometer is mainly to null out the radiation from the star, some $10^7$ times brighter than a planet. The planets are then seen as particular modulation patterns of the residual radiation as the device is rotated (see Angel and Woolf Fig. 3-5 or Beichman Fig. 5.4). Any variation in transmission of the star, will appear as noise, limiting the possibility of seeing planets. Those variations of transmission can result from pointing fluctuations of the device, phasing fluctuations of the device, variations in the amplitudes of signals at the beam combiners, and combinations of these.

In order to understand the phenomena, we shall consider the simplest version of the devices, 4-element interferometers. These have been proposed both as a diamond shape (DARWIN), see Leger et al. (1996) and Mennesson and Marriotti (1997) or a linear array (OASES) see Angel and Woolf. In Woolf and Angel (1997), paper I in this series, more complex linear arrays have been proposed to deal with the limitations of OASES, and the papers cited also propose a 5 mirror version for DARWIN. The analysis here however is helped by studying only a simpler system. In both of these devices, there are a pair of nulling arrays, and the signals are combined from the first nulls, also in nulling form so as to produce a broader minimum. The analysis here shows how to use a system that can be conformed as separate Bracewell nulling pairs for set up.

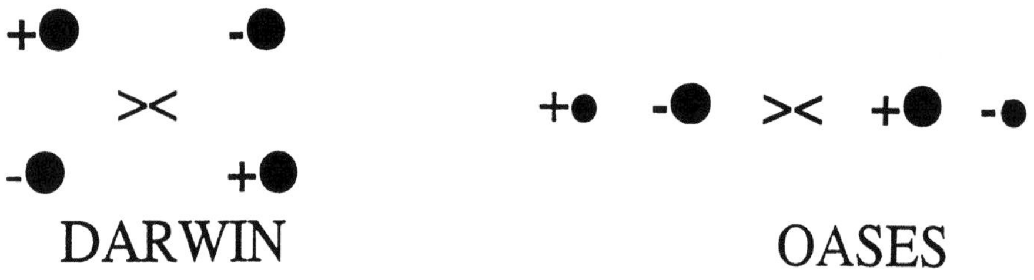


*Figure 1.* The arrangement of the mirrors in the two concepts. The phases are shown as + and - and refer to a 1/2 wave difference.

Changes in the amplitudes and phases from the four mirrors will change the shape and intensity of the minimum. Amplitude variations could occur from individual telescope pointing errors coupled to single mode operation. System mispointing, (effectively phase variations), also causes the observed signal from the star to fluctuate. If the star is not completely nulled, there will in addition be photon noise which will also make planets harder to detect.

There are two regimes to consider. In the first, the amplitude and phase wander within a certain range, and our task is to ensure that the fluctuating transmission of starlight is kept below a certain level. In the second regime, the amplitudes and phases are not exactly correct, but they are very stable. However, since photon noise will likely be dominated by exo-zodiacal dust emission (about 30,000 times fainter than the Sun in the solar system), photon noise from the star will not be important for a reasonably wide range of mis-setting. Originally attention has been focussed on the first regime. However, as later discussions here will show, amplitude matching may be difficult, while amplitude repeatability to high precision may be relatively easy. And phase repeatability may be much easier to achieve than exact phase matching.

Clearly the first, simplest and cleanest solution is to null the star so that it is fainter than any planet. Then if this is maintained, the noise contribution of the star is guaranteed to be negligible. Since the amplitude of the wave from the star is about 3,000 times as large as from the planet, an imbalance of amplitudes at beam combining of 0.03% is just barely tolerable. Similarly, since the amplitude in moving out of a minimum will vary as sin(z), and for small angles z is about equal to sin(z), the allowable variation in phase is also about $3{\cdot}10^{-4}$. Sample calculations for the OASES configuration as shown in Fig. 2 confirm these estimates.

Alternatively, if it is decided to allow the star signal to be larger, there needs to be a study of the monochromatic noise expected from the rate of variation of phase and amplitude errors permitted. The critical range of frequencies varies from about 500 per revolution for Saturn-like planets to about 10 per revolution for close-in planets, or from about 6 seconds to 36 minutes. If variations occur at a much faster rate than these, and are not correlated with rotation angle, the tolerances might be relaxed. But we are trying to measure an Earth-twin signal to a precision of 5%, and that corresponds to about 1 part in $10^8$ of the starlight.

An analysis below will suggest that phase fluctuations may indeed be controlled to very high precision. Then it becomes possible to allow a certain amplitude imbalance, up to the point where the leak-through signal from the star is almost equal to the exo-zodiacal light signal. However, any relatively rapid variation in amplitudes would then have a devastating effect on system noise. The tolerance for setting amplitudes and phases would become about 0.5% which may be the critical easing item to make focal plane design possible. But the tolerance for fluctuation of these during operation remains 0.03% as derived above.

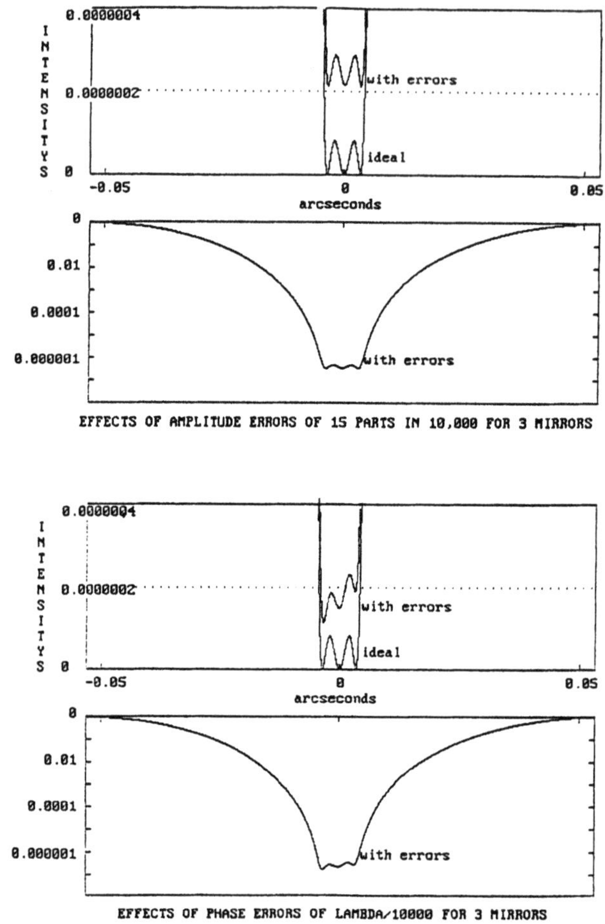


*Figure 2.* Effect of phase and amplitude errors on the null of OASES planet finder

## 5. Single Mode Operation, Beam Combining and Achromatic Path Length Control

We need single mode operation so that the aberrations and minor mispointing of the interferometer do not enter into the operation. The use of optical fibers in interferometry has been discussed by Mariotti et al. (1996). One advantage of using fibers is that an "X" junction between two fibers can act as a beam combiner. However, broadband beam combining needs to be done with correction of dispersion differences between the combining beams to about 0.5% in phase.

An alternative, passing the light through a diffraction limited hole is also available as a means of working with a single mode. In either case, the positioning of the beam on the hole or fiber needs to be both highly repetitive, and well centered. Suppose the beam is displaced by a fraction $\epsilon$ of the diffraction diameter. Then the fractional reduction in signal through the hole will be at most $\epsilon/\pi$. Thus $\epsilon$ should not exceed about $10^{-3}$. If we use the visual diffraction image of the star, and keep it centered to about $\pm$ 1/50 of its diameter, this should suffice, since the mid IR diffraction image is twenty times larger. Note however that there is a potential for the hole edge to behave somewhat like a knife edge, and there would be corresponding intensity fluctuations across the pupil. Then any non-uniformity of coatings etc. on the pupil would result in intensity fluctuations. Limiting such fluctuations may require tighter pointing control. This aspect needs further exploration.

The positioning could be by tilting the secondary or tertiary of each telescope. The visible radiation of the star could be sent to the star tracker, and the star tracker signals coupled back to the telescope mirror tilt. A beam displacer in the visible beam then allows the IR beam to be adjusted for maximum signal and minimum fluctuation, that is, to center it on the IR pinhole or fiber. The adequacy of this scheme will depend on the disturbance spectrum of the telescope. That is a topic of the mechanical configuration of the planet finder and will be dealt with in paper IV of this series.

If the beams are not sent through fiber beam combiners, some alternate beam combining is needed. Possibilities exist for cube-corner devices and beam combining systems as used in stellar multiplex spectroscopy. Indeed, because this is a multiplex device, many aspects have parallels with concerns and solutions in astronomical Fourier spectroscopy (Connes 1970). An interesting variant beam combiner has been proposed by Shao and Colavita (1992), in which the phase change at inclined totally internally reflecting surfaces, differing in the two linear polarization vectors is used to provide an achromatic half-wave difference between the beams.

As with the fiber beam combiners, the crucial problem is to provide precise dispersion correction, so that after beams from two telescopes have been combined, they have a similar variation of path-length with wavelength and can be combined within the phase tolerances discussed above. A special extra concern for the planet finder is that two combined beams differ by an achromatic  wave. In the more complex schemes with more mirrors, other achromatic relationships of 1/4 wave or 1/5 wave are needed.

We have investigated the possibilities of using pairs of plates of dielectric materials to achieve achromatic path differences. For the phase tolerances needed for the most precise nulling and using plates of CsI and ZnSe the IR band needs subdividing into four or five zones, each with its separate set of plates (Angel, Burge and Woolf 1996). The bandwidth can undoubtedly be increased if the phase setting precision is reduced as described above. Possibilities may exist for the entire spectrum to be workable with single set of wider band plates perhaps that use three different dielectrics. We have also calculated a system for a 1000Åwide bandpass in the visible, and a system could also be produced for near IR wavelengths.

Such phasing plates will also have associated absorption that varies across the spectrum. It may also be possible to use plates with three different materials to use the absorptive properties of the plates to get the spectral match of intensity.

## 6. Pointing

The pointing of an interferometer involves three parts. The first of these involves aligning the principal axes of the telescopes with the rotation axis of the spacecraft. Telescope mispointing brings in misalignment coma, and variable misalignment coma would make the single mode signal vary. This is an item for mechanical design.

The second aspect is as described above to tilt the secondaries of the telescopes so that the star diffraction pattern sits on the hole or fiber that defines single mode operation. The third aspect is to bring the interferometric axis into coincidence with the direction to the star by path length adjustment. If the spacecraft axis does not adequately coincide with this precisely, there will be a need for continual path length adjustments during spacecraft rotation.

While the internal configuration of an interferometer can be measured using laser interferometers, the orientation of the interferometric axis of the system is defined by the nulling at the output. Pairs of mirrors operating as Bracewell nulling interferometers define this axis and allow it to be set to high precision. In this case the optimum wavelength for monitoring will be at shorter wavelengths in the IR, probably $2\mu$m chosen so that the star diameter is small enough compared to fringe separation to give a good nulling measure, but also so that the photon rate is large. There will need to be commonality of path between this wavelength and mid-IR to the maximum possible extent.

The 2-mirror nulling precision will be set by the angular diameter of the star. The transmitted starlight will be about 4 times greater when the null is offset from the star by the star angular diameter, and the error will vary as the square of the mispointing/misphasing. For example, a Sun-like star at 10 pc distance has a diameter of 1 milliarcsec, and with precision of 0.1% in the adjustment of the intensity transmitted, the equivalent pointing error becomes about 2 microarcsec ($10^{-11}$ radian).

For the main nulls of interferometer elements spaced by 80m, this pointing, the pathlength error corresponding to 2 microarcsec would be 0.8nm, and for a spacing of 40m it would be 0.4nm. The corresponding mid-IR errors are $\lambda/12{,}000$, and $\lambda/24{,}000$, which are more than adequate. Consistency in nulling to 0.1% requires $10^6$ photons to be observed, and for a typical 4th. magnitude star, that is available in less than a millisecond. The corollary is that the needed instrumental errors in nulling (due to spacecraft pointing error or spacecraft vibration) should not change by more than 0.8nm in a millisecond or 800$\mu$m in 1000 seconds. Clearly there are two time scales involved. There is a mechanical time scale of phase slipping of the interferometer. The highest product of frequency and amplitude of this motion defines the needed response of the optical phasing servo system. That is limited by the photon output of the star, and so is a constraint in mechanical design.

Probably the simplest way to achieve the accuracy is to insert a rapidly varying path length of about 10nm in one 2$\mu$m arm, and use the phase sensitive signal from the detector to adjust the joint pathlengths on a fast timescale. If systematic changes are occurring, prediction may improve the result.

## 7. Amplitude Matching

As discussed above, amplitude matching to 0.5% or so is tolerable if amplitude fluctuations are kept below about 0.03%. It is critical to control the amplitudes so that the polarization planes and the different wavelengths for which the observation is made all satisfy the amplitude matching requirement. If it is not possible to keep the polarizations together so that when the amplitudes are adjusted in one plane, they are also adjusted in the other, then there will be a need to separate the polarization components, and there will be twice as much focal plane equipment needed.

The fundamental issue in this area is how to adjust amplitudes, how to deal with the need for wavelength variation of amplitude adjustment, and how to keep the efficiency of this process high. The first aspect that stands out, is that all schemes we have considered to date introduce a wavelength dependance of amplitude adjustment. Even diaphragming a mirror introduces wavelength errors because of the wavelength dependance of diffraction. It is possible to consider reflections and transmissions by dielectric coatings of variable thickness, or by metal meshes of variable spacing. It is possible to interfere a single beam with itself, and with a slight phase variation so that the transmitted radiation has a controlled reduction in amplitude. Amplitude matching is an area where both brainstorming and experiment will be needed. The likelihood is that many schemes will also introduce a wavelength dependent phase term. The simplest scheme at present would be to diaphragm pupils.

The test for amplitude matching is to use a steering mirror to mispoint one beam so that it does not go through to the detector, then to return it and mispoint the other. If the two detector signals appear identical in amplitude, to the required precision adjustment is complete. The mechanical design should ensure that the amplitudes vary very slowly with time, so that there is no appreciable component at the modulation frequency associated with planets. Consequently there should be rare corrections to the individual amplitudes, and these can be made at long intervals by observing a bright star.

## 8. Spectroscopic Separation

There are two different essential reasons for spectroscopic separation of signals. The first is that the interference pattern on the sky is wavelength dependant. The second is that we want to observe the spectrum of the planets. There is also a third possible need that may require a separate and different spectral divider. That is, if phase and/or amplitude cannot be kept within tolerance over the entire mid-IR spectral band, then the band will need to be divided into zones. The reason that these two tasks are separated is that it is very difficult to make the same precise band profile and center wavelength matching for the light from each telescope before it is interfered.

The largest pattern spectral differences show at the largest radial distances from the star. But the focus of efforts is on Earth-like planets, and these are expected to be relatively close to the star where the effect of changing wavelength is relatively small. Modeling of the interference patterns suggests that separation of the spectrum into bands $0.5\mu$m FWHM will not badly smear the fringe pattern records and will permit detection of planets out to 1 arcsec from the star. (In part because such planets will be cool, they are only expected to show at long wavelengths where tolerances are easier.) This bandwidth will just adequately detect the $9.7\mu$m ozone band for an Earth-like planet. On the other hand, because detector noise will be the likely limit to the operation of a planet finder, higher spectral resolution would make spectral study too time consuming.

The resolution is relatively easily obtained for a diffraction limited spectrograph. The spectrograph itself should be of the highest possible transmission. The number of optical elements in it should be minimized. However for the resolution needed, a simple prism spectrograph, perhaps employing a CsI prism appears adequate. The detectors could either be in the focal plane of the spectrograph or, because the detectors will need active cooling, the focal plane will probably be relayed into a Dewar containing the detectors.

The third possible need for spectral resolution is if it is found that either amplitude adjustment or phase adjustment cannot be performed to adequate precision with the full 7-17$\mu$m band. Then it becomes necessary to spectrally separate zones. But if this is done, the different zones must remain matched in amplitude at the edges! It is possible to perform this task with separate small spectrometers in the path from each telescope. Following the dispersion, the radiation must be approximately reversed in direction and de-dispersed to again be available for single mode operation. Any problems with band edges not matching to adequate tolerance would need to be dealt with in the final resolution, where small sections of spectrum might need to be cut out. Alternately, a stack of interference filters might well separate bands with less problems at the edges, but the efficiency would likely suffer and be unacceptable. It is concerns like these that redirect attention towards the possibility of using a single spectral zone.

## 9. The Sequence in the Light Path

There is no perfect answer to the problem of how to sequence operations, but some broad answers seem possible.

First, pointing and setting on a fiber or hole with commonality of both mid-IR and near-IR path with a visible startracker is needed. This path should contain the equalization adjustment from near-IR Bracewell nulling. Next, (if needed) the light must be separated into wide mid-IR zones. Following that, phase and amplitude adjustments are made for each zone.

Then, the signals go to a beam combiner for each Bracewell nulling pair. The outputs of the two Bracewell pairs have to be further adjusted in path length and amplitude to give the deep broad null when fed to a second beam combiner. It should be possible to have most of this part of the system very small and stable. The exception is the balancing of the pathlength between the two nulling systems. It should be possible to use the 2$\mu$m signals to servo this. Finally the mid-IR broad bands need feeding to one or more cold spectrographs where the nulled signals are divided into 0.5$\mu$m bands and fed to IR detectors.

## 10. Cooling

One aspect of some planet finder designs that needs attention is the cooling of the entire delay line/beam combining path, as is being discussed by Hawarden in this volume. One difference between the OASES and DARWIN designs is that OASES needs long delay lines with mirrors between the telescopes, whereas DARWIN could project beams across "uncooled" space. One can neither afford to have shields around telescope exteriors radiating into beam combining paths, nor shields around beam combining paths radiating into telescopes. This suggest a configuration where radiation shields are behind both. In consequence we are faced with all the adjustments described above needing to be made in equipment at a temperature of 30K or so, and using very little energy.

The time to reach a place at 3-5 AU from the Sun where the instrument is cold for planet observations seems to preclude a system spacecraft carrying cryogens. Since, as pointed out above, current detectors are marginal for a planet finder, it is not clear what temperature cooler is needed. There is concern that heat dissipated from a cooler should also not interfere with observations. One possibility for minimizing heat load is that the signals in the focal plane of the spectrograph could be fed by light pipes to detectors in a cryostat.

## 11. Conclusion

The focal plane of a planet finder is a complex system. It has a number of challenging aspects for which the form of a solution is quite uncertain, and laboratory studies are needed. One of the more challenging aspects of a solution will be to see how to integrate the issues discussed above with some mechanical design for the mission. This task will be considered in paper IV of this series.

**Acknowledgements:** This work has been supported by NASA under grant NASA7-1260

## References

Angel, J.R.P., Burge, J.H., Woolf, N.J.: 1996, *Proc. SPIE* **2871**, in press
Angel, J.R.P, Woolf, N.J.: 1997, *ApJ*, in press
Beichman, C.: 1996, A Road Map for the Exploration of Neighboring Planetary Systems (ExNPS), NASA: JPL
Connes, P.: 1970, *Advances in Astronomy and Astrophysics* **8**, 209

Leger, A., Mariotti, J.M., Mennesson, B., Ollivier, M., Puget, J.L. Rouan, D., Schneider, J.: 1996, *Icarus* **123**, 249
Mariotti, J-M., Coude de Foresto, V., Perrin, G., Peiquan, Z., Lena, P.: 1996, *A&AS* **116**, 381
Mennesson, B., Mariotti, J-M.: 1996, preprint
Shao, M., Colavita, M.M.: 1992, *Advances in Astronomy and Astrophysics* **30**, 457
Woolf, N. J., Angel, J.R.P.: 1997, in Planets Beyond the Solar System and the next generation of Space Missions, STScI, Baltimore, in press

# WORKSHOP SUMMARY

T. DE GRAAUW
*SRON, P.O. Box 800, 9700 AV Groningen, The Netherlands*

This international workshop on "Infrared Space Interferometry for Astrophysics and Detection of Earth-like Planets" has been, for several reasons, a very exciting one. First of all it has brought together over a hundred people with a wide variety of expertise: theoretical and observational astronomers, astrophysicists, laboratory physicists, chemists, space engineers, science journalists and science managers/administrators. Second, the meeting took place shortly after we learned from the first discoveries of planets orbiting solar type stars. This had a strong impact on the spirit of the workshop. Maybe that therefore most of the attention was directed to planet detection. Third, but not least, the interest in planet detection and infrared interferometry in the US, Japanese and European astronomy communities is rapidly increasing, which is reflected in a growing commitment by the corresponding space agencies to carry out the applicable space missions. See ESA's Horizon 2000+ plan and NASA's plans for the Exploration of Neighboring Planetary Systems (ExNPS road map). The resulting outlook is very stimulating for the research and development of the technology needed to accomplish these missions.

The interaction during this week has been very stimulating indeed. Thanks to excellent presentations and reviews on :

- theories on formation of planetary systems and on evolution of planetary atmospheres.
- results of planet detection programmes and related observations.
- astrophysics with IR interferometers.
- concepts for interferometric space missions and technical developments.
- precursor space missions and groundbased interferometry.

I will summarize the meeting along these topics.

## 1. Theories on formation and evolution of planetary systems.

Several reviews in this area were presented. The discoveries of several giant planets and of sub-stellar companions has put the theoretical



*C. Eiroa et al. (eds.),*
*Infrared Space Interferometry: Astrophysics & the Study of Earth-Like Planets*, 309–314.

work on the formation of planetary systems in a complete new perspective. The feedback for the models arising from these successful observations is very stimulating. For example, the very small separations between the star and its companion implies that the detected giant planet must have migrated inward, after its formation. Schemes how this could have taken place are being developed (see Boss) and IR spectra for the detected samples have already been calculated (see Guillot et al.). The evolution of planetary atmospheres in our solar system has been extensively reviewed by Gautier. He has made it clear that a thorough understanding of the evolution of solar system planets, and their oceans, atmospheres and solid component will be vital to model the evolution of exo-planets with potentially detectable signatures of life. One of the key questions he brought to our attention was whether we can predict the persistence of water on the surface for a period of several Gyrs, which is needed for the development of life.

In any case it seems that the solar system planetary research plays a very important role in the understanding of exo-planets, similar as the solar research had in the study of stellar evolution and stellar atmospheres.

How to detect life on exo-planets, once they have been discovered, has been discussed by Léger. His presentation was preceded by one from Fontana giving a calculus analogue for the chemistry-language description of life. The fascinating comparison is difficult to summarize. (It triggered me to read more about the origin and evolution of life and I found the book by Christian de Duve "Vital Dust; life as a cosmic imperative" very interesting). Léger argued that the simultaneous detection of $O_3$ and $H_2O$ is a good criterion for a Carbon based life. Simultaneous presence of $O_3$ and $CH_4$ seems to be another good criterion as well, but requires a much higher spectral resolution (R=500 in stead of 20) of the IR spectrometer. Although these criteria are important to discuss, I think that the spectral coverage of a space interferometer should, in any case, be wide enough to allow detection of $O_3$, $CH_4$, $H_2O$, $CO$, $CO_2$, $NH_3$, etc. and thus characterize terrestrial as well as giant planets..

## 2. Observational results.

The discovery of the Vega phenomenon by IRAS is at the origin of a more general interest in the studies of extra-solar planetary systems. The follow-up work by HST, ISO, KAO, mm-arrays, large optical telescopes operating in the visible and IR range, using very advanced techniques like the fiber optic recombination technique (Perrin et al.) and adaptive optics (Sandler) have been and will be crucial for the studies of evolution of planetary systems. With ISO one can expect not only to obtain a better insight in the presence of dust around candidate exo-planetary systems, but also important information about the chemical composition and evolution of the inter- and circumstellar dust in the various phases of star formation and stellar evolution. See also the paper by Wooden with a pre-ISO view on dust formation.

Astrometric and radial velocity techniques have long been used A.O. for studies of binaries and detection of brown dwarfs. Improved radial velocity equipment and methods have finally given the most direct proof of the existence of planets around solar-type stars. Although the announcement of the discovery of a Jupiter-size planet accompanying the solar-type star 51 Peg represents a major breakthrough, the confirmation by a second group and their announcement of the discovery of two more cases has put the field finally on a very firm base. For an overview of the results to date see the presentation by one of the discoverers, M. Mayor. Since the Toledo workshop several more planets have been added to the list, together with 5 new low-mass substellar companions. With the sample of companions around stars becoming larger and larger, one can start to attribute statistical significance to the distribution of derived orbital parameters such as radius and eccentricity and of the masses of companions.

The gravitational microlensing effect from a massive object near the line of sight towards a background source causes not only angular deflections but also an apparent magnification of the light. This latter effect is detectable and will show up as a smooth increase in the brightness of the background star. More than half a dozen groups are working with this method where the prime aim is to detect dark objects. Several tens of events have been observed. See Ferlet for the review. If the star is surrounded by planets the smooth curve will be changed and allows for the detection of a planet. The planet lensing time, depending on the mass, is of the order of a day for Jupiter-like planets and a few hours for Earth-like planets. It requires an on-line data reduction, to detect the ongoing lensing and to switch to a mode where one is able to measure small and rapid variations in the light curve. One team (PLANET) is preparing to be optimised for this method using a global network for continuous coverage. Model calculations indicate that the method is as powerful as the radial velocity method and a few Earth-mass planets could be detected per year.

There were two categories of observations presented, which are not directly aimed at detection of planets but which are extremely relevant for that work. These are the studies of the planetary atmospheres, comets, asteroids, meteorites, and studies of disks around Main-Sequence stars, YSO's, AGB's, etc. Th. Encrenaz discussed the pre-ISO status of IR spectra of planetary atmospheres and gave a detailed overview of the atmospheric compositions. She concludes with predictions of the fluxes of Jupiter- and Earth-like planets at a distance of 5 parsec, indicating that improvement of detector technology is needed. Waelkens demonstrated that the occurrence of dusty disks around main-sequence star seems to be a common phenomenon as it is the case for young stellar objects (Natta). This will seriously affect the interferometric detection schemes which have to discriminate between the IR contribution of the dusty disk and the planets. The likely presence of exo- zodiacal dust poses a similar and serious problem to the measurement and interpretation of the visibility function.

## 3. Astrophysics with IR space interferometers.

Although most of the attention of the workshop was for the planet detection issue, several speakers have given overviews of the science that can be done with space-based IR imaging interferometers. Quirrenbach and Eckart claimed that in all fields of astronomy major advances can be expected. Voit highlighted those for the studies of Active Galactic Nuclei and their host galaxies. Of course ground-based interferometry (see Paresce) plays an important role, not only in the development of techniques but particularly in defining the science objectives for a space interferometer.

## 4. Concepts for an IR space interferometer and technical developments.

Woolf and Mariotti, but also Bely, Hawarden and Menesson gave very interesting reviews and contributions on space based interferometer designs. The major challenge is to detect the radiation of the planet in a situation where the flux ratio between the star and the planet is of the order of $10^7$. The optimum optical scheme for planet detection is the nulling interferometer. In order to extend the scientific application to imaging, a different telescope configuration is needed. Both, Mariotti and Woolf, explored the feasibilities of combining both options. Hawarden discussed several design approaches for multi-aperture missions where he put the focus on the cooling of the telescopes. Bely explored the free-flyer version versus the moon-based version with individual telescopes. Although the free-flyer appears to have several technical advantages, the idea of free, not structually linked telescopes in space has certainly a psychologic arrearage. It seems that the necessary inclusion of spectroscopy will complicate the focal plane package. The technical development work presented covered a wide range of technologies, from cooling techniques to data transmision methods. All relevant for a space IR interferometer.

## 5. Precursor space missions and ground-based work

The development and construction of ground-based interferometers is of great importance for the space missions. Not only because of similarity of technology, but also from the scientific point of view. This was pointed out in various presentations and discussions. The earlier results from the Keck and VLT interferometers will be necessary to guide and optimize the designs of the space missions. For this reason NASA and ESA have shown an interest in the development of these projects. The Space Interferometer Mission (SIM; see SHAO's paper), planned to be started in 1998, will have a very important impact on the development of a planet finder project as it will demonstrate a number of key technologies and science objectives. The SIM is designed primarily as an astrometric one, with a wide-angle and high-througput and an accuracy of 1 microarcsec. Interferometric nulling will be possible down to a level of $10^{-4}$.

## 6. Conclusions.

The detection of extra-solar planets by several independent teams has given a strong boost to the field of IR interferometery, and in particular to the plans for an interferometer for space, which seems to be imperative to achieve the required sensitivity and suppression of the central star light (Woolf, Greenaway). In general, there appears to be much progress in the definition of interferometer concepts. And also the technology is advancing rapidly. Although the space misions appear to be of a complicated nature, there are no real serious technical road blocks. Maintaining coherence in the development and construction programmes and adequate system engineering will be the most difficult areas for these projects. Because of the complexity, costs and times scales are bound to be large. The very strong interest of the general public in the detection of exo-planets and the possibility of finding extra-terrestrial life have to be exploited to their limits, to obtain a steady support over a large period.

Several panel discussions highlighted the enthousiasm of the players involved. It showed the dynamics in the community. It gave me an impression of observing a multi-body system where the elements (theory, physics, chemistry, engineering, administration) are connected with springs and thereby pushing and pulling each other. The road to a stable and quiet multi-element space interferometer is slowly emerging, but a long one to go. This workshop has made a very positive contribution to this process.

## SUBJECT INDEX

# OBJECT INDEX

## AUTHOR INDEX

MIX
Papier aus verantwortungsvollen Quellen
Paper from responsible sources
FSC® C105338

If you have any concerns about our products,
you can contact us on
**ProductSafety@springernature.com**

In case Publisher is established outside the EU,
the EU authorized representative is:
**Springer Nature Customer Service Center GmbH**
**Europaplatz 3, 69115 Heidelberg, Germany**

Printed by Libri Plureos GmbH
in Hamburg, Germany